Buckminster Fuller

Anthology for the Millennium

Buckminster Fuller

This man is of the sort we now call original men . . . the first peculiarity is that they communicate with the universe at first hand.
—Thomas Carlyle

Buckminster Fuller

Anthology for the Millennium

Edited by Thomas T. K. Zung

Southern Illinois University Press • Carbondale

The Southern Illinois University edition of this book has been made possible through generous
donations from the RBF Dome NFP, Thomas T. K. Zung, and Southern Illinois University
Carbondale–Continuing Education.

Front cover design: Howard Collinge / Talley Carlson by Studio Zung

Cataloging data may be obtained from the Library of Congress.
Library of Congress Control Number: 2014942718

ISBN-13: 978-0-8093-3317-2 (paperback)
ISBN-10: 0-8093-3317-1 (paperback)
ISBN-13: 978-0-8093-3318-9 (ebook)
ISBN-10: 0-8093-3318-x (ebook)

Printed on recycled paper. ♻

The paper used in this publication meets the minimum requirements of
American National Standard for Information Sciences—Permanence of
Paper for Printed Library Materials, ANSI Z39.48-1992. ∞

Dedicated in Memoriam to
R. Buckminster Fuller and
Anne Hewlett Fuller

And to the generations of youths,
regardless of your age,
who will be touched by Bucky.

Contents

Foreword to the 2014 Edition

The numerous ideas R. Buckminster Fuller generated over the course of his lifetime were and continue to be exciting and influential. The structures, plans, and images created by Fuller and his principal associates, Thomas T. K. Zung and Shoji Sadao, have been significant to engineers, architects, designers, philosophers, and persons on the street worldwide. As voluminous as was Bucky Fuller's own output, he is among the few true innovators of the twentieth century whose work has in turn, even long after their death, inspired many more designs, structures, solutions, commentary, and especially ideas.

The Stanford University Libraries are privileged and honored to have the responsibility, passed to them by Fuller's family and immediate associates, of preserving and making accessible the archives of his work, in the form of designs, correspondence, reports, models, photos, videos, diaries, manuscripts, and a host of other artifacts arising from his most creative life. Bucky Fuller conceived of and articulated the notion of anticipatory design science. In creating and filling his own archive with its Dymaxion index, he foresaw the work such as we would do to permit and encourage successive generations of citizens and professionals in many fields to use his ideas and expressions for the improvement of humans and other denizens of Mother Earth.

Reissuing this anthology with some new material is definitely an aspect of keeping Bucky's great ideas in mind and considering their possible application to the grand challenges we all face, consciously or not. Bucky's ideas are here to serve humankind, and Stanford University is here to assure that they will continue to do so over the long run of time.

Michael A. Keller
The University Librarian
The Stanford University Libraries

Acknowledgments for the 2014 Edition

This revised anthology is dedicated to Buckminster Fuller and his wife, Lady Anne. To the Fuller family, Allegra, Jaime, and Alexandra, my thanks for your loving support over these many years, with acknowledgment to John Ferry. Without your help, this edition could not have been possible.

A special thanks to Lord Norman and Lady Elena Foster, Sir Harry Kroto, Calvin Tomkins, and Carl Solway, for their advice and friendship, and to the late Maria Stone, wife of architect Edward Durell Stone, Fuller's friend and my past New York architectural associate.

During Fuller's time in Cleveland, several organizations were supportive of his work, including the City Club of Cleveland, where Bucky and I presented at the City Club Forum, Fuller in October 1980 and me in July 2001. Kudos to past club director James Foster for assistance with the U.S. postage stamp honoring Fuller; the current club director is Dan Moulthrop. Thanks to Mort Mandel of the Mandel Foundation for support of the reissued Dymaxion map. Thanks to John Luteran, vice president of the Greater Cleveland Partnership; the David Ingalls Foundation; and Finnius Ingalls, who attended school with Fuller's godson Tommy Bates Zung. I am also grateful for the friendship of Richard Alt of Carnegie Capital; Steve Kadish of Kadish, Hinkel, Weibel; Peter Jefferson of Wright Gallery in Chicago; Nancy McCann; and Richard Pogue of Jones Day.

Many thanks to Stanford University librarian Michael A. Keller, Roberto Trujillo, Mattie Taormina, and Maria Grandinette. The portraits by Yousuf Karsh are printed with permission from the Karsh Estate, Estrellita Karsh, and Jerry Fielder, curator. The photo credit restrictions of the St. Martin's edition remain in place for this reissue; the images new to this edition are published with the permission of the Stanford University Archives, the Estate of R. Buckminster Fuller, BFI, SNEC, and Shirley Sharkey. Unless otherwise noted, other photos are from the office of Buckminster Fuller, Sadao & Zung Architects, LLC, and may not be reproduced without written permission from this office and Southern Illinois University (SIU) Press.

A special thank you to the late Herbert E. Strawbridge, known as Mr. Cleveland, and his late wife, Marie Strawbridge, of the John P. Murphy Foundation. Warmest remembrance for your friendship and guidance.

This 2014 edition was inspired by the Fuller dome home restoration project sparked by retired professor Bill Perk. I particularly need to thank my outstanding architectural partner Shoji Sadao, Shirley Swansen Sharkey, Robert J. Chin, and Michael Denneny for their support. Thanks to SIU Press editor in chief Karl Kageff; copyeditor Joyce Bond; and project editor Wayne Larsen and director Barbara Martin, both at SIU Press. Fuller lived in his dome home in Carbondale for twelve years while he taught at SIU, and some consider this one of his golden eras. Thus it is fitting that SIU Press has published this revised anthology. I also want to give my thanks to a special Fullerite person.

To my extended and beloved Zung family—Tommy Bates; dear Kaikoa; Joyce Burke-Jones; Yoko and her son; my late siblings, Max, Bill, and sister Louise; and all my nieces and nephews—this book is dedicated with much love to each of you.

Acknowledgments for the 2001 Edition

I express my gratitude to those whose patience and support made this book possible. My appreciation to the Estate of R. Buckminster Fuller, the Fuller family, and especially to Allegra Fuller Snyder and Jaime Snyder for granting me the permission to attempt this anthology.

A special thanks to Calvin Tomkins for his advice and friendship. My thanks to Bucky's and our firm's attorney, Robert D. Storey of Thompson, Hine & Flory, LLP, for his help. Heartfelt thanks to my sailing partner, Steve Kadish, Esq., and dear Joyce Burke-Jones for their overview, and to Herb Strawbridge, Dominic J. Guzzetta, and Steve Forbes for their early support. My acknowledgment to the late Ambassador and former Cleveland Mayor Carl B. Stokes. For her encouragement, thanks to Maria Durell Stone, the wife of the late architect Edward Durell Stone, Bucky's friend and my former New York architectural associate. I appreciate the permission to reprint Yousuf Karsh's photographic portraits of Bucky and me—Karsh records not just men, but history.

Comments from E. J. Applewhite, Shirley Sharkey, the late Kiyoshi Kuromiya, and Joe Clinton, with assistance from John Ferry at the Buckminster Fuller Institute, were helpful. And many thanks to Michael Denneny, my editor, and his assistant, Christina Prestia.

Sincerest gratitude to my office staff and Bucky's architectural associates, Shoji Sadao, Robert J. Chin, Kenneth L. Papes, Robert G. Arnold, and Joyce Burke-Jones, with help from Carl Solway, Kris Corradetti, and Linda Kenski. To my son, Tommy Bates Zung, who is Bucky's "grand-godson," and Carol Williams Zung, his mother, thank you. To a very special lady; and to my beloved late sister Lee-Fong Louise Lee, my late brother William W. K. Zung, M.D., Nina Zung Harman, Yoko Zung Chinen, and brother Max M. K. Zung, M.D.—this publication is also for you.

Thank you all.

An Introduction to Bucky

R. Buckminster Fuller over many years graced the cover of nearly every major publication in America and abroad. But since his death in 1983, he has been neglected, if not shamefully ignored. The need for the public, especially the young, to discover his thinking anew impelled me to take on this anthology. By training I am an architect; this publication is my first attempt at being an author/editor. Any criticism directed at me is probably deserved, and a small price to pay for introducing the young, regardless of age, to a man named Bucky.

Why publish now? Bucky would call the birth of a new millennium an auspicious occasion, a time line in history, a notch in the evolution of man's being on our "Spaceship Earth."

What better way to celebrate the new millennium than to let new readers discover a comprehensive anticipatory corporeal thinker, a nonconformist poet? I am convinced that if people understand him, they will be unwilling to let him go, and by adopting comprehensive planning, humanity will be significantly better off in this new millennium. As a student of Bucky's in the early 1950s and later as his architectural partner, I recall the host of highly creative people touched by him and note how their insights have dramatically affected this beautiful and elegant 8,000-mile-diameter—actually 7,926.42-mile-diameter—planet we call Earth.

In a 1964 cover story, *Time* magazine named R. Buckminster Fuller "the first poet of technology." Harvard University must have agreed, for in 1961 it had appointed Bucky, as he liked to be called, to the Charles Eliot Norton Chair in Poetry. In 1996, *Time* published *Great People of the 20th Century,* and included Bucky Fuller. As an inventor, architect, author, and forty other titles, he may be best remembered for his geodesic domes, but he is much more than just the notable U.S. Patent No. 2,682,235.

Time magazine called him an American original, a cranky genius, and an ingenious crank. Bucky Fuller's career started inauspiciously. Bucky had, in his words, "been fired twice from Harvard University." The first firing came after he expended his tuition funds by treating musical comedy star Marilyn Miller and the *George White's Scandals* cast to dinner at Churchill's. His family forced him to re-enroll at Harvard. The second firing: he was simply bored, and he felt that he was wasting his family's funds, which were quite limited since the death of his father.

In 1917, Fuller married Anne Hewlett, daughter of a prominent architect, and a year later their daughter Alexandra was born. Sadly the baby contracted infantile paralysis and spinal meningitis and died when she was four. Fuller felt tremendous responsibility and guilt.

In 1927, at age thirty-two, Bucky Fuller was to make a momentous decision. Considering himself a failure, he went to the shores of Lake Michigan, "to be a throwaway." In his own description, "I was an utter failure in business, I had lost all my money, my friends' investments, I was discredited and penniless." His second daughter, Allegra, had just been born, and he thought his small family would be better off without him.

Before jumping, he stopped himself, thinking: Instead of being a throwaway, could I be useful? Said Bucky, "Could I use myself as my scientific 'guinea pig': to see what a penniless, unknown human individual with a dependent wife and newborn child might be able to do effectively on behalf of all humanity?

"To make the world work for one hundred percent of humanity, in the shortest possible time, through spontaneous cooperation, without ecological offense or the disadvantage of anyone." It is here that Bucky Fuller starts his Guinea Pig "B" life experiment.

During the 1960s, Bucky Fuller was popular with the young, and he became a folk hero to the hippies, as some of them were known. Geodesic domes were multiplying like rabbits, and by the 1980s thousands had been built. Bucky believed in the young, and he urged young people to "reform the environment instead of trying to reform man." Bucky argued that "mankind has the capability through proper planning and use of natural resources to forever house himself and live in workless leisure." He put it very plainly: "Through technology, man can do anything he needs to do."

Sparks of rekindled interest in Bucky have recently emerged, reported a July 1999 *New York Times* article by James Sterngold. Currently, there is a major traveling museum show, Your Private Sky—R. Buckminster Fuller, produced by the Museum für Gestaltung, Zürich, with Curator Claude Lichtenstein and Dr. Joachim Krausse. The show started in Zürich and is touring Europe in 2000, with an Asia venue in 2001. The board of trustees of the Stanford University Libraries in California has acquired the massive archives of R. Buckminster Fuller from his family and the Buckminster Fuller Institute. Fuller's ideas are resurfacing, notes Michael A. Keller, Stanford University Librarian and Director of Academic Information Resources. In the fall of 2000 a play on Fuller, written by D. W. Jacobs, was performed by Ron Campbell to rave reviews on the West Coast.

The publication date for this book is set to coincide with the actual millennium year. I think Bucky would have been amused by the celebration of the millennium on December 31, 1999, when on New Year's Eve the ball dropped high above New York City's Times Square. The ball was actually a Phillips-made glass geodesic polyhedron. Fuller was a stickler for precise language and scientific fact. A century gets its name from its year. The first century was designated by the year 100; the second century is named for the year 200. The twentieth century's last year is 2000, and the century does not expire until December 31, 2000. January 1, 2001, begins the new millennium. At least that's what I learned at Stuyvesant High School, and in college I learned even more: that a millennium has 1,000 years and not 999.

Misinformation is something Bucky abhorred. It brings to mind Bucky's reaction when astronaut John Glenn was orbiting the earth for the first time and President John F. Kennedy contacted him to congratulate him on his successful mission. Remarked Fuller, "The president said, how is it up there?" At that precise moment, the astronaut was circling at the foot of the president, on the other side of the earth. The notion that the sun sets made him irate. "The sun never sets or goes down," Bucky would exclaim, "the sun eclipses, and the correct term would be that we sunsight in the morning and sunclipse in the evening." Seemingly harmless imprecision in language and idiom taints our understanding of scientific truths, and Fuller calls this to our attention.

In the nineteenth century, Thomas Malthus theorized that humankind was multiplying at a geometric rate but producing goods and food at only an arithmetic rate. Clearly, we were headed for failure. Following the Darwinian theory, Karl Marx then added that only the fittest class of men would survive, and that the fittest was the working class, because workers knew how to handle tools and all other people were parasites. Man would endure only because of the worker population.

It is reported that our human population reached the six billion mark in October 1999. In 1800 the population was just one billion. (Born in 1895, Bucky Fuller was himself part of this population explosion.) Projections are that in just fifty years, by the year 2050, our six billion population will nearly double to about 11.5 billion. If the population control agency predictions are accurate, in 2050, of the 11.5 billion people, 23 percent will live in Africa. Since Africa currently has only 8.8 percent of the world's population, many of whom are starving, these are troublesome numbers. In order to feed the world's enormous population, we need some innovative solutions. But neither Malthus nor Marx could look into a crystal ball and see that through science, mankind could accommodate a higher standard of living for all people. Man might be saved through a design revolution, one that Bucky would call anticipatory design science, an invisible revolution called technology.

Fuller predicted in *Critical Path* that by the end of the twentieth century the billionaires of the world would hold not physical property but information property. His prognostication about real wealth has become a reality, weaving new ways to measure "money," recorded now invisibly. He wrote in 1969, in *Utopia or Oblivion,* "Ephemeralization is accelerating by ever-increasing quantities of invisible energy events of the universe, it constantly does more with visibly less." Astonishingly, the current group of supermillionaires are very young—but this would not surprise Bucky. The richest men in the nineteenth century held vast wealth by owning tangible property like manufacturing plants, steel mills, tracts of land, railroads, shipping lines, oil wells, etc. Not so in the twentieth century.

Indeed, two of the richest men in the world, Bill Gates and Paul Allen of the Microsoft Company, have a combined wealth of over $125 billion via a computer chip that is weighted in nano ounces. They hold their power of wealth through innovative technology, the United States courts notwithstanding, and acquired this wealth in about a decade. Other recent millionaires have achieved their super-rich status within this same decade by the technology information network called the Web. And yet Bucky cautions us against "greed." Bucky once remarked on Instant Universe that while light's speed of 700 million miles an hour is very fast in relation to automobiles', it is very slow in relation to the "no time at all" of society's (obsolete) instant universe thinking. "Comprehensively the world is going from a Newtonian static norm to an Einsteinian all-motion norm. That is the biggest thing that is happening at this moment in history. We are becoming 'quick' and the graveyards of the dead become progressively less logical."

But even technology has traps. As the world prepared for the Y2K transition, the computer industry was panicked because some computer systems and some software would interpret the number 00, short for 2000, as 1900. A May 28, 1998, *New York Times* editorial commented: "It seems odd, somehow, that in an industry that is downright apostolic about the future—whose pundits make more prognostications than a Ouija board—no one saw the millennium coming until it was just this close." America's Y2K czar, the White House Y2K coordinator John Koskinen, estimated that the repair of this technical world glitch would cost our nation over $100 billion—more than the gross national product of over thirty of our least developed nations. And other nations collectively would pay an additional $500 billion for fixes. Technology can go awry, and the technicians, mostly our youth, must recognize their obligations and moral responsibilities to the world they helped create. My generation remembers Dr. Albert Einstein, who helped develop the atomic bomb via the Manhattan Project, and later expressed dismay and shame at the creation of such a hideous artifact and its use against his fellow men. Einstein, who studied at the Swiss National Polytechnic in Zürich, remarked after the bombing of Hiroshima and Nagasaki, "If only I had known,

I would have become a watchmaker." "God makes no mistakes," said Bucky. But man can make mistakes.

Bucky Fuller could see no sense in the leading industrial nations' spending huge portions of their budgets on military items like nuclear bombs. These weapons of mass destruction replaced the common sense of finding ways to house the homeless and feed the hungry. He called for "livingry instead of weaponry." Einstein said, "The unleashed power of the atom has changed everything save our modes of thinking, and we thus drift towards unparalleled catastrophes." We now have satellites that can shuttle destruction vehicles into space. These space weapons can launch "kinetic energy rods" and create havoc by hitting with pinpoint accuracy targets on Earth at speeds greater than five times the speed of sound. The United States has high-resolution satellites like Space Imaging's Ikonos, which from 24,500 miles in space can produce images of phenomenal detail like a six-inch animal. How productive is it to spy on a small mouse nibbling in a cornfield from outer space, while millions go hungry?

Society shifted its transportation thinking in the late fifties and early sixties as the delivery of Europe to the U.S. air passenger services outperformed in time, money, and energy the *Queen Mary* in crossing the Atlantic Ocean. Soon, air transportation, "doing more for less," rendered seaports obsolete. The docks along New York's Hudson River lay vacant, and so did the docks across the Atlantic at Southampton. Technology—in the form of the airplane—had changed their geographic importance. Nations were now building international airports with vast runways where jet planes could land.

Today, technology is having a dramatic effect in Brazil. Fuller would have been amused that an obscure location named Alcântara might be as well-known in the new millennium as Cape Canaveral and Vandenberg Air Force Base. At Alcântara, Brazil, which is 2 degrees south of the Equator, the launching of satellites is 30 percent more efficient than from Cape Canaveral. Why? Because the rotation of the Earth is at its maximum speed at the Equator, and its speed recedes in zones going north and south. The spin pushes the rockets into orbit faster. Alcântara can then launch rockets with heavier payloads than anywhere else on Earth. That means billions of dollars for Brazil's economy, says Ronaldo Sardenberg, the Science and Technology minister. What a rational economic activity, and one preferable to chopping down the rain forest and displacing scientific study in that invaluable virgin region. Fuller would call Alcântara an example of "anticipatory thinking" by Brazil, as that nation builds up the spaceport while waiting for Lockheed Martin, Rockwell International, and Boeing to queue up for a launch schedule.

Despite all our technical knowledge, we still strain our environment with dangerous nuclear energy plants. Accidents like that at Three Mile Island in the United States, the recent leaks at Japan's Tsuruga nuclear power plant, and the 1986 accident in Chernobyl are still not resolved. After years of flim-flam in Ukraine, the $443 million allocated in its 1999 budget for the Chernobyl disaster is only one-third of the amount necessary for cleanup and victim relief, according to reports by Ukraine's Emergency Situation minister. And today, fourteen years later, there are 3 million children who still require medical treatment for radiation exposure as a result of that accident. Despite worldwide warnings, on November 26, 1999, Chernobyl Reactor No. 3 was restarted.

Bucky said that the "sun is a nuclear energy plant, and Mother Nature put her safely 93 million miles away" and anything closer is pure folly. This generation, the next generation, and probably the generation after that, will need to study this catastrophic artifact that endangers our environment.

Fuller was not only concerned with intellectual abstractions; he would sometimes be so practical that he was delightfully amusing. During one of our trips to England, Bucky announced to me at breakfast that I was to arrange transportation from London to Greenwich. He wanted to show me the time-piece at the museum by which time is measured throughout the world, i.e., the Greenwich Mean Time piece. As we neared the artifact, a smiling Bucky saw that the timepiece was happily ticking away; no grinch had stolen it. Then he went into a half-hour talk on how, just twenty years ago, the sun never eclipsed over the British Empire. On another occasion, my son, Tommy Bates, watched "grand-godfather" Bucky negotiate, with a waitress at a four-star hotel a swap of five gauze-covered lemons, neatly tied with pink ribbon, for three additional stewed prunes for his breakfast order. He was very real and always very human, and he wrote with conviction about the options humanity faces.

In assembling this anthology for publication, I invited each essayist to contribute a piece on R. Buckminster Fuller. But as E. J. Applewhite notes, Bucky did not want to be misunderstood, so major portions of this book are Fuller's own words, taken from his many books. It's important to know the dates of publication for Fuller's books, and with one exception, each excerpt in this anthology follows those dates. This chronological order illustrates Fuller's thinking over the years. Like other innovative thinkers, Fuller would often come back to an idea. Thomas Carlyle called these thinkers "original men," noting that their "first peculiarity is that they communicate with the universe at first hand." Fuller often came back to his earlier thoughts on a subject, which surely accounts for his many versions of rethinking the Lord's Prayer.

Fuller's first book for general circulation, *Nine Chains to the Moon,* published in 1938, serves as this book's first chapter. It was reviewed by the architect Frank Lloyd Wright, and Wright's comments are printed in the appendix.

Some of the contributors knew Bucky personally, others knew him from his work. Each of the contributors' essays is paired with a selection from one of Bucky's books, but the essay is not paired with Fuller's work as a commentary. No conscious attempt is made to tie Fuller's work with the contributor's piece. Much of Bucky's work is universal and omnidirectional, so that first impressions can sometimes be deceiving. If there appears to be a connection to the reader, it is purely coincidental. There is, however, a sympathetic approach to the pairing. Many of Fuller's early writings are out of print, and I hope this millennium anthology publication will spark interest in reprinting some of his early books.

Reading Bucky is an exploration process. As the new reader ventures into Fuller's work, he may need a gestation period before he comes to an understanding of Fuller's words, as he talks about our universe and the nonsimultaneous scenario of omnirelated events, often in a lighthearted, Mozartian manner. Following Bucky's thought can sometimes be difficult, which is understandable, since he sometimes uses his own language. Just let the words flow, and you soon will be in the blend. Let Fuller's words flicker in your mind. He maintained that all children are born geniuses. Out of every 1,000, some 999 are swiftly and inadvertently degeniused by grown-ups. Motivated by love but also by fear for their children's future, parents in their ignorance act as though they know all the answers, and so they curtail the spontaneous exploratory acts of genius of their children. I remember Bucky's own words, "I see in a flash that the eyes of youth see that the world could be made to work for all humanity. And I see that they will settle for nothing less. And I see that they are impatient. And I see above all that they can and will soon make it so."

In 1998, *National Geographic* conducted an interactive Web questionnaire called Survey 2000, which gathered data from 80,000 people. Among young people thirteen to fifteen years old, 85 percent worry about the environment. And 60 percent of children disagree with the statement: "All in all the world population will be better off in the next hundred years." In fact, most would accept cuts in their standard of living to conserve global resources. Bucky would find these findings about the young an opportunity for humanity.

Civilization is engaged in the rapid acceleration of Fuller's design science revolution. Humanity now has the historically unprecedented option to produce a high standard of living for all people on an ecologically sustainable basis. Bucky said, "Think of it. We are blessed with technology that would be indescribable to our forefathers. We have the wherewithal, the know-it-all, to feed everybody, clothe everybody, and give every human on Earth a chance. We know now what we could never have known before—that we now have the option of all humanity to 'make it' successfully on this planet in this lifetime. To induce all humanity to realize full lasting economic and physical success, plus enjoyment of all the Earth, without one individual interfering with or being advantaged at the expense of another."

This collection of Bucky's writings and contributors' essays recalls the words of Sir Winston Churchill: "The further one looks back in history and delves, the further one will see ahead into the future." This book is an appetizer designed to encourage readers to explore the Dymaxion world of R. Buckminster Fuller, a publication guide for the new millennium for those who care about our future on Spaceship Earth.

THOMAS T. K. ZUNG, Architect

Chronology

1895—Born in Milton, Massachusetts, 12 July. Son of Caroline Wolcott and Richard Fuller. New England family since 1632. Great nephew of Margaret Fuller (1810–50), founder of American women's rights movement.

1904–13—Attended Milton Academy. Grandmother Matilda Walcott Andrews purchases Bear Island, Penobscot Bay (Maine), as summer residence.

1908—Death of father, Richard Fuller. RBF decides to keep all of the letters written to him. *Chronofile* develops out of this collection later.

1913–15—Graduates from Milton Academy. Attends Harvard College; expelled twice (he calls it "fired") and considered a failure by his family.

1914—Apprenticeship as machine fitter in Sherbrook, Quebec.

1915—Transportation work for the meat company Armour & Co., New York and New Jersey.

1917—Cadet at U.S. Naval Academy (spec.) Annapolis; marries Anne Hewlett.

1918—Birth of first daughter, Alexandra.

1922–26—Founds Stockade Building Company with his father-in-law, architect J. M. Hewlett. Death of Alexandra from polio.

1927—Birth of second daughter, Allegra. Loses his position as president of Stockade Building Company when company is sold. Serious personal crisis; loses all his money; considers being a "throwaway." Changes his life, dedicating himself to work for 100 percent of humanity, in shortest time possible, through spontaneous cooperation, without ecological offense or disadvantage of anyone.

1928—First 4D sketch, called Hexagonal House.

1929—Exhibition of 4D House in the Marshall Field Department Store, Chicago. Creation of the *Dymaxion* concept by the Marshall Field advertising staff. Meets the Japanese-American sculptor Isamu Noguchi, who becomes a close friend. Active in Romany Marie's Tavern in Greenwich Village; meets Alexander Calder, Martha Graham, and others.

1932—Purchases architectural magazine *T-Square,* renames it *Shelter.*

1933—Dymaxion Car No. 3 is introduced at the 1933 World's Fair in Chicago. H. G. Wells, Amelia Earhart, and Leopold Stokowski ride in car.

1934—Death of Caroline Wolcott Fuller, RBF's mother.

1935–36—Dymaxion Bathroom at the U.S. Bureau of Standards (1936–38).

1936—Explains to Noguchi the equation E=mc² in a telegram. Second book, *Nine Chains to the Moon,* written summer 1936 (published in 1938). Meets with Albert Einstein to discuss the manuscript of his book. Frank Lloyd Wright critiques book. Friendship with Christopher Morley.

1938–40—Technical advisor to *Fortune* magazine. Advisor to *Life* magazine.

1939—Dymaxion Bathroom exhibited at Museum of Modern Art, whose new building had been designed by his friend Edward Durell Stone.

1940—Writes *Untitled Epic Poem of Industrialization* (published in 1962). Model of the Dymaxion House destroyed while being transported to Rockefeller Center. Studies commissioned by the industrialist Henry J. Kaiser.

1940–42—D.D.U. (Dymaxion Deployment Units) prototype by Butler Manufacturing Company.

1942–43—*Dymaxion World Map* appears in *Fortune* and *Life* magazines. Geometric investigations, preliminary studies to energetic geometry.

1944–47—Development of the *Wichita Dwelling Machine, Dymaxion House,* in Wichita, Kansas. Production by the Beech Aircraft factory.

1946—Awarded patent for Dymaxion World Map. Founds Fuller Research Foundation. *Earth Inc.* (privately printed).

1947–48—Summer 1948: Constructs prototype of first geodesic dome together with students at Black Mountain College, North Carolina. Meets John Cage, Merce Cunningham, Josef Albers, Elaine and Willem de Kooning, Robert Rauschenberg, Arthur Penn, and Walter Gropius.

1949—Teaching at Institute of Design, Chicago; University of Illinois; Illinois Institute of Technology; and North Carolina State College. Visiting Critic at MIT. Seminars and visiting lectures at Bennington College, University of Michigan, Harvard University, and University of Toronto.

1949–53—Anticipation of global environmental problems becomes "anticipatory design science." Students and collaborators during the years 1949 to 1955 include T. C. Howard, James Fitzgibbon (Synergetics, Inc.), Don Richter, Kenneth Snelson, Jeffrey Lindsay, Shoji Sadao, Duncan Stewart, Thomas Zung, and others.

1950—Construction in Montreal of the first full-size geodesic structure, made of aluminum tubes, with a diameter of forty-nine feet.

1951—Fuller coins the term *Spaceship Earth*. Museum of Modern Art exhibits a geodesic dome.

1953—Ford Motor Company Dome H.Q. in Dearborn, Michigan.

1954—Awarded patent for the Geodesic Dome. First Radome installation at Mt. Washington. DEW (Distant Early Warning) Line. Grand Prize at the Triennale di Milano for the Two Cardboard Domes. Prizes from American Institute of Architects. Visiting Professor at the University of Michigan. Receives first honorary Doctor of Design from North Carolina State University, second from University of Michigan.

1955—Founding of Synergetics, Inc., and Geodesics, Inc.

1957–59—Construction of large geodesic domes, Kaiser Aluminum and Union Tank Car Company, Baton Rouge, Louisiana. Exhibition in the Museum of Modern Art (1959). Joins Temcor Board.

1959—Professorship at Southern Illinois University. Participates in first Delos Symposium, organized by Doxiadis. Memorial speech for Frank Lloyd Wright in Taliesin, Wisconsin. Conference on World Affairs, University of Colorado, Boulder.

1960—Project *Dome over Manhattan* two miles in diameter. RBF and Anne occupy their geodesic dome home in Carbondale, Illinois. Publication of *The Dymaxion World of Buckminster Fuller* by Robert Marks.

1961—Harvard University appoints RBF to Charles Eliot Norton Chair (Poetry Chair).

1962—Meets Marshall McLuhan at Delos Symposium.

1962–67—Initiative for a ten-year student program World Design Science Decade with the support of the Union of International Architects (UIA).

1963—Publishes *Ideas and Integrities* and *No More Secondhand God.* Advanced Structures Research for NASA. Advisor to NASA (1963–68).

1964—*Time* magazine cover story on RBF. Architectural office Fuller & Sadao in New York City.

1965—Monohex structures (Fly's Eye project). Inaugurates the World Game project at Southern Illinois University.

1966—Project Yomiuri Tower, tetrahedral tower two miles high (not realized). RBF profile by Calvin Tomkins in *The New Yorker.*

1965–67—Designs the U.S. pavilion for the World's Fair, Expo 67, in Montreal. Geodesic dome is a three-quarter sphere with a diameter of seventy-six meters.

1967—The Expo Dome makes Fuller and his geodesic domes world-famous. Jasper Johns's painting of the Dymaxion World Map is exhibited in the Expo Dome. RBF and Anne celebrate their Golden Anniversary.

1968—Receives the Gold Medal of the Royal Institute of British Architects from Queen Elizabeth II. With his Synergetics, Inc., constructs first elongated (caterpillar) geodesic dome, Cleveland, assisted by his former student, Thomas Zung (former design principal with RBF architect friend Edward Durell Stone).

1969—Delivers Nehru Memorial Lecture. Friendship with Indira Gandhi.

1970—Receives Gold Medal, American Institute of Architects, Washington, D.C.

1971—NBC television program *Buckminster Fuller on Spaceship Earth* (R. Snyder). Climatroffice project together with Norman Foster. Old Man River's City project for the renewal of a slum in East St.

Louis. Special editions of *Architectural Forum* and *Architectural Design* on RBF.

1972—Planning of *Spaceship Earth* at Disney World at Epcot Center. RBF is offered a position in Philadelphia as World Fellow in Residence. Moves residence and office to Philadelphia.

1973—Hugh Kenner publishes the book *Bucky: A Guided Tour of Buckminster Fuller.* Patent for tensegrity dome granted.

1975—*Everything I Know* is recorded on video (playing time is forty-two hours). Receives first license as an architect, from the State of New York (age seventy-nine). Fuller's magnum opus *Synergetics* (first volume), appears as a book; collaborator: E. J. Applewhite. Project Fly's Eye, a new geodesic dome. Meets John Warren with his "turtle dome."

1976—Publishes collection of verse *And It Came to Pass—Not to Stay.* *Tetrascroll* published as artist's book in limited edition. Hugh Kenner publishes *Geodesic Math and How to Use It.*

1977—Conferences at Campuan Society, Bali, and Penang, Malaysia, with Lim Chong Keat.

1979—Second volume of *Synergetics;* collaborator: E. J. Applewhite. RBF opens architectural office in Cleveland, as Buckminster Fuller, Sadao & Zung. Receives second license as architect from State of Ohio (age eighty-three). Arthur Loeb teaches geometry to students at Harvard University on the basis of RBF's *Synergetics.*

1980—New edition of the Dymaxion World Map. RBF's eighty-fifth birthday celebration features song written by friend John Denver, "What One Man Can Do."

1981—*Critical Path* appears. Receives forty-seventh honorary degree from Texas Wesleyan University. Starts two new books: *Grunch of Giants* (published 1983) and *Cosmography* (published 1992).

1983—Receives his twenty-eighth patent: Hang-It-All; with Adjuvant Thomas T. K. Zung. Produced by Thonet Company. Receives the Medal of Freedom from President Ronald Reagan at White House.

1983—R. Buckminster Fuller dies from a heart attack in Los Angeles, 1 July. His wife, Anne Hewlett Fuller, dies thirty-six hours later.

Buckminster Fuller

Anthology for the Millennium

The Pritzker Prize is by far the most prestigious award in architecture; it is often referred to as the Nobel Prize of architecture. Bucky's friend and colleague Norman Foster was the 1999 Pritzker Prize recipient. Laureate Foster says, "Every award is special, but there is only one Pritzker. It is a recognition of architecture itself." In 1999, Sir Norman Foster was elevated to the peerage. His new title is Lord Foster of Thames Bank.

Fuller and Foster were longtime associates, and their careers brought them some of the same honors. In different years, they received the highest honor the American Institute of Architects can bestow a member, the AIA Gold Medal. Their names are inscribed in marble at AIA headquarters, along with those of Frank Lloyd Wright, Louis Sullivan, Ludwig Mies van der Rohe, and Le Corbusier. And in Great Britain, Bucky and Norman received in separate years the Gold Medal of Architecture from Her Majesty the Queen.

Foster collaborated with Bucky in the design of the double deresonated dome, which Bucky had hoped to use in California. The model of this dome, along with that of Bucky's last dome, the Gigundo dome, is currently in the traveling exhibition Your Private Sky—R. Buckminster Fuller, produced by the Museum für Gestaltung, Zürich.

Norman Foster has been called a high-tech architect, but he himself says, "Since Stonehenge, architects have always been at the cutting edge of technology. You can't separate technology from the humanistic and spiritual content of a building." His work with Bucky incorporated exploration not only of architectural theory but of social, technological, and ecological theory as well.

Richard Buckminster Fuller

....

Lord Norman Foster

I can remember vividly my first meeting with Bucky, in 1971. He had been asked to design a theatre beneath the quadrangle of St. Peter's College, Oxford, and was looking for an architect to collaborate with him on the project. James Meller, a mutual friend, was helping by making some introductions. We met at the International Conference of Architects, which at that time had a paneled dining room overlooking the Mall. In this elegant setting, Bucky, James, and I talked through a long lunch. I had brought examples of our work to show, and the studio was on standby in the hope that Bucky would visit us. In the event, that was not necessary. Bucky decided on the spot that we should work together and headed off to the next engagement on his punishing schedule.

Only much later did I realize the extent to which Bucky was able to draw me out through that first conversation without my realizing it. He got me to reveal my attitudes to design, materials, research, and other issues, which ranged far and wide. Looking back over the twelve years of our collaboration and friendship, I realize that there are many papers that could be written on the insights that Bucky was to offer. But perhaps the themes of shelter, energy, and the environment—which go to the heart of contemporary architecture—best reflect Bucky's inheritance.

Bucky was a true master of technology, in the tradition of heroes such as Eiffel and Paxton. His many innovations—from the Dymaxion house to the geodesic dome—still surprise one with the audacity of their thinking. Yet while his public image may have been that of the cool technocrat, nothing could have been further from the truth. What was never discussed was his deeply spiritual dimension. For me, Bucky was the very essence of a moral conscience, forever warning about the fragility of the planet and man's responsibility to protect it. He was one of those rare individuals who fundamentally influence the way you come to view the world.

Bucky was the closing speaker on the occasion of my Royal Gold Medal Address at the Royal Institute of British Architects, in June 1983. He used that occasion to address issues of survival, a message that today seems even more pertinent, as some of his worst predictions are gradually coming true.

The world is changing rapidly around, but we are far from prepared for the consequences. The United Nations warned recently, in its report *Global Outlook 2000,* of a series of looming environmental crises sparked by water shortages, global warming, and pollution. It warned that these trends can be reversed only if the developed countries reduce their pattern of wasteful consumption of food, raw materials, and energy by as much as 90 percent.

An explosion in population growth is another crucial factor. Global population has doubled to six billion since 1960, and we are currently adding new humans to the plant at the rate of 78 million a year. That trend is expected to continue for at least the next decade. The UN predicts that by 2050 the developed world will have 1.16 billion people, slightly fewer than today. But in the developing world in the same period, the population will have nearly doubled from 4.52 billion in 1995 to 8.2 billion. As an illustration of what that means, the population of Africa in 1950 was half that of Europe; at the turn of the millennium it is equal; in fifty years it is expected to be three times that of Europe.

Alongside accelerating population growth is a shift toward living in cities. It is estimated that by 2030 two-thirds of the world's population will be urban. We can already see the growth of a new generation of mega-cities of unprecedented size, and urban conurbations in excess of 25 million people are predicted in the next fifteen years. This trend introduces us to new dangers.

In Latin America, where nearly 75 percent of the population is urban, serious problems have already surfaced. Throughout the region, in cities such as São Paulo and Rio de Janeiro, air pollution causes an estimated four thousand premature deaths a year. The reality behind these statistics, and the desperate state of our responses, was brought home to me when I was taken to see the Mexico City suburb of Chalco.

With a population of 3.5 million, Chalco is the size of many European cities. Yet there is a very significant difference. It is a place without transportation infrastructure, sewage or drainage systems, water mains, gas, or electricity. It has none of the basics that most of us take for granted. In one sense, however, the residents of Chalco are fortunate. One hundred million people around the world have no housing at all. This brings to life the estimate that in the developing world, two billion people have no access to energy other than burning natural materials or animal waste. Add to that the fact that just 25 percent of the world's population presently consumes 75 percent of the energy and the implications for energy and resources management are obvious.

In the developed world, buildings consume half the energy we generate; the remainder is divided between transport and industry, with all the associated problems of pollution. So what will happen as the rest of the world catches up? As architects—as a society—we cannot afford to sit on our hands: we have a responsibility to act. Bucky was fond of quoting Theodore Larson: "It is not to devise a better society so as to arrive at a finer architecture; it is to provide a better architecture in order to arrive at a more desirable society."

If those are not challenges enough to the design professions, then surely it is a paradox that we have "rapid responses" to war but no such responses to the social upheavals that follow. Certainly the needs of instant shelter for the victims of war, oppression, or natural disaster should be high on our collective agenda. The Kosovo conflict and the recent devastating earthquakes in Turkey demonstrated the degree to which entire societies can be overwhelmed by sudden housing crises. But still we remain unprepared.

Architects, of course, are only part of the equation. But how do we break down the boundaries between the design professions, the politicians, and industry; between conscience, provocation, and action? Bucky reminds us, "In architecture, 'form' is a verb." Architects and industrialists must be encouraged to work hand in hand, marrying innovation and production. But it is a partnership that has to be forged by political will.

However we might allocate the responsibilities, we must be able to do better than the tent cities that fill the pages of our newspapers, let alone the Chalcos of the future. Bucky himself said, "The proper goal of the architect-engineer is purposeful." By that he meant forcing the pace, challenging accepted conventions or the intellectual status quo. Asia has shown us the "can do" mentality in action. It presages a global shift that we will all soon face more out of necessity than by choice.

Hong Kong International Airport at Chek Lap Kok is just one example. Rather than expand an overcrowded city airport, you commit to building a new one. And when there is no remaining land on which to build it, you create its own island. And then, when buildings are complete, you make the entire move from the old one to the new one overnight. The bravery of this thinking demonstrates the way forward. And if it can be applied to the epic scale of an airport, surely it can be focused on a solution to the problems of shelter that can arise at any time, almost anywhere in the world.

As early as 1938, in *Nine Chains to the Moon,* Bucky said, "What is a house?" He responded with an industrialised solution to housing provision. As ever, he backed words with deeds. He was a master of the art of "technology transfer," harnessing new industries to produce pioneering solutions to old problems. The Dymaxion Deployment Unit is just one example. Commissioned at the outbreak of the war by the British War Relief Association, it anticipated the bombing of British cities and the need for an emergency housing unit. Characteristically, Bucky looked outside the housing industry for manufacturing expertise and approached a company that specialized in making corrugated metal grain silos—the Butler Company of Kansas City—to build a prototype. He drew on the strengths of that industry but pushed it to achieve the sophistication and speed of manufacture he required. It is a lesson that we can still benefit from today.

Allied to his willingness to explore new techniques was a concern for economy of means. Bucky spoke frequently, for example, about the relationship between weight, energy, and performance—about "doing the most with the least"—and that has consistently been the story of technological progress, from the earliest cathedrals to the latest cellular phones.

I remember, in 1978, showing him our Sainsbury Centre for the Visual Arts and being startled when he asked: "How much does your building weigh?" The question was far from rhetorical. He was challenging us to discover how efficient it was; to identify how many tonnes of material enclosed what volume. We did not know the answer, but we worked it out and wrote to him. We learned from the exercise as he predicted we would. The basement, which is only 8 percent of the volume of the main space, weighs 80 percent of the total, or about 3,600 tonnes. The main building weighs just over 900 tonnes—

less per cubic foot than a Boeing 747—and was built far more quickly than the basement and for half the unit cost.

Back in the 1970s we made that calculation in simple volumetric terms. Today our understanding is far more sophisticated. We are familiar, for example, with concepts such as embodied energy and sustainability; and we know that some systems of construction are inherently more energy-efficient and environmentally responsible than others. Furthermore, there is a universal acceptance that the planet's natural resources are not only finite but fast dwindling. Bucky was one of the first people to advocate the recycling of source materials. He proposed that major manufactured items be rented from industry—cars for eight years, ships for twenty years, and so on. In this way, he argued, the recycling process could be guaranteed. Only recently have major manufacturers taken steps in this direction—the automotive industry is a prime example—and begun to plan for recycling in a systematic way.

The pressure to "do the most with the least," which has long been felt in the context of manufacturing industry, applies just as powerfully to energy production and consumption. Long after Bucky first warned us, we have at last recognised that we must break the pattern of energy profligacy and pollution. We now acknowledge the fragility of the natural world and the destructive impact of our industrial installations. We know, for example, that power stations that burn fossil fuels to produce electricity are inherently wasteful and environmentally damaging. It is estimated that half the energy expended to generate electricity is lost in the form of waste heat that is dissipated into rivers and oceans, harming their natural ecology. These same power plants also deposit into the atmosphere huge amounts of carbon dioxide (CO_2)—a greenhouse gas—which has been a significant factor in global climate change.

The planet cannot naturally absorb the millions of tons of pollutants we currently tip into its oceans or pour into its atmosphere every year. As an illustration of the scale of the problem, it is calculated that a square kilometre of dense deciduous forest absorbs through photosynthesis approximately 570 tonnes of CO_2 per year. To throw this into sharper relief, you only have to take one of the coming mega-cities and consider its likely CO_2 emissions from burning fossil fuels alone. For a conurbation of 25 million people to be CO_2-neutral—that is, to absorb all the CO_2 emissions from buildings, vehicles, and industry at present levels—it would need to plant a forest of 400,000 square metres, equivalent to 114,000 times the area of Central Park, or fifteen times the entire metropolitan area of New York. That is clearly impossible, so something fundamental must change.

Alternative energy sources have an important role to play. For example, if we were to produce all our energy by alternative means—by burning renewable fuels or by using wind and water turbines and solar panels—global CO_2 emissions could be reduced by approximately a third. This, allied with a proactive approach to energy conservation, begins to provide us with a solution.

I am reminded of Bucky's exhortation to "think global, act local." Within my own practice, we have made significant steps in the direction of reduced energy dependency in the design of a new generation of ecologically sensitive projects. Among the most recent is our proposal for the London headquarters of Swiss Re—one of the world's leading reinsurance companies—which will be the capital's first ecological high-rise building.

Swiss Re is rooted in the thinking that Bucky first explored with us in the theoretical Climatroffice project, designed in 1971. The Climatroffice concept suggested a new rapport between nature and workspace in which the garden setting helped to create an interior microclimate sheltered by the most energy-conscious enclosures. The ovoid forms employed were selected for their ability to enclose the maximum volume within the minimum surface skin—analogies might be drawn with the naturally efficient forms of birds' eggs—while conventional walls and roof were dissolved into a continuous skin of triangulated elements. Similarly, the Swiss Re building is derived from a circular plan which, over forty stories, generates an elongated, beehive-like form that is fully glazed around a diagonally braced structure.

Successive floors are rotated, allowing voids at the edge of each floor plate to combine in a series of spiralling atria or "sky gardens" which wind up around the perimeter of the building. Socially, these green spaces help to break down the internal scale of the building, while externally they add variety and life to its facades. They also represent a key component in regulating the building's internal climate. The building's aerodynamic form generates large pressure differentials that greatly assist the natural flow of incoming and expelled air. Fresh air is drawn in at every floor via horizontal slots in the cladding and circulated through the gardens. This system is designed to be so effective that for the majority of the year, mechanical cooling and ventilating systems will not be required. As a result, energy consumption is reduced dramatically when compared with conventionally air-conditioned offices.

The rebuilt Reichstag in Berlin is equally progressive. It demonstrates the potential for a virtually nonpolluting, wholly sustainable public building. It makes extensive use of natural light and ventilation, together with combined systems of cogeneration and heat recovery, and eschews fossil fuels in favour of renewable "bio-diesel"—a refined vegetable oil derived from grape or sunflower seeds. The energy strategy for the building ensures that the minimum energy achieves the maximum effect at the lowest cost in use. In fact, because its own requirements are sufficiently modest, the Reichstag is able to perform as a local power station, supplying neighboring buildings in the new parliamentary quarter.

Refined vegetable oil can be considered as a form of solar energy, since the sun's energy is stored in the plants (the biomass). Furthermore, CO_2 emissions are considerably reduced in the long term, as the growing plant absorbs almost as much CO_2 in its lifetime as is released during combustion.

Heating and cooling the Reichstag by burning bio-diesel produces an estimated 440 tonnes of CO_2 per annum as opposed to the 7,000 tonnes generated annually by its previous installations, installed in the 1960s—a 94 percent reduction in emissions. As a further illustration, if the Reichstag were to burn natural gas instead of bio-diesel, its CO_2 emissions would be in the region of 1,450 tonnes per annum—more than three times the bio-diesel amount.

As well as forming the public focus of the building, the Reichstag's cupola, or "lantern," provides the key to our strategies for lighting and ventilating the assembly chamber. At its heart is a light-reflecting cone—a light "sculptor" and a sculpture in its own right. The cone is covered with faceted mirrors that together form a giant Fresnel lens just as you might find in a searchlight or lighthouse. In fact, the cone works as a lighthouse in reverse, reflecting daylight from a 360-degree horizon down into the chamber. An electronically controlled mobile sunshade tracks the path of the sun to block solar gain and glare, but is designed to allow a little sunlight to dapple the floor of the chamber. In ventilation terms the cone and chamber together perform as a solar chimney, drawing air up naturally through the chamber and expelling it via the open top of the cupola.

In ecological terms, the Reichstag has shown how public buildings can challenge the status quo: big buildings do not have to be big consumers of energy or big polluters. And although it represents a minuscule first step in terms of the journey yet remaining, imagine the impact these strategies could have if they were applied more widely around the world. If every new building—public or private—were to follow this lead, the energy equation could be stood on its head. Rather than consuming energy, these buildings would be net providers; rather than emitting CO_2, they would be broadly neutral. The savings in resources and running costs could be immense.

The cupola is the outward manifestation of these strategies, signaling a process of transformation. It represents the ultimate synthesis of old and new in the building and brings together all the elements that compose our program of renewal. Interestingly, it also carries more than a hint of the geodesic Autonomous House—an energy-self-sufficient dwelling with a rotating, sunscreening inner skin—that we developed with Bucky shortly before his death.

For me, with its environmental and democratic agenda, the cupola is certainly more closely related to Bucky's humanist vision of the future than it is to the symbolism of the past. And, as Bucky would surely want to know, its steel structure weights just 800 tonnes.

We Call It "Earth" from *Nine Chains to the Moon*

Of one planet, the earth, in one little star system (the sun's) in one relatively small galaxy, we know a little: its superficial geography, measurements, conditions and processes.

Of the nine planets in our solar system, the earth is the third nearest the sun, being 92 million miles distant from it. Mercury is nearest, with a minimum distance from the sun of 28 million miles, and Pluto is the farthest, being distant 3 billion 800 million miles. The ratio of distance from the sun to axial revolutions, or days, *per annum* is approximately of inverse proportion for all planets. Mercury has 88 days in a year, the earth 365, and Pluto 90,500.

We know that the earth is the densest body of the sun group, including the sun itself. If we were to call the density of the earth 1, Venus would be .88, the moon .60, and the sun .26.

The diameter of the earth is twice that of the smallest planet, Mercury, but is only one tenth that of Jupiter, the largest planet, the diameter of which, in turn, is but one tenth that of the sun. (The sun itself is a relatively small star, a new one having been discovered and measured in '37–'38 so large as almost to equal that of the whole solar system.)

Three fourths of the earth's surface is covered with a layer of moisture that is relatively thin (⅒ of 1%), its greatest depth being only 35,410 feet in the Mindinao Deep between the Philippine Islands and Japan, as compared to the earth's 8000 mile (42,000,000 ft.) diameter.

The remaining quarter of the surface of the earth consists of dry land concentrated within a relatively small sector. Indeed, one may so revolve a globular replica of the earth that 85% of all dry land is visible from one perspective point. When so revolved, there appears a "land hemisphere" in which the north pole is approximately one eighth of the way down from the top center. The center of the "land hemisphere" is the Spanish Riviera. In this "land hemisphere" two main continental bodies are apparent. One comprises all of Africa, Europe and Asia, penetrated by a small canal (the Mediterranean and the Red Sea); the other consists of the Americas and Greenland. These two great bodies are joined at their upper limits by Alaska and northern Siberia, with the Aleutian Islands re-enforcing the juncture. Bering Strait is scarcely discernible.

The Equator appears as a draped line girdling the globe one quarter of the distance between the bottom and the top of this view of the earth. It transits the center of Africa and what can be seen of South America. Above the line designating the Equator lies 85% of all the earth's dry land.

The "town plan" of an architect intent on devising a universal shelter service design is a fairly concentrated affair so far as the earth is concerned. This is emphasized by a study of population concentrations upon the earth's surface. Of the 2⅓ billion people currently on this earth-globe, only 13 million, or approximately ½ of 1%, are in the non-visible area of our "town plan." No teleologic designer, in view of the current world integration, can profess concern with building only within the "town plan" of Podunk when the materials, structures and tools he uses are so obviously derived from the entire surface of the earth. It is a different story from an early New England settler doing the best he could with the material at hand quarrying for himself a bit of granite for shelter construction. That was *architecture,* for he did the most with the least out of the available materials and tools. We cannot claim that we are doing the most with the least without carefully referring to our cosmic inventory and assertaining what is now most suitable and available.

Although the earth's land and water surface is protected by a blanket of atmospheric gas, which is frictionally cohesive to the earth in its fast rotation, the differential of speed of the earth's revolutions to that of the air and water produces a constant rotational current, in both its air and partial water covering, in a general direction of west to east. Scientifically, this is explainable as a slight rotational lag of the earth ball within its surface films of more mutably drawn liquid and gaseous elements,—drawn tidally by the electrical pull of the moon and possibly of the sun.

Were it not for the dry land projections into the gaseous and liquid films, these apparent currents would probably be true west to east currents. However, the continental projections, which, like three fingers, extend down from the north pole, set up turbulent back eddies in both the liquid and gaseous films around their southern extremities. Thus, warm equatorial waters are catch-basined into S-shaped depressions between the two main continental bodies, where they swirl about in such manner as to cause great vagaries in temperature in relation to the earth's theoretical parallels of latitude. These thermal conditions have a direct bearing on the areas favorable to human survival because the water content of the human is approximately 9 to 1, and water freezes at 32° F.

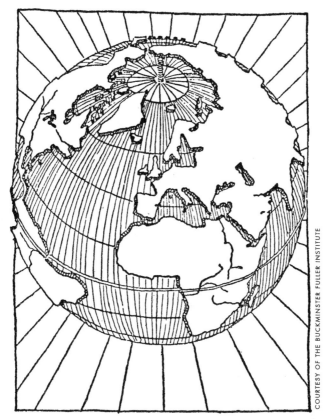

An SSA graph of universal architecture's prime "town limits" for an industrially emancipated human community.

The west-east currents of the gaseous film are further interrupted by great mountain ranges on the windward slopes of both main continental bodies. Molten snow and ice, flowing down the windward side of these ranges, have caused small alluvial plains to form like a shelf along their windward base. This windward lip is so slight as in no way to alter a general cross-sectional contour of the continent, similar to a cross-section of an aeroplane wing foil, as the continent tapers from its western mountainous edge to its eastward leeward flat-lands.

Whether or not there is significance in the coincidental streamline form of the continents (possibly so formed by air and oceans of yore passing over them from west to east), the fact remains that these westerly continental lead-edges cause a peculiar disturbance in their wake, to the east and over the hinterland. This still further affects the isotherm of average temperature, areas of moisture precipitation, and man-growth abetment conditions that must be heeded by the teleologic shelter designer.

The isotherm, or abstract temperature belt or zone, of an annual average variation of 48° F. and 32 ° F. mean low and 72° mean high, swirls from Alaska down the coast to Vancouver, B.C., thence to lower Kansas, after which it rises gradually to Lake Erie, Boston, Newfoundland, South Greenland, and mid-northern Russia. Then it veers back to and down through the Scandinavian Peninsula, centrally through Europe to the Black Sea, and, finally, passing through the Caspian Sea, Turkey, Persia, northern India and central China, rises again to traverse Japan and follow the coast of Siberia back to the southern tip of Alaska. *The significance of this isotherm is that it coincides with the central line of concentration of man population.*

One and one half billion or 70% of the total 2⅓ billion world population resides along this 48° range of temperature isotherm. This zone has an average rainfall of 40 inches annually and an average constant wind speed of 15 m.p.h. We are certainly getting down to specific conditions for the teleologic dwelling designer.

We now submit a new world map more suitable for our teleologist than the "land hemisphere" previously sketched. It centers on the North Pole and in it the whole dry land of the earth may be seen to be ONE CONTINENT instead of two as shown in the first sketch. There is one continent similar to a 3-bladed propeller with the hub at the North Pole. The winding dotted line is that of our population isotherm.

It will be noted from the following table that, if man were to be deployed over the whole surface of dry land, there would be but 40 persons to a square mile. At this rate, there would quite evidently be ample room on earth for man for a long time to come, this density being but one tenth that of the British Isles, or Rhode Island.

TABLE 1

Continent	Sq. Mile (Millions)	Population (Millions)	Population (per Sq. Mile)
North America	8.50	180.0	21
South America	6.8	81.5	12
Europe	3.7	550.0	149
Asia	17.0	1,155.0	67
Africa	11.5	150.0	13
All Other Land	10.0	197.0	—
Total	57.5	2,313.5	40 (average)

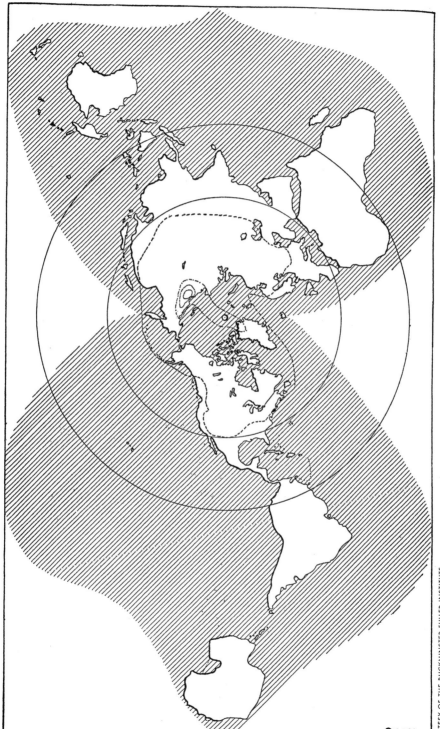

© 8-1-34
B. FULLER

Actually, however, population is highly concentrated in certain portions of the land area and sparsely existent in others. For instance, in Java the concentration is 804 to the square mile.

TABLE 2

| | Comparative Special Population Densities | | |
Country	Sq. Mile (Millions)	Population (Millions)	Population (per Sq. Mile)
Japan	.148	64.0	432
Australia	3.000	6.6	2
British Isles	.120	49.0	409
Java	.051	41.0	804

Of the dry land but approximately one half is arable and the population, were man theoretically deployed over all the pleasantly livable and arable land, would total 80 persons per square mile.

Reducing these figures further to comprehensible areas, on the basis of 80 inhabitants to the square mile of livable land, there would be 16 pleasant acres of land for each and every man, woman and child. However, the family unit is currently five and so, if mankind were completely deployed in family units over the "livable" dry land area, there would be 80 acres per family, or an area so large that, unless the family shelters were on hilltops, man need not be aware of the existence of others beyond the members of his immediate family group.

Man has evolved two unit measures of a mile: "statute" and "nautical."

The statute mile represents the earth's surface equivalent in feet to a longitudinal minute at the latitude of Greenwich, England,—Greenwich being, also, the starting point for the longitudinal reckoning of standard sun time. This statute or "legal" English mile was arbitrarily arrived at by England, and, with the zero longitude, was imposed by her on the world by means of her commercially dominant position as Queen of the Seas during the sailing era. It has been perpetuated by habit and the "necessity" of property title continuity. It must be irritating for the Japanese sailor-man always to find himself, as it were, in Row Z of the audience.

This lawyer's or "statute" mile was a "flop" at sea. The mathematical navigator had to have something much more actual and reliable to go to sea on. So he evolved the nautical mile, which is 6,080.20 feet in length and is equal to one sixtieth of a degree of a great circle of a sphere whose surface is equal in area to the area of the surface of the earth. This serves as an identical unit of measure on any great circle with but negligible error.

The nautical mile, which is 1.152 "statute" miles, was more scientifically determined, and is, therefore, more universally utilizable.

Although it may seem to be a digression at this point, it is nevertheless vitally significant to the comprehension of our book to state that a RATE of speed of *one nautical mile per hour* is known, at sea, as one *knot*. The name derives from the knots in the chip log line thrown over the stern of the old sailing ship whose run-out count, by means of a sand glass, determined the vessel's speed. It is improper, accordingly, to say "one knot per hour," which is like saying "one mile an hour per hour." The sailor says we are "making five knots at the present moment," or "we have covered ten nautical miles." We have no such speed or rate word relative to land miles, for land thinking has been relatively static and there has been no necessity for such a mobility word. On land, one measures in miles *or* hours. This has occasioned much mental confusion, in parlor talk about relativity, over the really very evident, from a rate viewpoint, integration of time and space. Among the first people on land to offer a rate word were the electrical engineers who evolved the DYNE (of the dynamo, dynamite, and dynamic family).

DYNE is a measure of force which, acting on a gram for 1 second, imparts to it a velocity of 1 centimeter per second. Result = 1 erg = 118 feet an hour. 1 dyne = the additional force exerted by a clock 30 ft. in diameter to carry a fly weighing $\frac{1}{453}$ lb. on the tip of its "minute" hand for one second. The exact additional *work* done by the clock to carry the fly completely around its circumference on the tip of the big hand would be an ERG hour. Three billionths of 1¢ is the hydraulic power cost of this erg-hour fly transportation. This is typical of the currently fine degree of appraisal of work costs, through these *rate* integrations.

The layman knows little about this except that he must pay for his electric light at some obscure rate, but the power sellers were quick to use it in their exploitation of this scientific phenomenon.

The earth's circumference at the equator is 24,903 nautical miles, and its total surface is approximately 197 million square miles, of which 57½ million are dry land. As there are 33 million square feet to a square nautical mile, or 189 billion cubic feet to a cubic mile, the earth has a volume of 260 billion cubic miles, and weighs 6½ thousand billion billion tons.

How does man stack up in size with all these volumes, weights and measures?

If all the earth's 2⅓ billion people were to stand on one another's heads, they would form a chain 2,423,000 statute miles long, or nine complete chains to the moon; that is, they would reach to the moon and back four and one-half times. It would require only 50 such chains to reach the sun. Man is, therefore, empowered to a sense of *personal* contact with all astronomical bodies of the universe in addition to his earth.

Were all the members of the human family to gather together in one spot with a density equivalent to that of people jammed in a New York subway car aisle, they would cover an area of 139 square miles, which is approximately that of the Virgin Islands or Bermuda. Compare this with the 31,820 square miles of Lake Superior, the 121,000 square miles of the British Isles, and the 472,000 square miles of Hudson Bay!

The whole of our present human family—only one-ninth of which would be required, standing on one another's shoulders, to reach to the not-so-far-away moon, could readily be housed overnight on the little speck of earth known as New York City, if the floor space of all the buildings were utilized for the purpose, although, of course, they could not be serviced with the city's present facilities.

There are 10 billion cubic feet of people on earth (⅟₁₉ of a cubic mile). They weigh 115 million tons and would fill 111 Empire State Buildings. Yet if put under a gigantic hydraulic wine press, so that all the water and gas might be squeezed out of them, they could be compressed into one Empire State Building.

There are approximately 50 Panama Canals to a cubic mile and there are 317 MILLION cubic miles of ocean. The "nine chains to the moon" would make but a small splash in those 317 million cubic miles of water, which, however, the moon lifts tidally twice daily from a few inches to fifty feet. This earth-moon force, compared with the minute muscle power of the human army, indicates well the vast excess of man's POTENTIAL mind power over his brawn potential, for he already CONTEMPLATES harnessing at least a portion of this vast earth-moon gravitational force.

The 115 million tons of people on the earth have an annual turnover, by birth, growth and death, of approximately 2⅓ million tons, with a net one million tons increase. This weight increase of human flesh and bones is equal to 14 S/S *Normandies* (the total weight of man being equal to 1400 *Normandies* or 287 Empire State Buildings). Although new human units weigh in at only 25,000 tons, they are increased by the conversion of energy from the sun and other stars (directly or indirectly, but from no other source) to twenty times their original weighing-in weight by the time they reach maturity.

The source of power for the operation of all man's instruments, inanimate as well as animate, is sun and other star energy, direct or indirect, primarily through latent storage depots of multitudinous forms. All people are nurtured and energized by the ultraviolet and gamma rays, as well as by most powerfully penetrating, highly energized cosmic rays. The latter are responsible, apparently, for all electrical polarity changes in integral-with-life mechanisms upon the earth's surface, these polarity changes or ionizations, in turn, sponsoring the birth of all new species in plants, known as "sports." The laws of chance, change and animate evolution are here involved.

Scientific shelter design, therefore, is linked to the stars far more directly than to the earth. STAR-GAZING? Admittedly. But it is essential to accentuate the real source of energy and change in contrast to the emphasis that has always been placed on keeping man "down to earth." The teleologic dwelling designer MUST visualize his little shelters upon the minutely thin dust surface of the earth-ball, dust which is a composite of inert rock erosion, star dust, and vegetable compost, all direct star (sun) energy resultants.

Man, living in shelters scattered over the earth's dust film and energized and nurtured by the stars, has evolved, through self-research and the comprehension of the dynamics of his own mechanism, the phantom captain's extension mechanisms. Through the leverage gained by his INANIMATE INSTRUMENT EXTENSIONS OF SELF, he has attained an extended mechanical ability far in excess of his own integral mechanical and energy content ability.

Utilizing sun-energized "fire" to work his metals and exercising intelligent selection and dynamic experience, man has harnessed inanimate power (of sun-star origin) to operate his extended depersonalized mechanisms. Thus he has "set" his environment under increasing control by the ceaseless operation of his depersonalized inanimate-powered mechanisms and, concomitantly, has broken down the original limitations of time and space (days, nights and seasons). We repeat: All of this has been done through radiant energy from the stars.

By means of his harnessed inanimate servant, power, and his extended mechanisms, man has now explored, measured, and "set" under control much of his earth's crust and his once-"outside" universe, entirely *despite* the inertia of vanities, superstitions, exploitation, humpty dumpty moralities, laws and destructive selfishness. He has flown in his imagination-conceived, intelligence-wrought, de-selfed mechanisms at 72,000 feet above the earth's surface, almost three times the height of the earth's highest mountain, and sixty times higher than the Empire State Building. Yet this is an insignificant feat compared with flights and heights to be attained in the NOT FAR AHEAD "NOW," in new intelligence-to-be-wrought mechanisms of flight.

Most extraordinary of all man's extension activities—and far superior to his extension physically into his physical universe by physical means—is his mental extension, on the basis of observations of the dynamic progressions involved in his tangible mechanisms, inferring progression continuity beyond the tangible bands, into an AWARENESS OF and EXPERIENCE IN the ABSTRACT. I do not mean the abstract of humpty dumpty, academic philosophizing or mysticism, but the mathematically rationalizable abstraction of electrical phenomena representing the 66 octaves or bands of radiation which he has discovered despite their non-sensorial nature. Not only has he explored much of the realm of RADIATION, but he is using it and ACTUALLY THINKING IN IT. The phantom captain's extension into participation in events of exterior mechanism occurrence has provided an "actual" sense in the realm of radio in our younger men.

In this sense-extension into radiation lies the promise of man's eventually understanding all the secrets of life-in-time, which, down through the ages, have evoked an intuitive, mystical and superstitious awe. Miracles, once irrational, will be continually rationalized and set under service to man by man.

Shoji Sadao was a student of Bucky's at Cornell University and worked with him on early geodesic domes. He was project architect on Fuller's Expo 67 geodesic dome, the United States World's Fair pavilion in Montreal, Canada. He also worked on the Dymaxion Airocean World Map that is currently used by World Game.

Bucky Fuller and Shoji Sadao established their architectural association first in Boston and continued it later in New York City. During the late 1960s, in New York's Long Island City, Sadao served as architect and planner with Bucky's lifelong friend Isamu Noguchi for Noguchi's sculpture and garden commissions. In the 1970s, R. Buckminster Fuller, Shoji Sadao, and Thomas T. K. Zung merged and formed their current architectural partnership with offices in Ohio and New York, with Fuller maintaining architectural registrations in both states.

Currently Sadao is Executive Director of the Isamu Noguchi Museum and Gardens and Foundation in Long Island City, New York.

A Brief History of Geodesic Domes

Shoji Sadao

The genesis of the geodesic dome can be attributed to Fuller's ambitious quest to find order in the universe. Fuller was convinced that the Cartesian, orthogonal view of the world was fundamentally inaccurate. His search for, as he put it, "nature's own coordinate system" led him on a lifetime journey of exploration of structure and process, one of whose paths led to his development of a three-way spherical grid and the invention of the geodesic dome. The geodesic dome, an icon of avant-garde architecture of the 1950s and 1960s, was issued U.S. Patent No. 2,682,235 on June 29, 1954. This brief history will attempt to familiarize the reader with Fuller's intellectual exploration and show how it led to the invention of the geodesic dome.

Despite, or possibly because of, his exposure to Harvard University (he dropped out twice), Fuller was an autodidact. He learned by thinking and acting on his own initiative. He was a great believer in what the individual human being, alone in the universe, could accomplish on his own without benefit of corporate largesse. After Harvard and a short stint in the U.S. Navy during World War I, he joined his father-in-law, architect James Monroe Hewlett, in the Stockdale Building Company, which manufactured and erected buildings using a block consisting of fibrous material, such as excelsior, and bonding it with magnesium oxychloride cement. Between 1922 and 1927, 240 buildings were erected. This exposure to the craft-oriented building industry with its inefficient use of materials and nonscientific analysis of the physical parameters affecting building design left an indelible impression on his mind that he was to refer to repeatedly in his criticism of architectural design and the piecemeal practices of the building industry.

Convinced of the futility of trying to solve the problem of buildings and providing shelter by conventional means, Fuller embarked in 1927 on what was to become his lifelong involvement with structures at the micro, macro, and human scale. His early search for space enclosures based on a more rational analysis of the forces of nature impinging on a structure led him to develop a rigorously outlined program of requirements. This list, which he called "Universal Requirements of a Dwelling Advantage—Teleological Schedule—A Checklist of Universal Design Requirements of a Scientific Dwelling Facility," succinctly stated, in tabular form, what he envisioned as essential to shelter design. Based on these precepts he designed the Dymaxion house in 1929.

The Dymaxion house was a radical, pole-suspended structure quite dramatic in appearance with many revolutionary features. The structure was to be made of aluminum; the mast was to house lenses to concentrate the heat and light of the sun and to direct it where needed; bathroom fixtures were to be manufactured in toto at the factory and merely hung in place (prefabricated); the floor deck was to be two layers of post-tensioned cable with pneumatic pillows sandwiched between and topped with solid decking.

Paralleling these developments was Fuller's utopian vision of integrating man's socioeconomic activities into a comprehensive Design Science that would deal effectively with the efficient and fair distribution of the world's resources. His 1927 "4D Time Lock" drawing (Fig. 1) illustrating the "one-town" Airocean World is the earliest example of his attempt to convey his concepts in graphic terms. It is a "moon's eye view" of the earth. This drawing does not deal with the fundamental problem of cartography—the representation of a three-dimensional surface on a two-dimensional plane with the minimum amount of distortion. However, by 1934 he is deep into this cartographic problem, as shown in a map (Fig. 2) which was reproduced in his book *Nine Chains to the Moon.*

His interest in cartography was the result of his dissatisfaction with existing world projections that distorted the size and shape of the continents, particularly the then-ubiquitous Mercator projection, which distorted the

Fig. 1

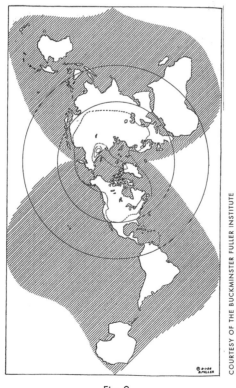

Fig. 2

landmasses near the north and south poles. He reasoned that anyone making an assessment of the world's resources would need an accurate world projection that portrayed the relative size and shape of the continental landmasses with minimum distortion. His investigation of this problem led to his filing with the U.S. Patent Office, in 1944, Patent No. 2,393,676, which was granted in 1946, the first patent issued on cartography in 150 years. In this patent he described the method of projecting data from the surface of the sphere (the earth) onto a planar surface (the two-dimensional map). The polyhedron he chose to demonstrate this transformation was the Dymaxion (cuboctahedron), consisting of six square faces and eight equilateral triangular faces. The spherical squares were subdivided by a two-way great-circle grid which, when transformed to two dimensions, became your familiar planar square with a rectilinear 90° grid (Fig. 3). The spherical triangles were subdivided by a three-way great-circle grid which when transformed to two dimensions became a planar equilateral triangle with a skewed three-way grid (Fig. 3). This three-way triangular grid, which Bucky chose to refer to as the "Regular grid," was his first geodesic grid. It was generated by constructing perpendiculars to the edges of the triangle from regular subdivisions along its edges (Fig. 4).

Before commencing the discussion of geodesic geometry it may be worthwhile to review a few basic points. Most readers will already be familiar with the five Platonic solids: the tetrahedron, the cube, the octahedron, the dodecahedron, and the icosahedron. These five regular solids were known to the ancient world and are treated in Euclid's great work, the *Elements*. Each is formed from regular figures, i.e., all the edges and face angles in each solid are identical. Geodesic geometry is the three-way gridding of a sphere using the spherical form of these solids (all vertices and edges lie on a common sphere)

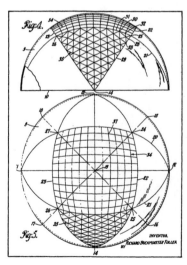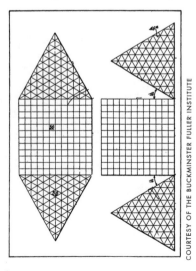

Fig. 3

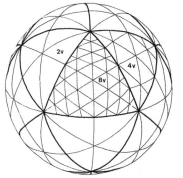

Fig. 4. Regular grid, first geodesic grid used for dome construction. Spherical icosahedron shown in bold lines. Numbers indicate frequency of modular subdivision of the icosa edge.

as the point of departure. It should be mentioned here that a three-way grid was selected for its inherent stability by virtue of its being omnitriangulated. A triangle is a structurally stable element independent of size. Squares, pentagons, hexagons, etc. are not stable configurations. Most of these spherical solids have been used in developing special-case applications, but the icosahedral grid (the largest number of identical faces: twenty) was used most frequently in the design of geodesic domes.

Fuller's early investigations yielded what now seems to be a cumbersome three-way gridding of great-circle arcs, the thirty-one-great-circle grid (Fig. 5). It was generated by successively spinning the spherical icosahedron on axes through vertices, mid-face and mid-edge. The breakdown is as follows:

6 great circles from spinning using the 12 vertices
10 great circles from spinning using the mid-face of the 20 faccs
15 *great circles from spinning using the mid-edge of the 30 edges*
31 great circles

This geometry was used to build one of the first geometric structures, the 48-foot venetian-blind dome at Black Mountain College. This dome was built from 2-inch-wide venetian-blind material. A second dome using the same geometry, the necklace structure, was made of sturdier tubular elements with a continuous internal cable net and was erected in the Pentagon Garden, Washington, D.C., in February 1949.

The thirty-one-great-circle grid's major drawback was the large difference in length between the longest and shortest member (2:1), which led to great inefficiencies in its design. A more efficient grid was required if geodesic domes were to become a viable alternative to other structural systems. Several structures were constructed using the "Regular grid" described earlier. They

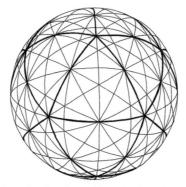

Fig. 5. Thirty-one-great-circle grid with spherical icosahedron shown in bold lines. Rotation on axes through midpoints of edges define fifteen great circles; on illustration, bold lines of icosahedron and their extension to midpoint of opposite edge. Rotation on axes through vertices define six equatorial great circles that do not pass through any vertices; on illustration, lines connecting midpoints of icosahedron edges. Rotation on axes through faces define ten great circles that do not pass through any vertices; remaining set of lines on illustration.

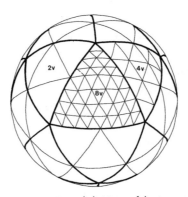

Fig. 6. Alternate grid showing successive subdivisions of the icosa triangle. Spherical icosahedron shown in bold lines. Numbers indicate frequency of modular subdivision of the icosa edge.

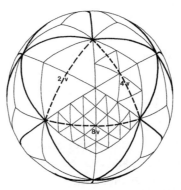

Fig. 7. Triacon grid discovered by Duncan Stuart in the spring of 1951. Spherical icosahedron shown in bold lines. Numbers indicate frequency of modular subdivision of the icosa edge.

appear in the August 1951 issue of *Architectural Forum*—an aluminum dome built by Jeffrey Lindsay and a wooden geodesic frame by Zane Yost. Another variation that was developed was the "Alternate grid," whereby the basic icosa triangle and each successively formed triangle was divided at mid-edge and joined to congruent points on adjacent edges (Fig. 6). This geometry yielded a simpler mathematical routine for calculating the grid and concomitantly generated fewer types of parts than the Regular grid.

The real breakthrough in geodesic grids was discovered by Duncan Stuart in the spring of 1951 with the development of the "Triacon grid." He was a professor at North Carolina State College in Raleigh, North Carolina, where Fuller had one of his two offices of Geodesics, Inc. (the other office was in Cambridge, Massachusetts). Stuart was the house mathematician at Geodesics, Inc., and was an invaluable contributor to many of the projects going through the office. His discovery of the Triacon grid, like many discoveries, was a serendipitous occurrence. There had always been a nagging problem with the Regular grid—it had windows. That is to say, at certain vertices the three great circles did not go through a common point. Calculations were checked and rechecked, but the windows persisted to the point where Fuller thought it was a message from God that the trigonometric tables had an error in them! A copy of *Tables of Sines and Cosines to Fifteen Decimal Places at Hundredths of a Degree*, published by the U.S. Department of Commerce, National Bureau of Standards, was used, to no avail. Stuart, who taught at the School of Design at North Carolina State, was in addition to being a painter an accomplished mathematician. Fuller had many discussions with him regarding this problem, and Stuart's brilliant solution was the Triacon grid, a simple yet elegant solution reducing the number of types of parts and keeping the difference between longest and shortest members to a minimum. What Stuart did was to use the spherical diamond generated by the thirty-one-great-circle grid rather than the spherical icosahedron as the basic element to be subdivided (Fig. 7). By subdividing the icosa edge, which is the long axis of the diamond, he was able to generate a grid with no windows. This grid was used on almost all subsequent large geodesic domes designed by Geodesics, Inc., in Raleigh.

The office of Geodesics, Inc., in Cambridge, Massachusetts, discovered another significant grid that was developed in response to the problem of the base condition of domes. The Cambridge office was working at that time with Western Electric Company to develop a series of fiberglass-reinforced plastic radomes for the Arctic DEW line (Distant Early Warning line) ranging from 30 to 55 feet in diameter. To enclose the rotating radar antenna, a portion of the sphere larger than a hemisphere was required. If one used any of the grids mentioned previously, the intersection of the grid with a base plane below the equator created base vertices that did not fall on the grid.

Special-length struts and special triangles are generated that increase the number of types of parts to be fabricated and hence the cost of the structure. Bill Wainwright of the Cambridge office discovered that for three-, four-, and five-frequency alternate grids, vertices could be plotted such that the lesser-circle base truncation of the sphere could be accommodated within the new geometry. This geometry was named the "truncatable" or "parallel" grid (Fig. 8).

The 1953 Ford Rotunda dome (Fig. 9), a 93-foot-diameter, 8½-ton aluminum space frame structure, was Fuller's first major commission. Structurally, it was an impressive demonstration of the lightweight, high-tech construction philosophy Fuller had been espousing for twenty-five years. But the problem of finding a watertight skin for this structure, and for that matter all subsequent structures, had to wait for sealant technology to catch up with the complicated demands of these multifaceted, multijointed structures.

In 1958 the Union Tank Car Company ordered a 384-foot-diameter geodesic dome (Fig. 10), the largest clear-span structure of its time. It was constructed at its Baton Rouge facility. This structure solved its waterproofing problem by having an all-welded 11-gauge sheet steel skin suspended from its exoskeleton space frame.

Probably the best-known geodesic structure is the United States Pavilion in Montreal, Canada, designed for Expo 67 (Fig. 11). This 250-foot-diameter

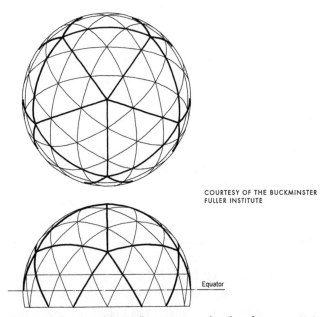

COURTESY OF THE BUCKMINSTER
FULLER INSTITUTE

Equator

Fig. 8. Parallel (Truncatable) grid discovered by William Wainwright. Three-frequency (3v) plan and elevation shown of 6/10 sphere. Spherical icosahedron shown in bold lines. Four-frequency (4v) and five-frequency (5v) cases also exist.

Fig. 9. Ford Rotunda Building geodesic dome skylight. View of ceiling showing workmen and delicate tracery of octet truss. Dome spanned 93 feet and weighed 8.5 tons. Completed June 1953.

Fig. 10. Union Tank Car Company geodesic dome, Baton Rouge, Louisiana, October 1958, was the largest clear-span structure ever built at that time.

diaphanous, silvery sphere caught the imagination of all who visited Expo and became the symbolic icon of all subsequent world's fairs and visionary urban construction. Every Expo after 1967 had its spherical exhibition structure; every city of the future had its spherical building prominently positioned in its urban fabric.

Geodesic domes are but one aspect of the multifaceted "random element" that Fuller often compared himself to. His contribution to molecular structures was acknowledged in 1985 when the stable C_{60} molecule (derived from the 3 frequency hex-pent configuration of the icosahedron) discovered by Nobel laureates Kroto, Curl, and Smalley, was named "buckminsterfullerene," his fifty-odd years of structural exploration has gained solid scientific accreditation. Geodesic domes are here to stay. Their patents having long expired, the system is in the public domain for all to use. This is what Fuller wished. He often referred to his role as one of providing the instruments with which others could play lovely music. It is now up to mankind to make the most of this exquisite instrument.

Fig. 11. United States Pavilion (250-foot-diameter steel pipe and acrylic panels), Expo 67, Montreal, Canada: R. Buckminster Fuller/Fuller and Sadao, Inc./Geometric Inc. Associated Architects.

Geodesic Structures from *The Dymaxion World of Buckminster Fuller*

The Octet Truss

It was indicated earlier that the Vector Equilibrium could be subdivided into tetrahedrons (four-sided pyramids) and octahedrons (eight-sided "solids"); actually it is composed of eight tetrahedrons and six half octahedrons.

A complex of Vector Equilibriums joined together form a matrix of alternating tetrahedrons and octahedrons. Such structures form what Fuller calls the Octet Truss. A frame built of tetrahedron-octahedron combinations provides an omnidirectional and equal dispersion of load pressures, with no member of the truss duplicating the function of any other. For this reason the truss has an enormous load-carrying ability; and its strength to weight ratio increases as the truss grows in size.

In 1953, at the University of Michigan, Fuller load tested an Octet Truss made of 170 slim 33" aluminum struts, each weighing one-third of a pound. The entire truss, when riveted, weighed 65 pounds. This frame, no heavier than an ordinary canoe, supported a total load of six tons, the weight of a small army tank.

Fuller himself did not expect such a performance from the tetrahedron-octahedron combination. It was a surprise—the behavior of a whole not predicted by its parts. Describing the Octet Truss in a letter to his patent attorney, Donald W. Robertson, Fuller wrote, almost apologetically, "I am sorry that my whole family of inventions tends, by rational acceleration, to sneak up on you and press you for attention. But isn't this the nature of invention? Invention is always a surprise."

Tensegrity

The Wichita Dymaxion House had been designed to be delivered across the Pacific in a single DC-4. The 1927 house was to be delivered by dirigible. The passing of two decades had been marked by such improvements in technology that shelter delivery was now possible by heavier-than-air aircraft.

After the Wichita house, Fuller concentrated on the problem of air delivery. He had never departed from his 1927 4D assumption that the air is our ultimate ocean, and that man's "mobilizing, recirculating, design-regenerating technology" will eventually evolve into gossamer. But evolution toward gossamer depends on radical weight reduction; a spider's web can float in hurricanes only because of its high strength-to-weight ratio. For new design strategies aimed at radical weight reductions and strength intensifications, Fuller once again scanned the premises of his Energetic Geometry; he explored possibilities of intertwining the geometry with the inventory of war-developed technical advances.

In the planned 1927 4D house, Fuller had minimized weight by separating compression members from tension members. The central mast was a compression unit, around which hung a multiple-rimmed tensionally-cohered, horizontal wire-wheel house structure. Guy wires supporting the mast provided the balancing tension. In developing the Wichita house, however, he discovered that as he increased the diameter of the mast-and-guy-wire complex, the over-all mast complex weight grew less. And ultimately, at its dimension of least weight, the mast complex structure was congruent with the outside shell of the house.

When this "congruent phase" had been reached, the inner wall of the shell (the "mast" complex) would be in compression, the outer structure would be in tension. Although to the viewer there would be no visible separation of compression and tension elements, there was nevertheless a universal, comprehensive tension system in operation; this system laced the entire structure into a single, finite, energetic embrace.

The universal comprehensive tension system could be interspersed locally with islands of compression, in the form of struts, in such a manner that the islanded compression struts would not touch one another. Yet these struts would force the tension network into outward patterning from the center of the total structural system in precisely the same way that molecules of gas inside a balloon press the balloon bag outwardly from its center.

Fuller saw that the gas molecules in balloons were not exploding in a radial pattern from the center of the system, but were bouncing around the inside circumference of the balloon, as sounds bounce around the wall of a circular structure. But if the skin of a plastic balloon is viewed with a microscope, it is found to be full of holes. Therefore it was clear to him that an accurate description of a balloon is a "network"—but one in which the holes in the network are smaller than the molecules of gas. These molecules, coursing independently of one another like so many herrings inside a weir, impinge upon the weir net repeatedly, thus forcing it to balloon outwardly. Thus the action is not the result of a consolidated, group effort of the herrings in a shoulder-to-shoulder radial attack outwardly, in all directions, against the net, but rather of the high frequency ricocheting impingements of each herring.

This theoretical consideration of balloons and fish nets, herrings and molecules, suggested that the comprehensive tension network of his structural system could be patterned in such a manner that the individual compression struts would not touch one another, yet would hold the tension network outwardly in firm spherical patterning. That is, Fuller saw that he might be inventing a spherical building in which the bricks, or compression members, did not touch one another. Thus there would be a spherical building of bricks, in which the bricks would be interlaced with "rubber bands"; each brick would be in effect restrained from escaping from the pattern only by the rubber bands for no brick would be in direct contact with any other brick.

Fuller later concluded, after he had developed and successfully demonstrated a variety of discontinuous-compression, continuous-tension structures, that it was only the habitual tendency to think of all forms of matter in terms of brick-on-brick structuring that led to the assumption that the structure of the atom's nucleus could not be represented by a model—"even though the nuclear physicists had discovered certain geometric system pattern relationships with respect to the nuclear coherence."

Fuller called this special discontinuous-compression, continuous-tension system the *Tensegrity*.

What is startling about the conception is its pertinence to fields which ordinarily seem to be unrelated. Tensegrity supplied a generalized approach to the most economic forms of "man-occupiable" structures. And again, as nuclear physicists have suggested, it might provide in fact a true model of the atom's nuclear structure.

To understand how his compression struts could be successfully islanded from one another while thrusting the net outward, it is only necessary to think of a large number of pairs of live herrings, with the members of each pair so close to each other that the two appear as a unit. Each of these unit-couples are approximately evenly spaced away from the other couples but all of these evenly dispersed couples are within a complete spherical fishnet dropped into the ocean by a trawler (the neck of the net has closed after the herring have swum inside and the connecting line to the trawler has been inadvertently severed).

Imagine the herring pairs setting up a patterned herring dance; each member of each pair takes a position facing away from the other; then swims away from its partner, and continues in a straight line until it strikes the net—even if only with a glancing blow—thus pushing the net outward. After making a racing swimmer's turn, each herring races swiftly in a straight line back again to its mate, joins the mate momentarily, and then repeats this out-to-the-net-and-back linear darting, over and over again. Thus, we have a piscine ballet pushing the net outwardly in all directions.

Let us substitute for each pair of herrings, one round rod, whose two ends represent the two members of the couple; and arrange a pattern of these rods, acting as chords within a sphere, pushing at an acute angle in the opposite directions against the net in such a manner that the sum total of chordal patterns provides an omni-triangulated grid wherein the point of impingement of one rod is congruent with the mid arc of the chorded action of the next rod. It will be seen that such triangulated outward-pushing can be independently accomplished by the positive and negative chordal impingements on the net; yet the chords' ends will not be in continuous array.

This Tensegrity network principle could also be demonstrated in a linear manner—as Fuller, enlightened by a linear Tensegrity discovery of his student colleague, Kenneth Snelson, showed by developing a series of Tensegrity masts. From 1949 to 1952 his Tensegrity masts were exhibited on the campuses of many universities, including the Massachusetts Institute of Technology, the University of Oregon, the University of Michigan, and North Carolina State College.

The Tensegrity principle in its spherical omni-triangulation intensifies the structural integrity of Fuller's Geodesic structures.

It can be seen that Fuller's Tensegrity geodesics, like fishnets or balloons, could result in highly flexible structures. When it is desirable to have a Geodesic integrity with a non-mushy exterior, Fuller provides concentric Tensegrity spheres, one of lesser radius than the other, and the inner one of one modular frequency less than the other. He interlaces the inner and outer spheres respective omni-triangulated point patterns. Each of the inner points connects outward to three of the outer points; and each of the outer points, as a result, is found to be interconnected to three inner points. The resulting intertriangulating of the concentric Tensegrity spheres provides an Octet Truss.

The Octet Truss, in this spherical arrangement, will be seen to be the same finite omni-triangulated patterning of Fuller's energetic-synergetic geometry, closest-packing Vector Equilibrium layers of any modular radius and frequency.

Compression columns have a limit slenderness ratio (the ratio of column length to cross-section diameter). If this ratio is exceeded, the column (strut) will buckle. (The slenderness ratio of a column of ordinary steel is approximately 33 to 1.) On the other hand, tension cables have no inherent limit ratio of section diameter to length. The "pulling strength" of a cable is the same in lengths of two feet or two miles. Thus it can be said that compression is limited and tension unlimited in relative slenderness ratio magnitudes, and their respective structural applications.

It followed from this that structures developed according to Tensegrity principles, with discontinuous compression, continuous tension, have no size limit. Theoretically it is possible to dome the entire earth in a Tensegrity envelope. Unlike other structures, Tensegrity domes increase in strength by a factor greater than that governing their growth in dimensions; the larger they are made, the stronger they become. It is only at the toy-size level that their strength-area relation does not show up to dramatic advantage.

Even at this writing, Fuller has plans developed for structures now feasible which could dome in all of lower Manhattan, or the site of an entire town. Such a dome, erected in the Antarctic, would give colonizers a temperate environment long before actual living and industrial facilities were installed.

Geodesic Structures

The vertexes of the geometric figures which form Fuller's "systems" are points which determine great circles on the surface of a sphere. In modern geometry, as we have seen, any arc of a great circle is called a "geodesic."

When Fuller began to construct domes that were essentially networks of spherical triangles formed by the intercrossing great circles, he called these structures "Geodesic."

The three sides of a spherical triangle are formed by three great circles. A complete over-all network of great circles can be defined as a "grid"; since to form triangles a grid must have lines extending in three directions, Fuller regarded the Geodesic dome as a three-way grid of great circles.

It is not practical to catalog the thousand or more Geodesic domes constructed between 1948 and 1959 by Fuller, his associates, his companies, his licensee corporations, and his university students. It suffices to note that once the Geodesic idea got going it began—as Eugene Field once predicted of Chicago—to make culture hum.

In 1952 the Ford Motor Company became the first industrial organization to be licensed under Fuller's patents. Under this license they had constructed the 93-foot aluminum and plastic dome over the Dearborn Rotunda Building. Fuller considers the Ford Geodesic Dome as the fulfillment of his 1927 prediction of a quarter-century gestation period for his Dymaxion enterprise. The dome arrived on schedule. Fuller refers to his first customer as "Mr. Industry himself."

Of great importance were the Geodesic radomes Fuller began producing in 1955 for the frozen tundra and icy hills of the U.S. Air Force's DEW (Distant Early Warning) line—the 3,000 mile strip of radar installations which clings to the northern rim of Alaska and Canada. Because of the violent and uncertain weather along the Arctic Circle, the Air Force required a structure that could be flown knocked-down to site, and then set up in the 20-hour margin of predictable good weather. The installation, when completed, was required to withstand a 210-mile-per-hour wind, and to be fabricated from materials which would be invisible to the radar's microwave beam. A radar beam is reflected by metal.

Fuller responded with domes 55 feet in diameter, made of fiberglass plastic. Standing 40 feet high, these domes in 1954 were the largest plastic structures that had ever been built. They were assembled on delivery, not in 20 but 14 hours; and they withstood static load testing for wind velocities in excess of 220 miles per hour.

By the time the Air Force radomes were constructed across the DEW line, the Marines had about 300 Fuller domes in use, some in the Antarctic, some around the equator. Fuller's 4D "anticipatory" realism of 1927 had at last begun to orbit; his structures, delivered in the air ocean, had spiralled the earth.

Another innovation was Fuller's paper dome; two such domes, manufactured by the Container Corporation of America, were sent, on invitation, to Italy—to be shown at Milan's international design exhibition, the Tenth Triennale, in 1954.

The domes were awarded the Triennale's *Gran Premio,* the highest prize given to any participating country. The award was ironic since the United States had no official entry at the Triennale; Fuller's exhibit was a consequence of his exuberance, his dedicated belief in Dymaxion-Geodesic values, and the fact that he was able to muster enough support to lob his structures across the Atlantic.

In Fuller's opinion, however, the paper Geodesic domes were anticipatory rather than actual; they bear the same relation to the corrugated paper available today as the 4D house bore to the soft aluminum which was the only aluminum available in 1927. Even in 1954 Kraft paper having exceptional "wet tensile strength" had been developed—"wet strength" meaning the ability of the paper to retain its structural quality when saturated. But in 1954 corrugated paper board with good wet compressive strength had not yet been developed. When wet, corrugated paper board folded up like an accordion. To avoid the collapse of the Triennale and other paperboard domes, Fuller covered them with vinyl "bathing caps," aluminum foil, and other water impervious materials. He has delayed, however, any production enterprise in this area. High wet compression strength papers have already been demonstrated successfully in the laboratory, but they are not yet industrially available. When they become so, Fuller proposes to license the paperboard domes for mass production.

Large paper manufacturing mills have the capacity to produce 3000 domes per day, each dome with a floor area of 1000 square feet. Fuller estimates that domes of this type could be retailed in the $500 price range, that is, at approximately 50¢ per square foot. A concrete floor would cost about $200. The autonomous "mechanical package" for the domes—sanitary facilities, cooking and heating units—could be rolled in under such a dome for another $2000 purchase price or rented on a trail-it-yourself basis for a dollar per day. The conclusion Fuller draws is that with this type of structure, people, in time, may be able to enjoy high standard dwelling advantages at costs readily met out of a single year's income.

The U.S. Department of Commerce decided to set up a Geodesic dome as its Pavilion in the 1956 International Trade Fair, at Kabul, Afghanistan. What followed was perhaps an historical speed record for engineering planning, manufacture, and construction. The project contract was signed May 23rd. Seven days later all designs, calculations, engineering plans had been completed. By the end of June, the entire dome had been manufactured and packaged, ready for air shipment to Kabul in the company of a single engineer. The dome was light enough and compact enough to be flown from America to Afghanistan in one DC-4 plane. It was designed to be erected anywhere, by workmen speaking any language, who were in no way trained or briefed for the operation. Directed only by one Geodesic engineer, the Afghans fastened blue-ended dome parts to other parts whose ends were blue. Red ends were matched to red ends. And forty-eight hours after the arrival of the air shipment, the Afghans found that they had erected a great dome. A stranger, ambling innocently into Kabul, might reasonably have concluded that the Afghans were the most skilled craftsmen.

The Kabul dome, like Fuller's Air Force radomes, established another historical "first"; it was, in 1956, the world's largest Geodesic structure, 100 feet in diameter, 35 feet high at the center. It provided a clear-span, entirely uninterrupted floor area of approximately 8,000 square feet. The dome frame was formed by 480 aluminum tubes, three inches in diameter. The frame weighed 9,200 pounds; the nylon skin, 1,300 pounds.

A signal feature of the dome's "informational" value, at Kabul, was the fact that the dome attracted far greater attention, and attendance than all other exhibits including the Russian and the Chinese Communist. Both groups had spent months and many times the cost of the U.S. exhibit, in the preparation of their special pavilions.

Capitalizing on this success, the United States government arranged to have Geodesic domes set up at other international trade fairs. The Department of Commerce had now become interested in the kudos value of Geodesic domes. The Geodesics, it was argued, dramatized American ingenuity, vision, and technological dynamism; as structures to house American trade exhibits they would be tangible symbols of progress. Fuller's three-way grids were better propaganda than double-meaning speeches broadcast to regions in which radios were scarce. Domes as large as the Kabul dome, and larger, were flown from country to country, girdling the globe; and many of these also set attendance records. Within a short space, Fuller's domes were seen in Poznan, Casablanca, Tunis, Salonika, Istanbul, Madras, Delhi, Bombay, Rangoon, Bangkok, Tokyo, and Osaka.

The breakthrough to large-scale industrial marketing of the Geodesic idea began in the latter part of 1956. Donald Richter, a former student and associate of Fuller's, had gone to work for Henry J. Kaiser. Like many of Fuller's students, Richter had become an avid constructor of Geodesic models and had installed one miniscule dome in his office. Kaiser strode through the office one day and saw the model. "What's this?" Kaiser asked, with reasonable interest. Richter explained; and the consequence of this seemingly accidental event was that Kaiser's metal-fabricating customers geared up to mass-produce quarter-million-dollar domes.

The Geodesic "building construction," as it is called in the patent application, was covered fully by a U.S. patent (No. 2,682,235) issued to Fuller in June, 1954; and from this time on, all users of the system were required to be licensed by Fuller. Kaiser Aluminum became one of the early licensees. The initial Kaiser project was an aluminum-skinned, 145-foot-diameter auditorium for Henry Kaiser's Hawaiian Village, in Honolulu. The project construction men were startled by the speed with which the Geodesic dome went up. For Kaiser the speed had almost a shock effect. He wanted to see the dome rise; and the day the workmen started on the structure he hopped a plane from San Francisco, intending to be on hand during the first week's construction. By the time his plane reached Honolulu, however, the dome was finished. As a dramatic fillip, the Kaiser promotion men arranged to have it formally opened the same night, when it housed an audience of 1,832 and a symphony orchestra.

By the end of 1958, Kaiser Aluminum's fabricator customers had produced eight domes, including one used as a theatre in Fort Worth, Texas, one as a bank in Oklahoma City. The Kaiser organization assumed that there was a probable market for at least one dome to every American town large enough to use a community center. The domes are now available in a number of sizes, at prices ranging from $50,000 to $190,000, exclusive of foundation and interior detail. The most publicized Kaiser-erected Fuller dome of 1959, however, was the Geodesic dome housing the United States exhibit at the World's Fair in Moscow. It was on seeing this dome that Nikita Khrushchev exclaimed, "I would like to have R. Buckminster Fuller come to Russia and teach our engineers."

The largest clear-span enclosure ever to be erected anywhere in the world is the steel-skinned Geodesic structure Fuller's own company, Synergetics, Inc., designed for the Union Tank Car Company, and which was completed and put into operation at Baton Rouge, La., in October, 1958. This dome is about 23 times the volumetric size of the dome on St. Peter's Church in Rome. It has a total clear span (i.e., without posts or obstructions of any kind) of 384 feet; it rises 128 feet at the center. The relation of dimensions to weight and to cost is extraordinary: the dome has a floor area of 115,558 square feet and encloses 15,000,000 cubic feet, yet its total weight is only 1,200 tons. In simple units, this is two ounces of structural weight for every cubic foot the dome encloses. Total cost was less than $10 per square foot.

A similar dome, planned of the same dimensions was under construction in Wood River, Illinois, by Union Tank Car Company's Graver Tank Division, and was to be completed by December 1959.

The Union Tank Car projects were born when the company was scouting for some economic way to construct a railroad car rebuilding and reconditioning plant large enough to accommodate trainloads of cars at a time. It was important to have an enormous span of clear space to permit shuttling of engines and the swing of cars around a central turntable.

Union Tank is now a licensee of Fuller's, and, through its Graver Tank and Manufacturing Company Division, is offering all-steel Geodesic domes at a probable cost of $10 per square foot, or less, in competition with the Kaiser aluminum domes.

At the close of 1959, there were more than a hundred licensees operating under Fuller's cumulative array of patents. He now has patent coverage in many foreign countries, and a number of foreign licensees. And his experiences in past operations have led him to a personal philosophy of patents. In the craft equation, he holds, the patrons have the design initiative. Professional architects and engineers are retained, and rewarded for services rendered, only at the express command of the patron initiator. In the industrial equation, by contrast, a "comprehensive, anticipatory, design scientist" not only takes the initiative in development, but holds it—years in advance of any awareness on the part of industry, the government, or the public, that there is such an initiative to be taken and held.

In the industrial equation, as Fuller views it, the designer never renders his service under patronage command. Because of the complexity of industry, and of the economic accounting by industry and the government, the only possible control the individual designer can exercise over the economic inhibition by society of the technical advantages he anticipates on behalf of society, is through the patent. The patent safeguards the designer's right to protect the future from the inertias of the past. Society, like the guppy, devours its offspring. "Future comprehensive designers," says Fuller, "will have to be masters of patent law as well as their other fundamental disciplines—if they are going to be able to preserve the regenerative advantage innate in the individual."

Of the hundred Fuller industrial licensees, the largest, at this writing, is North American Aviation—a company whose total gross is on the billion dollar level. North American, in 1959, constructed a 250-foot-diameter aluminum Geodesic dome for the American Society for Metals, the official organization of metallurgical scientists. This dome, at the headquarters of the metals society, in Cleveland, was designed by John Kelly, and is a delicate, open structure functioning as a gossamer net arching over the society's buildings, gardens, and pools. Kelly looks on this dome as a forthright statement of the advances made in the alloying sciences, and as a realization of Fuller's concept of advancing technology's "over-all trend to invisibility."

Advancing technology, Fuller reasons, crossed the threshold to invisibility in World War I, advancing from wire communications to wireless, from tracked transportation to trackless. Technology's alloying evolution developed invisible solutions to problems of strength; and these invisible solutions indirectly, in 40 years, have shrunken the world to a one-town community. The real Magic Carpet is an alloy web.

Significance of the Geodesic Breakthrough

For most of the three decades, following 1927 and the days of 4D, the Dymaxion house, and the Dymaxion car, architectural and national news magazines made frequent reference to Fuller as "failure prone." Condescending accolades were heaped on his ideas, but the assumption was that nothing would come of them; nothing, it was held, came of the house, the car, the bathroom, and the host of other early prototype developments. Today the picture is quite different. Fuller has suddenly become the conservative industrialists' ideal of the pioneer scientist. Pictures of his latest projects appear regularly on the front cover of the magazines which symbolize the tycoon press, and both business and the Armed Services have deluged him with construction projects.

To keep pace with the demands for his ideas, his technological knowledge, and his computations, and to keep in order his rapidly expanding bookkeeping chores, Fuller organized several corporations wholly owned by him, which channel the licenses for the use of his patents. Geodesics, Inc., handles all government and Armed Services developments; Synergetics, Inc., deals with design and research for all private industrial operations; Plydome, Inc., is one of Fuller's private research and development companies.

There are several explanations for the sudden change-about in the world's attitude toward Fuller. Long ago Fuller observed that conservatism is part of the normal social process, and that—according to the timetable then in effect—about 25 years were required to bring about general acceptance of an important new idea. Fuller waited out his quarter-century. The praise which is now generally heaped on his head can be attributed in part to another factor: industry, which recently awoke to the vision of the great economies and profit possibilities of Geodesic structures, has tried to side-step Fuller's patent and found the evasion impossible. Fuller has a hammerlock hold on the construction principles.

Fuller attributes his sudden success to the fact that technological developments have caught up with him. His early designs were "anticipatory," not actual; they required materials which were to come but which were not then in existence, particularly the extremely strong light alloys, and strong, transparent, weather-resistant plastics. "All you need now is the knowledge of what you want to do—the billion dollars' worth of anticipatorily scheduled research has been done," he claims, referring to the estimate he made of the cost of producing the true prototype Dymaxion house in 1927. "But society did it the easy and slow way, which partially accounts for the 300 billion dollar national debt."

He regards domes as basic environment valves, differentiating human ecological patterns from all other patterns, microcosm from macrocosm, yet permitting a controlled interchange of energy (including heat and light) between the two separated pattern regions. As an environment valve, the Geodesic dome is not limited in size; its span can be anything from a few yards to a few miles; it can envelop living quarters, gardens, lawns, acres, or cities. As an environment valve it can make possible cities of temperate climate domed over in the Arctic, the Antarctic, or at the bottom of the sea. It can shelter the lawns, gardens, and grounds in the midst of which a house is customarily established, thus causing the conventional house to become, if not obsolete, at least increasingly superfluous. To erect an expensive house, with rugged foundations and solid walls, under a highly efficient and relatively inexpensive environment valve would be equivalent to wearing a mink coat in an apartment with central heating.

Geodesic domes of sufficient size, covered with a transparent plastic skin, tend to become invisible; the permitted extreme slenderness of supporting struts enable them to escape detection when the radius of the sphere increases beyond a certain limit. The domes can be geared to rise from the ground, or to hug the earth, at the instance of control devices operating pneumatic or hydraulic jacks. Air vents and light-regulating louvres can be introduced at will. Winter heat can be effected locally, with radiant coils coupled to heat-exchange pumps. Privacy and space division, even room and room divisions, can be established in a variety of ways without requiring an architectural imitation of an Italian palazzo, a Norman villa, or the peristyle of a Greek temple. Some alternate possibilities are suggested in the latter part of the book.

Yet Fuller puts no undue emphasis on his domes. They are steps in a progression, not an end in themselves. What is important to him in the domes is their Pythagorean overtones—the fact that they are tangible, measurable illustrations of laws fundamental to the nature of the universe, of the spread and temper of energy patterns. He finds a measure of satisfaction in that the domes perform according to the predictions of Energetic Geometry, and that they function as evolving forms in a comprehensive design science.

The possibility of the good life for any man depends on the possibility of realizing it for all men; Fuller holds to this credo today as intensely as in 1927, when he organized the first decisive postulates of his synergetic cosmology and its consequent philosophy. And the full life, which encompasses the elements without which neither freedom nor higher social expressions are possible, is a function of society's ability to turn the energies of the universe to human advantage.

All we have to work with, in our span of life, is the energy system of the universe—the system which determines the dynamic structuring of the 92 elements found in nature, and the secondary, tertiary and sequitor phase structures (molecules, crystals, alloys, shelters, vehicles) into which these elemental dynamic patternings can be formed. The universe is what is given to us in experience; it is to be found as an integrated whole, not an assortment of parts. It is a Gestalt.

The problem of science, more particularly of a "comprehensive design science," is to separate out local eddies from the universe as it is experienced, directly or conceptually; to isolate specific instances of the behavior patterns of a general, cosmic energy system, and to turn these to human use. "I am not a creator," Fuller once said. "I am a swimmer and a dismisser of irrelevancies. Everything we need to work with is around us, although most of it is initially confusing. To find order in what we experience we must first inventory the total experiences, then temporarily set aside all irrelevancies. I do not invent my thoughts. I merely separate out some local patterns from a confusing whole. The act is a dismissal of pressures. Flight was the discovery of the lift—not the push."

At the birth of the twentieth century, the architect Louis Sullivan observed that production steel, which men were insinuating within the stone faces of buildings, was permitting "stone" buildings to assume shapes grotesquely alien to the nature of stone. Sullivan, in Fuller's view, pioneered a revolution of integrity. He sought to make honest and unashamed statements in materials that expressed society's new industrial capabilities. He inspired corps of esthetic disciples and emulators. Yet in the stampede of subsequent design exploitation, both his integrity of conception and his philosophic message were lost. The exploiters sidestepped the essence of Sullivan's phrase, "Form follows function." They made the words read "The ends justify the means," ergo, "Do business at any price."

In spite of Sullivan's recognition of the industrial equation—whose myriad patterns are invisible—the building arts, until now, have been pre-empted by the non-industrial, non-priority, catch-as-catch-can crafts. And in this stream-lined chaos, architects have become as increasingly marginal as journeymen, tinkers, and drivers of hansom cabs. Like patients who diagnose their own ail-ments and sketch for the surgeon the operation they want, clients design their own buildings, and then demand of the architect his blueprints for action. The creative architect is hamstrung. Not only do his clients tell him what kind of building they want, and how much it should cost, but community codes, building laws, and bank mortgage biases have become instruments to the tyranny. Architects have left to them little more than the privilege of being exterior-interior decorators to skeletons prefabricated by the major steel com-panies.

Yet Sullivan's slogan held as a justification for all the late architectural stereotypes. The more glass and shiny metal used in the decorative ensemble, the more it was claimed that form was following function. The functions were not techniques for doing more with less; the functions were shine and gleam. In contrast with this distortion of the significant virtue of form in the build-ing field, where form is conceived only as obvious structure, the industrial equation, Fuller points out, was creating decisive advantages in invisible structures. The Model T Ford is a case in point. Henry Ford's apparent doggedness in continuing to produce the Model T over a period of years, con-cealed the fact that the Model T was improving functionally, while competi-tive cars were improving only in cushioning and external styling. Before Ford finished with the Model T, he had introduced 54 different alloys of steel into it. It was these alloys which gave the car its service durability and pioneered Ford's success. Ford was improving his cars more rapidly than his competi-tors, but the improvements were invisible. Visible form could no longer fol-low the subvisible functions.

Fuller, today, sees a new industrial world forming—one that is a decisive step forward in progression to "a second derivative and surprisingly satisfac-tory world era." It is symbolized by the Geodesic dome of the American So-ciety for Metals; for here the notion of doing more with less, as expressed in the trend toward invisibility, is dramatized by the dome's open structure which is pure system integrity. And he takes it as a straw in the new wind that the dome was fabricated by the most powerful of the aircraft corporations.

When Sputnik rocketed successfully into orbit, Fuller maintains, it shot down the military airplane. This signal act closed the half century in which the world's larger nations put behind the airplane weapon a subsidy adding up, in capital enterprise, to more than two trillion dollars. The new controlled, un-manned missiles made the airplane, by comparison, virtually stand still in the air; as a weapon it was finished. The immediate consequence of this military reality was that the two trillion dollar air frame and air power plant industry was roughly thrown out of its kept-mistress luxury quarters. It was con-strained to seek a living on its own.

To Fuller, this event was not a catastrophe, but an opportunity to begin "the fundamental reorientation of the whole vital economic patterning of man." This was the day he had foreseen, some 32 years earlier—the day when man's highest knowledge and comprehensive resources could be applied di-rectly to his living needs, instead of being assigned exclusively to negative functions.

With the two-trillion-dollar subsidy of high technical capabilities now ten-tatively available for living rather than military problems, Fuller believes that this reorientation is about to become a reality. The touchstone is the aircraft industry. In 1946, North American Aviation, together with Douglas, Boeing, Grumman, and others, had looked on Fuller's Dymaxion house as a possible, if not probable, post-war field for their respective enterprises. But the Cold War's cumulative half-trillion defense budgeting—which produced a jet age—temporarily shunted the industry from the building arts. It postponed the last great slum clearance project of technology.

But Sputnik destroyed the airplane weapon. The aircraft industry, paced by North American, is in a position to inaugurate a world-circling building and building mechanics service industry which can fly whole cities into position overnight, as great fleets sail into great harbors, fundamentally in grace with a vast environment. And as the great fleets can sail on, to continue their useful-ness wherever they are needed, Fuller holds, so may the environmental facili-ties of man be repositioned about the earth, giving him access to the dwellings of yesterday, the productive resources of tomorrow, and a vaster reach of the universe gained without political revolution or panacea.

Bucky's association with Philadelphia was a result of his ties with Martin Meyerson, then the President of the University of Pennsylvania, and Harris Wofford, then President of Bryn Mawr (later a U.S. Senator). They were determined to relocate Bucky to Philadelphia, and Fuller moved there in 1972, living in that city until his death on July 1, 1983. Meyerson describes Fuller as "a Leonardo-like character." A University Professor at the University of Pennsylvania, Bucky was also a World Fellow in Residence with the University City Science Center, which was headed by Dr. Randall Whaley. He was pleased that the consortium of three Philadelphia-area Quaker colleges, Bryn Mawr, Haverford, and Swarthmore, sponsored him.

Bucky was so much in demand that it took a taskmaster to manage his schedule, which looked like several airline schedules pushed together. His world cable address was the single word "BUCKY." He liked that. A sample of a typical Bucky Fuller diary schedule is reproduced for the reader's interest.

Bucky as a Leonardo-Like World Fellow in Residence

Professor Martin Meyerson

Bucky was a Leonardo-like character. Knowing most of his architectural, urban, and design concepts, I also had the privilege of being exposed to the automobile he built, the poetry he wrote, the mathematics works he authored, the boats he conceived and loved.

When I was in my early twenties and in my first job, which was at a policy research center adjacent to the University of Chicago campus, I met Bucky Fuller. In the evenings I would go on occasion to the New Bauhaus and be tutored by its head, the great design educator Moholy-Nagy. We would explore the planning of new communities, or we would speculate about a postwar Europe. And then one evening Bucky joined us, and I was enchanted by him ever since that initial meeting. I had come to Chicago from graduate-student days at Harvard, and Bucky reminisced about his short-lived Harvard days, and about much else of his life and work.

SEPTEMBER (Continued)

Sept. 28	Sun.		Pacific Palisades
Sept. 29	Mon.	10:30 a.m.	Appointment with Dr. Charles DiPilla, podiatrist 321 N. Larchmont, Suite 607, LA (213) 464-5188
			Stay: Pacific Palisades
Sept. 30	Tues.		Pacific Palisades

OCTOBER

Oct. 1	Wed.	9:55 a.m.	Fly from Los Angeles, CA United # 70
		5:05 p.m.	Arrive Cleveland, OH (met)
			Dine at hotel with Thomas Zung, Cyrus Eaton
			Stay: Stouffer's Inn on the Square (216) 696-560
Oct. 2	Thurs.	9:10 a.m.	Interviewed for TV Channel 5 "Morning Exchange" on CRITICAL PATH and Jitterbug
			Contact: Joel Rizor (216) 431-5555
		12:00 noon 12:30 p.m.- 1:00 p.m.	Luncheon at the City Club Speak on "Humans in the Future" for ½ hour ½ hour Q & A follows
			Contact: Alan Davis (216) 621-0082
		5:30 p.m.	Dine at hotel with Chancellor and Mrs. Nolan Ellison, Dr. and Mrs. David Mitchell, Dr. and Mr Alan Davis, JoAnn Liebow, Thomas and Carol Zung
		7:30 p.m.	Reception
		8:00 p.m.	Speak at Cuyahoga Community College "Philosophy and Architecture"
			Contact: JoAnn Liebow (216) 464-1450 x 3001
		9:30 p.m.	Meet with department heads from Cuyahoga Communi College
			Stay: Stouffer's Inn on the Square

Not too long after, I was on the University of Chicago faculty and Margy, my wife, and I would on occasion house-sit for David Riesman, like Bucky a most imaginative mind: social scientist, lawyer, theoretician, utopian. In the Riesman home, the Meyersons assembled a group of up-and-coming professionals and scholars for dinner, with Bucky as our special guest. We listened and reacted, listened and reacted, and soon it was eight in the morning. We went off to our tasks. No one who was at that evening ever forgot it. Bucky's extraordinary energy throughout his life was one of his special characteristics.

Later, at the Joint Center for Urban Studies of MIT and Harvard (I was Williams Professor at Harvard and first head of the center), and then at Berkeley (where I was a professor, dean, and interim chancellor), Bucky would come and delight us as well as instruct us with his world views and

OCTOBER (Continued)

Oct. 3	Fri.	8:30 a.m.	Breakfast at City Club with members of Cleveland Exposition Group
		12:00 noon	Luncheon with Tom Vail, editor of CLEVELAND PLAIN DEALER
			Rosamund Fuller Kenison dies
			Stay: Sheraton Hopkins (216) 267-1500
Oct. 4	Sat.	10:00 a.m.	Fly from Cleveland, OH NW # 292
		11:27 a.m.	Arrive Boston, MA
		12:51 p.m.	Fly from Boston, MA Delta # 130
		1:25 p.m.	Arrive Portland, ME (met)
			Stay: Wolcott Andrews Residence (207) 882-5578
Oct. 5	Sun.		Attend Rosy Kenison's funeral, Bear Island, ME
			Stay: Andrews Residence
Oct. 6-7	Mon.-Tues		Stay: Andrews Residence
Oct. 8	Wed.		Attend Memorial Service for Rosy Kenison, Wiscasset, ME
			Stay: Andrews Residence
Oct. 9	Thurs.	2:45 p.m.	Fly from Portland, ME Delta # 129
		6:32 p.m.	Arrive Atlanta, GA
		7:15 p.m.	Fly from Atlanta, GA Delta # 1647
		8:45 p.m.	Arrive Daytona, FL (met)
			Stay: University Inn (904) 734-5711
Oct. 10-12	Fri.-Sun.		Meet with Marshall Thurber and Jim Patterson Deland, FL
			Stay: University Inn

insights. I suggested that he move to Berkeley, but it was southern California which for family reasons attracted him as a base.

Bucky and I were together on as many occasions as possible. For example, we were among the fifteen or sixteen "regulars" in the yearly Greek island boat trips, with their discussions/seminars on human settlement stimulated, organized, and hosted by Constantine Doxiadis, who coined the term "ekistics." Bucky was a star.

I will not forget the times Bucky was with us in the Frank Lloyd Wright house in Buffalo, which we had restored and lived in when I was president of the state university center there. He would visit us there, and our four-year-old son adored Bucky—even now, when he is an assistant professor at the Harvard Medical School, he has kept the tetrahedrons and other mementos

Bucky gave him. When Bucky left after those visits with us, Matthew would cry. Over the many years we knew each other, when parting company, I did not cry, but I always longed to be with him again soon.

During my presidency of the University of Pennsylvania (1970–1981), Ed Schlossberg (who had gotten to know Bucky) mentioned to me and Harris Wofford (then president of Bryn Mawr, later to be a U.S. senator) that Bucky might like to be based in Philadelphia. Harris and I were determined to make that happen. We developed an arrangement in which the three Quaker colleges (Bryn Mawr, Haverford, and Swarthmore) joined with the University of Pennsylvania, with which they had an affiliation, to create a professorship for Bucky. Soon he also became a university professor at Penn.

He had the appealing notion of putting a dome over the entire University of Pennsylvania. I tried more modestly to get a dome of Bucky's design built on land next to the university museum. Penn has in the Lower Egyptian gallery of the museum the outstanding Egyptian temple in the United States. Unfortunately, it was not possible to place it in the beautiful Upper Egyptian gallery, because of weight. Thus, the temple columns are in pieces in the lower level of the museum. A Bucky dome would have permitted it to achieve its full height and glory and would have been an attractive focal point for Philadelphia as well as the university, just as the more modest Egyptian temple at the Metropolitan Museum in New York is. One day that still may happen.

Bucky in addition was attached to the University City Science Center, which had been established earlier by the University of Pennsylvania and other institutions. The Science Center, headed by Dr. Randall Whaley, was and is an "incubator" for various kinds of small and medium-size research enterprises. There Bucky was a resource for the varied activities going on. Those in the Science Center would ask advice from Bucky on particular projects; sometimes the requests were modest, sometimes complex. Either way, Bucky would have important contributions to make. And Bucky was practical as well as theoretical. In the World Game which he and his associates devised at the Science Center, his judgments on energy, food, and all the resources required by civilization were quantified and interlinked. He was mindful of the requirements to actualize such prospects and not confine them to conceptualization alone.

The Science Center was a fine base in many ways for Bucky, but by then he was attached to many institutions throughout the country and the world. He maintained a hectic activities schedule, traveling constantly; only Shirley Sharkey, his executive administrator, could keep his appointments in order. Incidentally, Bucky always had his chronofile, but it was at the Science Center that Bucky's archives first got established. And here in Philadelphia, with the help of the Bell System, we prepared a video record, about forty hours in length, of Bucky's life, work, and findings.

Philadelphia does use a slogan these days: "Philadelphia, the city that loves you back." It did love Bucky back, and I kept trying to think of settings which would help amplify Bucky's varied brilliant qualities so that we could gain even more from him locally.

In that same period, I was a very involved board member of the Aspen Institute. Neva Rockefeller Kaiser and I enticed Bucky to Aspen, to the satisfaction of all.

On occasion we would accidentally encounter each other in places such as the Tokyo airport, and would readily resume our most recent conversation. Bucky would expand on his interest in quantitative analysis and on his continuing studies on form, structures, artifacts, and all the world. How the Internet would have pleased Bucky!

It was fitting that he and Anne, his wife of half a century, died within a couple days of each other. They had been mutually supportive and unconditionally accepting of each other. Bucky greatly loved his daughter and grandchildren, and had a very special relationship with young people of all ages.

At Bucky's funeral at Mount Auburn Cemetery, it was clear that, as he was returning to his New England soil, his impact on individuals and on the world would continue to influence all of us in the new era ahead.

Education from *Education Automation*

The headache of a president of a great university is today probably the next biggest headache to that of a quasi-nation's president. Take the problem of how to get the funds for this enormous educational undertaking. You educators are uniquely associated with people who are well educated and who have a great feeling of responsibility toward the new life. There is an enormous task to be done, and the budget gets to be formidable. How do you raise the funds? The now world-populated state universities have to keep raising funds from a political base which as constituted is inherently static, operating exclusively in terms of Illinois or Ohio or whichever state it may be.

The point is that we—both as individuals and as society—are quite rapidly uprooting ourselves. We never were trees and never had roots, but due to shortsightedness we believed blindly and behaved as though we did. Today we are extraordinarily mobile. In this last election, 10 per cent of the national electorate were unable to vote because they hadn't been in their new places long enough. The accelerating mobility curve that I just gave you indicates that by the next election 25 per cent of America will not be able to vote due to recentness of moving, and in the following election possibly less than the majority will be able to vote. We are simply going to have to change our political basis. We are now at the point where the concept of our geographically-based representation—which assumes that it realistically represents the *human* beings—is no longer valid. The political machine alone will continue to stay local. It sees the people as statically local. So those who are politically ambitious just stay put while society moves on, and, therefore, the static politicians become invisible to the swiftly moving body politic, which cannot keep track of their static machinations since society does not stay long enough in any one place to be effective in reviewing the local political initiations. The political machines soon will have no one to challenge realistically their existence validity except the local newspapers, whose purely local political news becomes progressively of less interest to a world-mobilizing society.

Comprehensively, the world is going from a Newtonian static norm to an Einsteinian all-motion norm. That is the biggest thing that is happening at this moment in history. We are becoming "quick" and the graveyards of the dead become progressively less logical. I would say, then, that your educational planners are going to have your worst headaches because you will have political machines that are less and less visible to the people because the people are more and more mobile. You will have to be serving the children of the mobile people who really, in a sense, don't have a base, and you will have to justify it with very hard-boiled local political exploitation. I am not particularly optimistic about the kind of results you are going to get. Therefore, when I begin to talk about the educational revolution ahead I see that the old system is probably going to become paralyzed. That is why your headache will get worse and worse until nature just evolutes and makes enormous emergency adjustments. President Morris, I not only recognize that your job is fabulously challenging, I recognize you as an extraordinarily able man. Yet I see that you are going to have a harder and harder time, and nobody could care more than you do about the good results you might get. What I am saying, then, is realistic. It is also going to be obvious to you, I am sure, that the kind of changes I will talk about next are probably going to have to take place.

We know that our world population is increasing incomprehendibly swiftly. There are enormous numbers to be educated. We are going to develop very new attitudes about our crossbreeding and our reversion to universal pigmentation. That is going to be slow, but it is going to be a great and inevitable event. In the end we are going to recognize that there are no different species of living man, and we will get over that kind of color class-distinction.

The big question is how are we, as educators, going to handle the enormous increase in the new life. How do we make available to these new students what we have been able to discover fairly accurately about the universe and the way it is operating? How are we going to be able to get them the true net value won blindly through the long tradition of ignorant dedications and hard-won lessons of all the unknown mothers and all the other invisibly heroic people who have given hopefully to the new life, such as, for instance, the fabulous heritage of men's stoic capacity to carry on despite immense hardships?

The new life needs to be inspired with the realization that it has all kinds of new advantages that have been gained through great dedications of unknown, unsung heroes of intellectual exploration and great intuitively faithful integrities of men groping in the dark. Unless the new life is highly appreciative of those who have gone before, it won't be able to take effective advantage of its heritage. It will not be as regenerated and inspired as it might be if it appreciated the comprehensive love invested in that heritage.

The old political way of looking at things is such that the political machine says we first must get a "school house" for our constituents, and it must look like Harvard University, or it must be Georgian and a whole big pile of it. "We see that the rich kids went to school in automobiles; so let's get beautiful buses for our kids." "Harvard and Yale have long had football; our school is going to have football." There is nothing boys used to have that they are not going to "get" from their politicians, who, above all, know best how to exploit the inferiority complex which they understand so well as handed down from the ages and ages of 99 per cent have-not-ness of mankind. There is a sort of class inferiority amelioration battle that goes on with the politicos in seeking the favor of their constituents to get into or back into office, and little if any attention is paid to the real educational problems at hand.

In thinking about these problems, I have thought a lot about what I have learned that may be useful as proven by experiments in my own self-disciplining. I have met some powerful thinkers. I met Dr. Einstein. I wrote three chapters in a book about Dr. Einstein, and my publishers said that they wouldn't publish it because I wasn't on the list of people who understood Einstein. I asked them to send the typescript to Einstein, and they did. He then said he approved of it—that I had interpreted him properly—and so the chapters did get published. When Einstein approved of my typescript he asked me to come and meet him and talk about my book. I am quite confident that I can say with authority that Einstein, when he wanted to study, didn't sit in the middle of a school room. That is probably the poorest place he could have gone to study. When an individual is really thinking, he is tremendously isolated. He may manage to isolate himself in Grand Central Station, but it is *despite* the environment rather than because of it. The place to study is not in a school room.

Parents quite clearly love their children; that is a safe general observation. We don't say parents send their children to school to get rid of them. The fact is, however, that it is very convenient for mothers, in order to be able to clean the house for the family, to have the children out of the way for a little while. The little red school house was not entirely motivated by educational ambitions.

There is also a general baby-sitting function which is called school. While the children are being "baby sat," they might as well be given something to read. We find that they get along pretty well with the game of "reading"; so we give them more to read, and we add writing and arithmetic. Very seriously, much of what goes on in our schools is strictly related to social experiences, and that is fine—that's good for the kids. But I would say we are going to add much more in the very near future by taking advantage of the children's ability to show us what they need.

I have taken photographs of my grandchildren looking at television. Without consideration of the "value," the actual concentration of a child on the message which is coming to him is fabulous. They really "latch on." Given the chance to get accurate, logical, and lucid information at the time when they want and need to get it, they will go after it and inhibit it in a most effective manner. I am quite certain that we are soon going to begin to do the following: At our universities we will take the men who are the faculty leaders in research or in teaching. We are not going to ask them to give the same lectures over and over each year from their curriculum cards, finding themselves confronted with another roomful of people and asking themselves, "What was it I said last year?" This is a routine which deadens the faculty member. We are going to select, instead, the people who are authorities on various subjects—the men who are most respected by other men within their respective departments and fields. They will give their basic lecture course just once to a group of human beings, including both the experts in their own subject and bright children and adults without special training in their field. This lecture will be recorded as Southern Illinois University did my last lecture series of fifty-two hours in October 1960. They will make moving picture footage of the lecture as well as hi-fi tape recording. Then the professor and his faculty associates will listen to this recording time and again.

"What you say is very good," his associates may comment, "but we have heard you say it a little better at other times." The professor then dubs in a better statement. Thus begins complete reworking of the tape, cleaned up, and cleaned up some more, as in the moving picture cutting, and new illustrative "footage" will be added on. The whole of a university department will work on improving the message and conceptioning of a picture for many months, sometimes for years. The graduate students who want to be present in the university and who also qualify to be with the men who have great powers and intellectual capability together with the faculty may spend a year getting a documentary ready. They will not even depend upon the diction of the original lecturer, because the diction of that person may be very inadequate to his really fundamental conceptioning and information, which should be superb. His knowledge may be very great, but he may be a poor lecturer because of poor speaking habits or false teeth. Another voice will take over the task of getting his exact words across. Others will gradually process the tape and moving picture footage, using communications specialists, psychologists, etc.

For instance, I am quite certain that some day we will take a subject such as Einstein's Theory of Relativity, and with the "Einstein" of the subject and his colleagues working on it for a year, we will finally get it reduced down to what is "net" in the subject and enthusiastically approved by the "Einstein" who gave the original lecture. What is *net* will become communicated so well that any child can turn on a documentary device, a TV, and get the Einstein lucidity of thinking and get it quickly and firmly. I am quite sure that we are going to get research and development laboratories of education where the faculty will become producers of extraordinary moving-picture documentaries. That is going to be the big, new educational trend.

The documentaries will be distributed by various means. One of the ways by which I am sure they will be distributed eventually has very much to do with an important evolution in communications history which will take a little describing. First, I point out to you that since the inauguration of the United States and adoption of its Constitution some very severe alterations have happened in the evolution of democracy's stimulation and response patterning and the velocity and frequency rates of that patterning's event-transformations.

At the time we founded our country, men were elected in small local areas out of communities wherein all the people were familiar with all the faces. Everybody knew Mr. Forbes or whatever his name was, and they trusted him and elected him to represent them in their federal assembly meetings. These "well known" representatives of the eighteenth and nineteenth centuries had to go to the Congress by foot or horse, for those were the means of travel. For instance, they went from some place in Massachusetts to Philadelphia or Washington, wherever the Congress was convening, and it took them a week or so to get there. They stopped along the way, meeting many friends and other folk and finding out what the aspirations of the different people's localities were.

Let us hypothetically consider how they conferred at their Congress on their individual needs and requirements; how they found certain things that were of general pertinence to all of them and found some things that were relevant only to individual areas. While they were meeting they received a letter from France, and they were very excited because France, who had helped them in the Revolution, now critically needed some help from the new United States of America. They talked about what they might do about that letter. All of these men then went back by foot or horse to their different homes and conferred face to face with their townspeople. They told their constituents what they had found out about the various things, and they said: "Here's a letter from France; this is what the various representatives at the Congress thought about it—what do you think about it?" Then they went back to the central meeting place again and acted on that letter and other pertinent matters in view of their direct knowledge of their constituents' thoughts and ambitions. The term of office that we gave representatives was predicated upon this ecological pattern of on-foot and horseback traveling. It took about four years to complete the two trips just outlined to effect a basic democratic stimulation and response cycle. The velocity rates of stimulation and response were in a one-to-one correspondence.

Suddenly new industrial technology made scientific harvesting available through invention. Lincoln became the first "wired" president—the first head of a state to be able to talk directly by telegraph to his generals at the front. This was the first time generals no longer needed to be sovereignly autonomous, because now the head of state became practically available for the highest policy decisions right at the front. World War I brought in the radio, and in World War II, for the first time, the admirals at sea were hooked up directly to Washington. They didn't need the autonomy they had to have when they took the fleet away for a year with no way to communicate with the president other than by a messenger sailing ship. Now "we the people" have radio and TV, and we obtain world-around event information from the telegraph, newspaper, and broadcast. With world-around news broadcast to us in seconds, there is no way we can respond directly to their problem-content stimuli.

We no longer have the one-to-one velocity and frequency correspondence between stimulation and response that we had in the early formative days of the U.S.A. We now have enormous numbers of stimulations and no way to say effectively what we think about them or what we would like to do about each of them. By the time that presidential voting comes around every four years we have accumulated ten thousand unvented, world-around emanating stimulations, and usually we are no longer in the same town with the representatives that we previously elected.

Automobiles move through the streets with pictures of political candidates' faces on their sides, and we try to pick out the candidates whom we think least offensive. We rarely know them or whether we may trust them. So we vote superficially for the "least offensive" ones, depending primarily on the major party selections. That is about the best we can do.

Because all this is so, those now doing the representing, wishing to be returned to office, wish to know what people are thinking about all the important issues. So the surveys of public opinion have developed, and congressional investigations of many phenomena have increased. We have to have a kind of anticipatory political reconnaissance going on all the time. Even then, when the elected man comes in he knows that it is only as the result of indirect effects of total psychological moods; so he pays little attention to any specific "mandates," and he begins to work right away on the psychological culturing of his next election. He is not really sure that there are any true mandates. He doesn't really know what the people think. That is one large reason why democracy is in great trouble today, because of the vacillation and compromise arising from the lack of one-to-one correspondence between stimulation and response of the electorate. The Communists and dictatorships scoff at democracy—saying it doesn't work. I am sure that democracy is inherently more powerful and capable and appropriate to man's needs than any other form of government, but it needs proper updated implementation to a one-to-one velocity correspondence in respect to each and every stimulation-and-response, and then democracy can work—magnificently.

I have talked to you about solving problems by design competence instead of by political reform. It is possible to get one-to-one correspondence of action and reaction without political revolution, warfare, and reform. I find it possible today with very short electromagnetic waves to make small reflectors by which modulated signals can be beamed. After World War II, we began to beam our TV messages from city to city. One reason television didn't get going before World War II was because of the difficulty in distributing signals over long distances from central sources on long waves or mildly short waves. We were working on coaxial cables between cities, but during the war we found new short ranges of electromagnetic frequencies. We worked practically with very much higher frequencies, very much shorter wave lengths. We found that we could beam these short waves from city to city. Television programs are brought into the small city now by beam from a few big cities and then *rebroadcast* locally to the home sets. That is the existing TV distribution pattern. My invention finds it is now possible to utilize the local TV masts in any community in a new way. Going up to, say, two hundred, three hundred, or four hundred feet and looking down on a community you see the houses individually in the middle of their respective land plots. Therefore, with a few high masts having a number of tiny massers, lassers, or reflectors, each beam aimed accurately at a specific house, the entire community could be directly "hooked up" by beams, instead of being broadcast to. This means a great energy saving, for less than 1 per cent of the omnidirectionally *broadcast* pattern ever hits a receiving antenna. The beaming makes for very sharp, clear, frequency-modulated signals.

In the beaming system, you also have a reflector at the house that picks up the signal. It corresponds directly to the one on the mast and is aimed right back to the specific beaming cup on the mast from which it is receiving. This means that with beam casting you are able to send individual messages to each of those houses. There is a direct, fixed, wireless connection, an actual direct linkage to individuals; and it works in both directions. Therefore, the receiving individual can beam back, "I don't like it." He may and can say "yes" or "no." This "yes" or "no" is the basis of a binary mathematical system, and immediately brings in the "language" of the modern electronic computers. With two-way TV, constant referendum of democracy will be manifest, and democracy will become the most practical form of industrial and space-age government by all people, for all people.

It will be possible not only for an individual to say, "I don't like it," on his two-way TV but he can also beam-dial (without having to know mathematics), "I want number so and so." It is also possible with this kind of two-way TV linkage with individuals' homes to send out many different programs simultaneously; in fact, as many as there are two-way beamed-up receiving sets and programs. It would be possible to have large central storages of documentaries—great libraries. A child could call for a special program information locally over the TV set.

With two-way TV we will develop selecting dials for the children which will not be primarily an alphabetical but a visual *species* and *chronological category* selecting device with secondary alphabetical subdivisions. The child will be able to call up any kind of information he wants about any subject and get his latest authoritative TV documentary, the production of which I have already described to you. The answers to his questions and probings will be *the best information* that man has available up to that minute in history.

All this will bring a profound change in education. We will stop training individuals to be "teachers," when all that most young girl "education" students really want to know is how they are going to earn a living in case they don't get married. Much of the educational system today is aimed at answering: "How am I going to survive? How am I going to get a job? I must earn a living." That is the priority item under which we are working all the time—the idea of *having to earn a living*. That problem of "how are we going to earn a living?" is going to go out the historical window, forever, in the next decade, and education is going to be disembarrassed of the unseen "practical" priority bogeyman. Education will then be concerned primarily with exploring to discover not only more about the universe and its history but about what the universe is trying to do, about why man is part of it, and about how can, and may man best function in universal evolution.

Automation is with us. There is no question about it. Automation was inevitable to intellect. Intellect was found to differentiate out experience continually and to articulate and develop new tools to do physically repeated tasks. Man is now no longer *essential* as a worker in the fabulously complex industrial equation. Marx's *worker* is soon to become utterly obsolete. Automation is coming in Russia just as it is here. The word *worker* describing man as a muscle-and-reflex machine will not have its current 1961 meaning a decade hence. Therefore, if man is no longer essential as a worker we ask: "How can he live? How does he acquire the money or credits with which to purchase what he needs or what he wants that is available beyond immediate needs?" At the present time we are making all kinds of economic pretenses at covering up this overwhelming automation problem because we don't realize adequately the larger significance of the truly fundamental change that is taking place in respect to man-in-universe. As automation advanced man began to create secondary or nonproductive jobs to make himself look busy so that he could rationalize a necessity for himself by virtue of which he could "earn" his living. Take all of our bankers, for example. They are all fixtures; these men don't have anything to do that a counting machine couldn't do; a punch button box would suffice. They have no basic banking authority whatsoever today. They do not loan you their own wealth. They loan you your own wealth. But man has a sense of vanity and has to invent these things that make him look important.

I am trying to keep at the realities with you. Approximately total automation is coming. Men will be essential to the industrial equation but not as workers. People are going to be utterly essential as consumers—what I call *regenerative consumers,* however, not just swill pails.

The vast industrial complex undertakings and associated capital investments are today so enormous and take so long to inaugurate that they require concomitantly rapid regenerative economics to support them. The enterprise must pay off very rapidly in order to be able to refund itself and obtain the economic advantage to inaugurate solution of the next task with still higher technical advantage. In that regenerative cycle of events, the more consumers there are the more the costs are divided and the lower the individual prices. The higher the frequency of the consuming the more quickly the capital cost can be refunded, and the sooner the system is ready for the next wave of better technology. So man is essential in the industrial equation as a consumer— as a regenerative consumer, a critical consumer, a man who tasting wants to taste better and who viewing realizes what he views can be accomplished more efficiently and more interestingly. The consumer thus becomes a highly critical regenerative function, requiring an educational system that fosters the consumer's regenerative capacity and capability.

At present, world economics is such that Russia and China work under an integrated socialist planning in competition with our literally disorganized economic world (for our anti-trust laws will not permit organization on a comprehensive basis). The Communists have high efficiency advantage because of their authoritarianism. We have very little centralized authority, save in "defense." The Communists now have the industrial equation, too, in large scale, and soon complete automation will be with them. They are very much aware of the fact that the more customers there are, the more successful the operation will be, because the unit costs are progressively lower. This is why the Soviets were historically lucky in getting China as customers. They would like also to have, exclusively, India and Africa as customers. If Russia acquires the most customers, we will not be able to compete. They will always have the lower costs on any given level of technology. We are going to have to meet this possibility and meet it vigorously, swiftly, and intelligently. Within the next decade, if we survive at all as an organized set of crossbreeding men on the American continent it will be because we will have suddenly developed a completely new attitude on all these matters. In case you are apprehensive that social and political economics are to be so laggard as to impede your advanced educational programming, it is well to remember that the comprehensive world economics are going to force vast economic reforms of industries and nations, which incidentally will require utter modernization of the educational processes in order to be able to compete and survive.

Every time we educate a man, we as educators have a regenerative experience, and we ought to learn from that experience how to do it much better the next time. The more educated our population the more effective it becomes as an integral of regenerative consumer individuals. We are going to have to invest in our whole population to accelerate its consumer regeneration. We are going to be completely unemployed as muscle-working machines. *We as economic society are going to have to pay our whole population to go to school and pay it to stay at school.* That is, we are going to have to put our whole population into the educational process and get *everybody* realistically literate in many directions. Quite clearly, *the new political word* is going to be *investment.* It is not going to be *dole,* or socialism, or the idea of people hanging around in bread lines. The new popular *regenerative investment* idea is actually that of making people more familiar with the patterns of the universe, that is, with what man has learned about universe to date, and that of getting everybody intercommunicative at ever higher levels of literacy. People are then going to stay in the education process. They are going to populate ever increasing numbers of research laboratories and universities.

As we now disemploy men as muscle and reflex machines, the one area where employment is gaining abnormally fast is the research and development area. Research and development are a part of the educational process itself. We are going to have to invest in our people and make available to them participation in the great educational process of research and development in order to learn more. When we learn more, we are able to do more with our given opportunities. We can rate federally paid-for education as a high return, mutual benefit investment. When we plant a seed and give it the opportunity to grow its fruits pay us back many fold. Man is going to "improve" rapidly in the same way by new federally underwritten educational "seeding" by new tools and processes.

Our educational processes are in fact the upcoming major world industry. This is *it;* this is the essence of today's educational facilities meeting. You are caught in that new educational upward draughting process. The cost of education will be funded regeneratively right out of earnings of the technology, the industrial equation, because we can only afford to reinvest continually in humanity's ability to go back and turn out a better job. As a result of the new educational processes our consuming costs will be progressively lower as we also gain ever higher performance per units of invested resources, which means that our wealth actually will be increasing at all times rather than "exhausted by spending." It is the "capability" wealth that really counts. It is very good that there is an international competitive system now operating, otherwise men would tend to stagnate, particularly in large group undertakings. They would otherwise be afraid to venture in this great intellectual integrity regeneration.

I would say, then, that you are faced with a future in which education is going to be number one amongst the great world industries, within which will flourish an educational machine technology that will provide tools such as the individually selected and articulated two-way TV and an intercontinentally net-worked, documentaries call-up system, operative over any home two-way TV set.

The new educational technology will probably provide also an invention of mine called the Geoscope—a large two-hundred-foot diameter (or more) lightweight geodesic sphere hung hoveringly at one hundred feet above midcampus by approximately invisible cables from three remote masts. This giant sphere is a miniature earth. Its entire exterior and interior surfaces will be covered with closely-packed electric bulbs, each with variable intensity controls. The lighting of the bulbs is scanningly controlled through an electric computer. The number of the bulbs and their minimum distance of one hundred feet from viewing eyes, either at the center of the sphere or on the ground outside and below the sphere, will produce the visual effect and resolution of a fine-screen halftone cut or that of an excellent television tube picture. The two-hundred-foot geoscope will cost about fifteen million dollars. It will make possible communication of phenomena that are not at present communicable to man's conceptual understanding. There are many motion patterns such as those of the hands of the clock or of the solar system planets or of the molecules of gas in a pneumatic ball or of atoms or the earth's annual weather that cannot be seen or comprehended by the human eye and brain relay and are therefore inadequately comprehended and dealt with by the human mind.

The Geoscope may be illuminated to picture the earth and the motion of its complete cloud-cover history for years run off on its surface in minutes so that man may comprehend the cyclic patterning and predict. The complete census-by-census of world population history changes could be run off in minutes, giving a clear picture of the demological patterning and its clear trending. The total history of transportation and of world resource discovery, development, distribution, and redistribution could become comprehendible to the human mind, which would thus be able to forecast and plan in vastly greater magnitude than heretofore. The consequences of various world plans could be computed and projected. All world data would be dynamically viewable and picturable and relayable by radio to all the world, so that common consideration in a most educated manner of all world problems by all world people would become a practical event.

The universities are going to be wonderful places. Scholars will stay there for a long, long time—the rest of their lives—while they are developing more and more knowledge about the whole experience of man. All men will be going around the world in due process as everyday routine search and exploration, and the world experiencing patterning will be everywhere—all students from everywhere all over the world. That is all part of the new pattern that is rushing upon us. We will accelerate as rapidly into "yesterday" through archaeology as we do into "tomorrow." Archaeology both on land and under the seas will flourish equally with astronautics.

Glenn Olds served as U.S. Ambassador to the United Nations Economic and Social Council. He has been president of three universities, including Ohio's Kent State University after the tragic shooting of students on May 4, 1970, and has served as special assistant to four U.S. presidents. He is currently chairman of the board of the World Federalist Association.

Olds, E. J. Applewhite, and Neva Goodwin Rockefeller are best remembered by Fullerites as the nucleus of the Fuller International Design Science Institute. Olds was president of the institute as Bucky was developing the Wichita Dymaxion house. The Wichita house, when completed, would weigh only 6,000 pounds, with two bedrooms, two Dymaxion bathrooms, a kitchen, and a 28-foot diamond-shaped living room. Although designed more than half a century ago, this unique structure defies time and is today still a *tour de force*, an affordable experimental house for the average man or woman. It is regrettable that Bucky never saw the Wichita house go into full production. Currently the Henry Ford Museum at Greenfield Village in Dearborn, Michigan, is restoring the Wichita Dymaxion house.

Glenn Olds remembers Bucky as a "Cosmic Surfer."

R. Buckminster Fuller: Cosmic Surfer

Dr. Glenn A. Olds

My first and last meeting with Bucky was fittingly on a university campus. He was then, first and last, inviting us to enroll in the University of the Universe, from which there is no graduation.

"Dare to be naive," he was telling us. "In cosmic calculus, you are its highest number." Had he lived more fully in the lingo of our computer age, he would have said we are its indispensable Web site, the "personal regenerative energy" through which and by which the universe works. To know this is to be wise; to do this is to be successful in the cosmic venture; to share it with others is the highest form of freedom and fulfillment.

Such wisdom comes not merely by opening your eyes, though seeing is essential, but by knitting your brows, which is to exercise your true nature, your spiritual birthright, your authentic and unique place in the scheme of things. Such wisdom is open to every person on the same condition. It cannot be commanded or coerced. It is never secondhand. It is first person singular, experimental, universal, the way the world works.

"Look," he would say, "at who you are, and where you are." The law of entropy may be a foundational building block of physics, but not for the human mind. Indeed, he would remind us, as had Thomas Huxley, father of Julian and Aldous, nearly a century before, that the human mind can reverse that law. It *is* regenerative, not only bringing order out of chaos, inventive creativity, or in Bucky's trenchant phrase, "Doing more with less." This anchored Bucky's optimism, rallied a confused and diffused student generation across the campuses and world to an inspired empowerment, a new sense of self-education, a new "World Game" designed for win-win strategies, resisting the counsel of scarcity, fear-ridden violence and possessiveness, and in Bucky's cosmic classroom, full participating partners in discovering the "operational manual" for Spaceship Earth.

Surfing

One whose life and early listening were set in the surf surroundings of his beloved Bear Island in the rugged contour of the Maine coast would recognize surfing as a fit symbol for the regenerative integrity of the pulsating phases of Bucky's extraordinary life. Bucky was schooled in that surfside sea. Anyone who ever visited him there, as I have done, sensed the strength and wisdom he drew from that place. Whether under the silent stars, his navigation chart, with his "summer of children visitors," or standing almost awestruck on the shore with him marveling at the unusual energy of the tidal lift of the moon or undulating tidal comings and goings; or listening to the wind whisper through the tall trees, arm stretched, skyward, en route from boat to house, one knew this was Bucky's "home." Like Thoreau's Walden Pond, Muir's Wilderness, or Kant's belfry beacon, it was where all the intricate webbing of a marvelous mind knit together. It was the netscape of his horizons. It anchored his soul.

Bucky knew, at the crest of the wave, nature marshaled its propelling power. An article I did on his remarkable life, "At the Crest of the Wave," in the resurrected *Saturday Evening Post* of March–April 1973, made this telling point. The agility, flexibility, and strength required to negotiate the balance, movement, and forward thrust at the crest of the wave was Bucky's mind dancing with innocent delight at the naked edge of nature's design and power. This was why even children, as well as the world's seminal thinkers, found such fun in the frolicking of Bucky's mind.

It was Bucky's genius to see the extravagance of Nature's unused and abundant energy, the raw material for human inventiveness. His early proposals for connecting the energy grids of The World (see the remarkable strategy of GENI—Global Energy Network International—for its deployment) to use the idle power at night on one side of the earth to light and lift the load of the other was but one of the multitude of his insights on this issue. Another was his early plan to create a new sea-city in Tokyo Bay energized by harnessing the unused power of the tidal pull of the moon and thermal differentiations of that remarkable sea.

It was this pondering of our human capacity to harness what nature lavishly provided, and navigate its most troubled waters by the delicate orchestration of its powerful and provident resources, that was at the root of Bucky's confidence in harnessing technology and resources to make the all-around human venture truly successful for all its players. He thought every person, if properly "schooled," could sail through life at the crest of the wave.

Significance of the Sea

It is not difficult to understand why for Bucky it was easy to see the sea as the metaphorical substitute for space in our cosmic setting. In his enduring and endearing description of "Spaceship earth," "ship" is as important as "space" in localizing our habitat.

His speculation that our origins may more reasonably be located in the Polynesian culture of the Pacific than in Africa is not the work of an empirical anthropologist. His reasons are not dug from the earth but from the human mind, and its inferences form what we know about how the earth turns, winds blow, prevailing tides propel, and stars guide. Though Bucky never had a James Michener to tell *this* story, he was intrigued by its potential scenario, only partially portrayed in Michener's *Hawaii*.

Bucky's Dymaxion map, which hangs on my study wall, reminds one of how he saw the world as connected islands in a sea of water. Even as he believed the lunar cycle, its tidal counterpart, and woman's menstrual periods had some cosmic connection not yet found, so he believed it was not sheer chance that so much of our earthly embodiment was water. He did not go so far as Aldous Huxley in proposing that we are amphibious, but the metaphor was not lost on him. That the basic elements of earlier cultures seemed reducible to earth, air, fire, and water left Bucky puzzling over man's Piscean nature in the Age of Aquarius. His sense of the role of astronomy, astrology, and numerology in the evaluation of mathematical and geometric models in his almost daily doodling on napkins and scraps of paper over dinner reflected his preoccupation with sailing the seas, navigating in unknown and uncertain waters, and how we can use wind, tides, and stars in all of nature's abundance to awaken us to ways of invention and partnership to take us places we had never been before. Somewhere, in all this, he felt sure was a remarkable cosmic design.

Design

Few people grasped so immediately, intrinsically, and profoundly the factor of design in all that is and is yet to be. That is why even his beloved Harvard, which threw him out twice, then welcomed him finally as deserving of its honor, classifying him as poet and architect, saw this incredible talent for discerning pattern in all possibilities.

It was this sense of pattern-making, whether in the poetry of words, the "geometry of thinking," (as he characterized his *Synergistics*), or the artistry of construction which characterized his inventive architecture, that made *Design* so central to his *ideas* and the *integrities* of his life. (See his book *Ideas and Integrities*.) And it was this factor, I believe, which moved Bucky, a real "loner" who was suspicious of all forms of power, political or organizational, to reluctantly concede to the formation in 1972, at the Carnegie Center for Peace, across from the United Nations, the International Design Science Institute.

The Institute, launched in midsummer, to which the *New York Times* devoted a full page in surprise and speculation, met with interest and puzzlement. Its credentials were impressive enough. The World Advisory Council was composed of Arthur C. Clarke, Constantinos A. Doxiadis, Walter J. Hickel, Margaret Mead, Jonas E. Salk, U Thant, and Jerome B. Wiesner. The organizing and management body included some of Bucky's oldest and most devoted friends: Fritz Close, former chairman of Alcoa; Gerald Dickler, Bucky's lawyer and distinguished New York publishing legal expert; Edgar Applewhite, family member and Bucky's coauthor and editor; Norman Cousins, founding editor of the *Saturday Review;* Charles M. Harr, professor of law at Harvard and onetime assistant secretary of housing and urban development; Neva Kaiser, one of Bucky's brilliant young campus enthusiasts, later to head one of Tuft's distinguished institutes; Martin Meyerson, president of the University of Pennsylvania and former dean of the College of Environmental Design at Berkeley; John S. Rendleman, president of Southern Illinois University; M. W. Whitlow, Bucky's longtime banker and personal friend; and William M. Wolf, president of Wolf Computer Corporation, early involved in the U.S. space program. I myself, just leaving my post as U.S. Ambassador to the United Nations Economic and Social Council and assuming the presidency of Kent State University after the tragic shooting there, was the institute's president.

It was our conviction that the impressive range of Bucky's inventions, patents, and designs should be more widely known and, where possible and practicable, converted into useful artifacts and made humanly available for what Bucky called "the World around success of every human being." And by converting weaponry *to* livingry, to quote another of his powerful images, help a then deeply troubled world, paralyzed by the Cold War and Vietnam and on the eve of Watergate, to find its way "home."

The institute's modest influence was a commentary on the resistances of the times and what Whitehead called the inertia of empty categories and dead ideas. With minor instances of support from the Rockefeller Brothers Fund and isolated friends, the Institute was never able to rally the financial resources to implement its mission. In some ways, alas, it confirmed Bucky's deepest conviction that the evolutionary design of the universe rested on solitary, particular, individual inventiveness and sustaining power, and the universe could not be forced or bargained with and was indifferent to the games we play to control or subordinate others. This conviction was sufficient to bear Bucky up and carry him forward. But, of course, it could not sustain the Institute.

The final deathblow to its effort came after a last valiant effort to enlist Ambassador Phil Klutznick of Chicago, my predecessor at the UN, confidant of presidents, brilliant civic leader of Chicago, and most successful businessman, who had known Bucky in the brief period in Wichita after World War II when he had been asked to mass-produce his geodesic-dome house to provide low-cost, environmentally sound, and replicable housing for returning servicemen. Ambassador Klutznick was persuaded of the institute's purpose, the relevance of Bucky's powerful ideas, and the need for sustainable financing for the institute's functions and Bucky's future.

One never knows why the final negotiations of so delicate a deal—negotiations between a man whose second nature was privacy but whose ideas were genuinely universal and invited implementation and an exceedingly successful public servant, business entrepreneur, and social engineer—should fail at the last moment, but they did.

It seemed enough to Bucky to say "I told you so" and to convert the institute into an advisory council primarily concerned with preserving and perpetuating Bucky's considerable archival repository over a lifetime. Happily, Norman Cousins was willing to succeed me in that new venture, and Bucky's dedicated grandson and daughter together with other close friends have kept that vision of the Design Science Institute's appropriate role vital and growing.

Key Concepts

If one were to characterize Bucky's range and depth of ideas over his productive lifetime as a system, it would have to be a different kind of system from those we are generally used to describing. Though for him the universe was finite, that did not mean it was finished, bounded, or final. Indeed it was Bucky's intent to show that Einstein's genius in characterizing nature's behavior in his famous equation $E = mc^2$, did more than show the transformational relation of energy and mass, gravity and radiation, space, time, and motion. It provided Bucky a structure to show that form itself is a verb, and that at work in the universe is this incredible design that can remain orderly but changing, substantive but transformational, binding but never finally bonding, that its pattern is evident but never repeatable. This deep discernment runs through each of his generic concepts.

Comprehensivity

Here is a characteristic Bucky term. He wants to say more than "comprehensive"; he means whole, inclusive, all-comprehending. He wants to include more than the formal, rational, mathematical, or abstract. He wants to suggest a combination of sensitivity and compassion, a dynamic grasp of the aesthetic and technical, the appreciative and descriptive, the pattern of the whole and its process.

For Bucky, the tragedy of the specialization that dominates our educational and knowledge model is not its analytic clarity or simplicity, or even indeed its insular isolation, limiting as they are. The tragedy is to miss the whole. It is to mistake simplicity this side of complexity for simplicity on the other side (to use Oliver Wendell Holmes's trenchant description). It is to miss the marvel of the total design, in which orchestration of the parts must be ultimately understood. Miss this, and you miss its essential meaning, its mirroring of cosmic design, the transformational character of pattern, the telic character of all true knowing. To invert Marshall McLuhan's maxim "The medium is the message," the message is always more than the medium. The more, to use Bucky's characteristic language, is weightless. It does not add to physical properties. It is the formless form of the half hitch, which can inform any material (hemp, cotton, steel, twine, etc.). It is the way we were meant to think, to exercise our transformational propensities, to *be*, which is always a verb.

Synergy

This follows comprehensivity. This is, to use Bergson's term, the *élan vital* of all our forms of structure, life, and understanding. It is the manifestation of non-entropic energy where the whole is always more than the sum of its parts. It gives the lie to all reductionistic forms of logic, driving through the rearview mirror, perpetuating the past as a vital and viable part of the unformed tomorrow.

At work in this concept is a new meaning for causality. It faintly echoes Aristotle's formal and final causes, but is not reducible to either. It is not simply push or pull, measures of linear progression, but can more readily be characterized as the artist's idea that permeates the product, the living oak that celebrates its form in the tree, the manner in which a dream informs the deed. It is why secondhand concepts of experience are already shadowy in substance. It is why so much of our learning is boring, tedious, or dumb, and why our images of ourselves are so frequently blinding, binding, or deadening of potential. Synergy is the style through which energy dances through space and time. It is the way we become what we are though perpetually differing. It is what sustains and propels our spaceship and informs the design of all our courses. It is the right relationship for sustaining community through every civil or political design.

Integrity

Integrity is more than making the outer and inner man fast friends, as Plato affirmed. The category is more than moral or mathematical, though it is both. It is metaphysical. It is an appropriate characterization of the trustworthiness of God. It makes the concept of covenant, the constant in all equality, the patterned integrity of all diversity, the one in many, understandable, dependable, authentic, and demonstrable.

Without integrity, neither mathematics nor morality could stand. Diversity would dissolve into anarchy, and difference would beget enmity and violence. It was this understanding, so central to Bucky's thought and life, that endeared him to thousands of students, worldwide. They were sick of duplicity, deception, and contradiction, and their attendant manipulations and exploitive power. They welcomed Bucky's conviction as beliefs buried within them. His credibility was rooted in integrity, his own and that of his ideas.

Generosity

Though Bucky speaks little of this virtue in his writings, it was eloquent in his life. One of the pages of the book Bucky kept at Bear Island for his young visitors every summer, in which they were invited to write what they had learned from their visit, leaps out at me:

"Dear Bucky, Thanks. I learned a little about sailing and a lot about loving."

His name I have forgotten, but the age was ten, and the sentiment ageless. Bucky felt if the universe was so provident and generous, could we be less as the local manifestation of its regenerative design? For him this was no burdensome obligation or societal imperative. It was the natural expression of who we are and were intended to be. We were lent to be spent. We pay the rent for our space and place, not for possessing, grasping, or holding, but by giving, channeling, releasing in regenerative form our unique inventiveness. His favorite illustration was how the half hitch can inform and pass through any rope like string, so long as it remains loose; but tighten the loop, knot it, and that's the end of the form. Though I never knew him to say as much, I believe his view of immortality lay in the imperishability of the way in which we let life work through us, giving out and away the energy, and synergy, and integrity, of who we are and what the cosmos intended.

Anticipatory

Bucky had a marvelous sense of "fast forward." He seemed always "ahead of his time." Any biography of Bucky will disclose the remarkable way in which he "anticipated" future events. They were so clear to him he did not think of them as imagination or speculation. They were real, awaiting only actualization. Nor was anticipation purely visionary. His views were grounded in fact, pattern, and cumulative history. His archives, kept personally over the years, augmented here and there by volunteers and staff, were a montage of meaning mirroring nearly a century. His "World Game," so successfully developed in Philadelphia when Bucky served as a distinguished world scholar/statesman under President Meyerson's creative support, was soundly rooted in this aggregation of fact-event-resources from which patterned trends and consequences could be seen and anticipated.

He saw wealth as created by the regenerative power of the human mind in partnership with resources ever inventive of that technology capable of doing more with less. His Dymaxion revolution and revelation root in this anticipatory turning of problems into opportunity, limited resources into wealth.

The future was not a place in space or time, but the actualization of anticipatory events through human invention, and application of design sciences. This is why he remained to the end, with the innocence and wonder of a child, open to the discovery of ever new linkage that held the promise of regenerative principles and their applicability.

Doability

"Only the impossible is possible," Bucky never tired of saying, driving home his conviction that ideas were meant to inform action. He was not a pragmatist. He did not believe that the test of the truth of an idea was its doability. On the contrary, in an inverted way, he believed that if you could not think it, you would not do it. Its doability was its thinkability. This sprang from his conviction that it was the grasp of principles, and their intrinsically integrative interrelatedness, that made both knowledge and dependable, transformational action possible. Creative action was not merely the result of its grasp or thinkability, it was part of the dynamic of both grasping and doing.

As with discovery of the principle of the lever, after human stumbling over, walking on, encountering the principle at work in fallen trees, load-lifting propensities of nature, to finally grasp its significance carried intrinsically its adaptability and fruitfulness in a wide range of applications for human good. Its adaptation with the wheel to the waterwheel effected a giant revolution in agriculture, in construction, in levered tools of every kind and description.

This made Bucky so magical on campuses so devoted to the simonizing of secondhand symbols. As Whitehead had said, "In the garden of Eden we saw the animals before we named them." The tragedy of modern education is that it learns the names and rarely sees them.

The impossible had its fascination for Bucky because of its open-ended invitation to discovery. If an idea was only possible, it was already disposed to become actual, and it remained but to trace out that disposition of probability. But if it was impossible, there were no barriers of probability to deter the free play of anticipatory rehearsal, naive wonder, or unpredictable consequences, or the surprise and delight of finding what is not, or even cannot be, suddenly and surprisingly appear. It made one wonder at Tagore's wisdom from the East, "To exist is a perpetual surprise, which is the joy of life." What cannot be through its own self explanation, or any form of necessity, should, indeed, be, is the ultimate surprise. It is not surprising it anchored one of the classical and endurable arguments for God, the cosmological. It further inferred, such an impossible surprise, since not necessitated or anticipated, must be a free act, to *be* must be an act, not a fact, a verb not a noun!

In the end, I suppose, one could say Bucky was the best illustration of both his talk and walk. "What he was," as Emerson once eloquently remarked, "shouted so loudly I could not hear a word he said." His very being anchored this thinking, and was the final vindication of his ideas and his life.

Bucky's life was full of "jitterbug transformation," a kind of dance with pattern but spontaneous, anchored in individuality, never secondhand, yet embracing a whole ever more than the sum of any step or succession of steps. It kept a beat both uniquely its own and yet in some strange way synergistic with the integrity of the others, and generous in affording all others the space, incentive, and opportunity for their own thing. This perpetual movement of pattern and form, extravagant of energy yet inclusive of synergy, to be anticipated but never slavishly copied, generous yet mindful of the integrity of personal responsibility and choice, was how he lived out his days.

His call to me was never loud or raucous, never commanding or coercive. It was ever invitational, a call to being, ever becoming. It was, as I said at the beginning, Bucky's invitation to the University of the Universe. He expected no graduation, nor do I.

Comprehensive Designing from *Ideas and Integrities*

The Comprehensive Designer emerges as the answer to the greatest problem ever addressed by mankind: The Human Family now numbering three billion is increasing at an annual rate of three per cent and is trending toward seven billion expected by the end of the second half of this century. Of this number, sixty-five per cent are chronically undernourished, and one third are doomed to early demise due to conditions which could be altered or eliminated within the present scope of technology; specifically, that area of technology comprising the full ramifications of the building arts, which now contains the negatives or blanks which match the lethal factors. Relative to this premise, Jawaharlal Nehru once said in Chicago: "It is folly to attribute the disquietude of the Orient to ideological pressures." Nehru went on to point out that the de-energized and doomed are prey to any political shift of the wind that might promise arrest of their fate.

At present all the world's industrial, or surfaced, processed and reprocessed functional tonnage (the *Industrial Logistics*) is preoccupied in the service of four tenths of the world's population, though one hundred per cent are directly or indirectly involved in its procurement, processing and transportation.

All the politician can do regarding the problem is to take a fraction of that inadequate ratio of supply from one group and apply it to another without changing the over-all ratio. The politician can, of course, recognize and accept the trend rather than oppose it, but this does not accelerate it in adequate degree to arrive at a solution in our day and generation, and, more importantly, before the deadline of the doomed.

All that money can do is shower paper bills of digits on the conflagration. Relative denominations neither decrease nor increase the velocity of combustion.

How and by whom, if at all, may the problem be solved? Scientists are often charged with the task, but scientists as a class (irrespective of their proclivities as individuals) do not function in the *comprehensive* capacity, they function as specialists in taking the universe apart to isolate and inventory its simplest behavior relationships. Engineers function as invoked specialists in reproducing satisfactory interactions of factors ascertained as "satisfactory" by past experience and a wealth of behavior measurement. Both engineers and politicians would lose their credit from society if they incorporated the unprecedented in wholesale manner.

We hear and read frequently in scientific and philosophical journals of the desirability of ways in which problems of the universe may once more be approached by comprehensive and scientific principles.

A New Social Initiative

There emerges the need for a new social initiative which is not another function of specialization but is an integral of the sum of the produce of all specializations, that is, the Comprehensive Designer.

The Comprehensive Designer is preoccupied with anticipation of all men's needs by translation of the latest inventory of their potentials. Thus he may quickly effect the upping of the performance-per-pound of the world's industrial logistics in fourfold magnitude through the institution of comprehensive redesign, incorporating all of the present scientific potentials that would otherwise be tapped only for purposes of warfaring, defensively or offensively.

In view of our myriad of performance-per-pound-advances of multifold degree (in contrast to percentage degrees) typified by pounds of rubber tire upped in performance from one thousand miles to thirty thousand miles expectancy without poundage increase (yet with complete chemical, though invisible, transformation), or of communication advance from one message to two-hundred-fifty concurrent messages per unit of cross section of copper wire (and both of these multifold advances have been accomplished within a quarter of a century), it is seen as a meager technical problem to consider advancing the over-all efficiency of worldwide industrial and service logistics fourfold (to serve one hundred per cent of the population).

Some may tend to underestimate the comprehensive nature of the problem, saying the people are thus starving and we have the land capacity to raise the food. This conception voiced by the theoretical specialist or casual observer is without benefit of logistic experience. It is not just a matter of raising food but getting food to people, anywhere from zero to twenty-five thousand miles distant. And then it is not just a matter of getting food to people zero to twenty-five thousand miles away—it is a matter of getting it there at certain velocities; and it is not just a matter of getting it there at certain velocities, but it is a matter of getting it there on schedules in certain conditions, conditions of nourishing content, palatability and vital preservation.

And even then it is not a matter of success concerning all the preceding conditions, for the dumping of a year's food supply in front of a helpless family huddled on the street-curb is but an unthinkable tragedy. The maggots appear in hours. And once again the continuing energy controls providing progressive freezes, heatings, etc. cannot be effected by refrigerators and stoves dumped in the street along with a year's tonnage of food. Obviously, a world continuity of scientific-industrial controls resultant from comprehensive and technical redesign is spelled out as the irreducible minimum of solution.

For those who think that this minimum can be obtained through legislative enactment by the politician or by the establishment of new dollar credits, and who are forgetful that the total world tonnage is already preoccupied with service of only forty-four per cent of the world-people, it is to be noted that the economic-statistical approach has been voiced by the press in conjunction with water shortages in the great American cities, such as New York and Los Angeles.

These are not problems unique to those cities, but symptomatic of the trend of the great industrial interactions. The economic-statistical solution, voiced by the politicians and the news, proposes further encroachments of the watershed origins through the rerouting of waters otherwise destined to lesser centers.

A typical question asked by the Comprehensive Designer is: What do people want the water for? They are using two hundred gallons per day per capita, consuming only one gallon for their vital processes while employing one hundred ninety-nine gallons to dunk themselves, and gadgets, and to act as a liquid conveyor system of specks of dirt to the sea. We note that scientists do not need water to dunk their instruments in, nor industrialists water to soak their machinery in. Are there not superior ways to effect many of the end purposes involving no water at all, and where water is found to be essential, can it not be separated out after its combining functions and systematically recirculated as chemically pure, sterilized, "sweet" and clear, and with low energy expense or even an improved energy balance sheet as a result of comprehensive redesign?

The specialist in comprehensive design is an emerging synthesis of artist, inventor, mechanic, objective economist and evolutionary strategist. He bears the same relationship to society in the new interactive continuities of worldwide industrialization that the architect bore to the respective remote independencies of feudal society.

The architect of four hundred years ago was the comprehensive harvester of the potentials of the realm. The last four hundred years have witnessed the gradual fade-out of feudalism and the gradual looming of what will eventually be full of world-industrialization—when all people will produce for all people in an infinity of interacting specialized continuities. The more people served by industrialization, the more efficient it becomes.

Positive Constituents of Industry

In contrast to the many negative factors inherent in feudalism (such as debt, fear, ignorance and an infinite variety of breakdowns and failures inevitable to dependence on the vagaries of Nature), industrialization trends to "accentuate the positive and eliminate the negative," first by measuring Nature and converting the principles discovered in the measurements to mastery and anticipation of the vagaries. Day and night, winter and summer, fair weather or bad, time and distance are mastered. Productive continuities may be maintained and forwardly scheduled. There are three fundamental constituents of industry; all are positive.

The first consists of the aspect of energy as *mass,* inventoried as the ninety-two regenerative chemical elements which constitute earth and its enclosing film of alternating liquid-gaseous sequence.

The second fundamental component of industry consists of energy but in a second and two-fold aspect—that is: (a) energy as *radiation* and (b) energy as *gravitation,* both of which we are in constant receipt of from the infinite cosmic fund. The third and most important component of the industrial equation is the intellect factor, which secrets a continually amplifying advantage in experience-won knowledge.

Complex-component No. 1 cannot wear out. The original chaotic disposition of its ninety-two regenerative chemical elements is gradually being converted by the industrial principle to orderly separation and systematic distribution over the face of earth in structural or mechanical arrangements of active or potential leverage-augmentation. Component No. 2, cosmic energy, cannot be exhausted.

Constituent No. 3 not only improves with use but is interactively self-augmenting.

Summarizing, components No. 1 and No. 2 cannot be lost or diminished and No. 3 increases, with the net result inherent gain. Inherent gain is realized in physical advantage of forward potential (it cannot be articulated backwards; it is mathematically irreversible). Thus, industrial potential is schematically directional and not randomly omni-directional. Thus, the "life" activity, as especially demonstrated by man, represents an anti-entropic phase of the transformations of non-losable universal energy.

The all-positive principle of industry paradoxically is being assimilated by man only through emergent expedients, and only in emergency because of his preponderant fixation in the direction of tradition. Backing up into his future, man romantically appraises the emergent dorsal sensations in the negatively parroted terms of his ancestors' misadventures.

The essence of the principle of industry is the principle of synergy, which I have explained in an earlier chapter. This principle is manifested both in the inorganic and organic. The alloying of chrome and nickel and steel provides greater tensile strength than that possessed by any of its constituents or by the constituents in proportional addition. Three or more persons by specialized teamwork can do work far in excess of that of three independently operating men. Surprisingly, and most contradictory to the concept of feudal ignorance, the industrial chain's strength is not predicated on its weakest link. So strong is the principle that it grows despite a myriad of superficially failing links! In fact, there are no continuous "links" in industry or elsewhere in the universe because the atomic components are, interiorly, spatially discontinuous.

The strength of "industry" as with the strength of the "alloy" occurs through the concentric enmeshment of the respective atoms. It is as if two non-identical constellations of approximately the same number of stars each were inserted into the same space, making approximately twice as many stars, but none touching due to the difference in patterns. The distances between stars would be approximately halved. It is the same with alloyed atoms whose combined energetic cohesion increases as the second power of the relative linear proximities of the component parts. Though the parts do not "touch," their mass cohesive dynamic attraction follows the gravitational law of proportionment to second power of the distance apart of centers. Therefore, alloying strength is not additive arithmetically but is advantaged by gravity, which, as Newton discovered, is inversely proportional to the "square" of the distance apart.

Man has now completed the plumbing and has installed all the valves to turn on infinite cosmic wealth. Looking to the past he wails, "How can I afford to turn on the valve? If I turn it on, somebody's going to have to pay for it!" He forgets that the bill has been prepaid by men through all time, especially by their faithfully productive investments of initiative. The plumbing could not have been realized except through absolute prepayment of intellectually organized physical work, invested in the inherent potentials of Nature.

Epochal Transformations

Not only is man continually doing more with less—which is a principle of trend which we will call ephemeralization, a corollary of the principle of synergy—but he is also demonstrating certain other visible trends of an epochal nature. Not only does he continue to increase in literacy but he affords more years of more advanced study to more people. As man becomes master of the machine—and machines are introduced to carry on every kind of physical work with increased precision, effectiveness and velocity—his skilled crafts, formerly intermittently patronized, graduate from labor status to continuity of employment as research and development technicians. As man is progressively disemployed as a quantity-production muscle-and-reflex machine, he becomes progressively re-employed in the rapidly increasing army of research and development—or of production-inaugurating engineering—or of educational and recreational extension, as a plowed-back increment of industrialization.

Product and service production of any one item of industry trends to manipulation by one man for the many through push-button and dial systems. While man trends to increasing specialized function in anticipatory and positive occupations of production, he also trends to comprehensive function as consumer. Because the principle of industry improves as the number of people it serves is increased, it also improves in terms of the increase of the number of functions of the individual to which it is applied. It also improves in terms of its accelerated use.

Throughout the whole history of industrialization to date, man has taken with alacrity to the preoccupation of the specialist on the production side of the ledger; but the amplifications of the functions of the individual as a comprehensive consumer have been wrenched and jerked and suffered into tentative and awkward adoption in the mumbo-jumbo and failure complex of obsolete feudal economics. Up to yesterday man was unaware of his legacy of infinite cosmic wealth. Somewhere along the line society was convinced that wealth was emanating from especially ordained mortals, to whom it should be returned periodically for mystical amplification. Also with feudal fixation man has looked to the leaders of the commercial or political states for their socioeconomic readjustments—to the increasingly frequent "emergencies."

Throughout these centuries of predominant ignorance and vanity, the inherently comprehensive-thinking artist has been so competent as to realize that his comprehensive thoughts would only alienate him from the economic patronage of those who successfully exploited each backing up into the future. The exploiters, successively successful, have attempted in vain to anchor or freeze the dynamic expansion at the particular phase of wealth generation which they had come to monopolize.

The foolhardy inventors and the forthright prospectors in humble tappings of greater potentials have been accounted the notable failures. Every industrial success of man has been built on a foundation of vindictive denouncement of the founders.

Thus the comprehending artist has learned to sublimate his comprehensive proclivities and his heretic forward-looking, toward engagement of the obviously ripening potentials on behalf of the commonwealth. The most successful among the artists are those who have effected the comprehensive ends by indirection and progressive disassociations. So skillful have the artists of the last centuries been that even their aspiring apprentices have been constrained to celebrate only the non-utilitarian aspects of the obvious vehicles adroitly employed by the effective artists to convey their not-so-obvious but all-important burden.

Thus the legend and tradition of a pure art or a pure science as accredited preoccupations have grown to generally accepted proportion. The seemingly irrelevant doings of the pure scientist of recent decades exploded in the face of the tradition of pure mathematical abstraction at Alamogordo. No one could have been more surprised than the rank and file of professional pure scientists. The results were implicit in the undertakings of artist-scientists whose names are in the dim forefront or are anonymous in the limbo of real beginnings. How great and exultant their secret conceptioning must have been!

The Time Has Come

The time has arrived for the artist to come out from behind his protective coloring of adopted abstractions and indirections. World society, frustrated in its reliance upon the leaders of might, is ready to be about-faced to step wide-eyed into the obvious advantages of its trending. We will soon see the emergence of comprehensive training for specialists in the husbandry of specialists and the harvesting of the infinite commonwealth.

Will the Comprehensive Designer, forthrightly emergent, be as forthrightly accepted by the authorities of industrialization and state? If they are accepted, what are the first-things-first to which they must attend?

The answer to the first question is yes. They will be accepted by the industrial authority because the latter has recently shifted from major preoccupation with exploiting original resource to preoccupation with keeping the "wheels," which they manage turning, now that the original inventory of wheels, or tools in general, has been realized from original resource. Though original resource-exploiters still have great power, that power will diminish as the mines now existing above grade, in highly concentrated use forms (yet in rapidly obsoleting original design), become the preponderant source of the annual need. Severe acceleration in the trend to increase of performance per pound of invested material now characterizes all world-industry. With no important increase in the rate of annual receipt from original mines, the full array of mechanics and structure requisite to amplifying the industrial complex, from its present service to approximately one-fourth the world's population to serve all the world's population, may be accomplished by the scrap "mined" from the progressively obsoleting structures and mechanics.

World-industrial management will be progressively dependent upon the Comprehensive Designer to accelerate the turning of his wheels by design acceleration. Each time the wheels go round, the infinite energy wealth of cosmos is impounded within the greater receptive capacities of the ninety-two regenerative-element inventory of earth, and those who manage the wheels can make original entry on their books of the new and expanding wealth increments even as the farmer gains cosmic energy wealth in his seasoned cycles.

An answer as to whether the designer will be accredited by political leadership has been made. Political leadership in both world camps has announced to the world of potential consumers their respective intents to up the standard of living of all world peoples by "converting the high technical potential to account through design."

Only the designer can accomplish this objective. Legislative mandate and dollar diplomacy cannot buy the realization.

As first of the first things, the designer must provide new and advanced standards of living for all peoples of the world. He must progressively house and re-house three billion people in establishments of advanced physical control. The mechanically serviced sheltering must be a continuity of roofs, stationary and mobile, sufficient to allow for man's increasing convergent-divergent interactions of transciency or residence, of work, play or development, interconnecting every center of the world and penetrating to autonomous dwelling facility of the most advanced standard, even in the remotest of geography. The logistics of this greatest phase of industrialization must impound cosmic-energy wealth, within the inventory of ninety-two chemical elements, to magnitudes, not only undreamed of, but far more importantly, adequate to the advancing needs of all men. Implicit is man's emancipation from indebtedness to all else but intellect.

The Very Reverend James Parks Morton was until his recent retirement the Dean of the Cathedral of St. John the Divine in New York City, the largest (though unfinished) Gothic church in the world. It is significant that he hosted the celestial as well as the cerebral. He took the whole world into his arms, welcoming ideas from many, and literally opened up the church to all the city's ethnic and social groups, creating a vital church where God is more than a hyperbole.

Morton has hosted such "livingry" as the famous Poet's Corner, major art exhibitions, and performances of portions of the musicals *Godspell* and *Jesus Christ Superstar*. And he hosted Philippe Petit, the man who walks on air, as an artist in residence at St. John's. Petit is the man who tightroped-walked the Eiffel Tower and performed on a cable stretched between the Twin Towers of the World Trade Center in New York City, to the delight of jaded New Yorkers.

The unfinished towers of the Cathedral of St. John the Divine challenged Bucky Fuller to design an icosahedron tower for the crossing of the neo-Gothic structure in 1978. Morton presented a special program, held at the cathedral, for the Centennial Celebration of the life of R. Buckminster Fuller in 1995. God is omnipresent and omnidirectional; somewhere there is a halo. Morton presents his memories of Bucky as a series of epiphanies.

Bucky Fuller—Nine Epiphanies

The Very Reverend James Parks Morton

So far I have witnessed nine significant Bucky epiphanies in my brief seventy years on spaceship earth: in February, June, November, and December 1978, February 1979, September 1983, October 1991, November 1995, and June 1996. I hope this book will be the tenth, and I expect more to come.

Epiphany I

Bucky entered my life forcefully in February of 1978 on a sun-drenched atoll in the Seychelles. Prince Chahram Palavi of Iran had assembled fifteen environmental activists on his tiny island of D'Arros and given them a serious task: to develop a strategy for preserving the earth! Bucky on this subject, morning, noon, and night for seven straight days was overwhelming for everyone, and to me transforming. In addition to seeing him at our group gatherings, which Bucky dominated, I met with him privately several times— extraordinary meetings. He asked me to read his new rewrite of the Lord's Prayer to the group, and I asked him how we should continue building New York's Cathedral of St. John the Divine for the solar age.

By the time we left D'Arros, Bucky had agreed to preach at the Cathedral the first Sunday in June. More: Bucky most astonishingly volunteered to plan with me over the next few months how to complete the Cathedral as a vast solar bioshelter "to export solar energy to the surrounding community."

Epiphany II

Bucky's cathedral sermon at the eleven-o'clock service on June 4, 1978, was an hour-long dialogue in which I related the appointed scripture—doing the truth—with Bucky's insistence that things work because they are true. I called our dialogue sermon "Technology and Theology" and began by pitching this question: "Bucky, your second book was called *No More Secondhand God.* So I ask you: what does firsthand experience with God, not secondhand experience, really mean?"

Bucky: "At the age of thirty-two, having grown up on a world in which I was continually being told, 'Never mind what you think—pay attention to what we're trying to teach you,' . . . I came to the conclusion that I must do my own thinking and give up all the things that I've been taught to believe— that is, accepting the explanation of physical phenomena without experimental evidence. I said, if I'm going to have to go on direct experience in doing my own thinking, I think the first question you must ask yourself is 'Do you have any experiential necessity to assume a greater intellect to be operating in the universe than that manifested by humans?' And I said I'm overwhelmed by the intellect and the intelligence and the order manifest in the universe, which is completely transcendental to any human capability."

Bucky then spoke of his early discovery that the universe is both eternally regenerative and accelerating—the one and only system that is 100 percent regenerative—"meaning that all of the familiar design alternatives as such are intercomplementary. That is what is meant by ecology."

After reflecting on the acceleration of events in his early years—when he was three, the electron was discovered; at eight, the Wright brothers flew; at twelve, there was the first wireless SOS—Bucky expanded on the new relationship of humans to the universe.

"Human beings are given mind and not muscle. So I say the most important of all the truths that we are being confronted with is that we are here for our mind functioning. Why are we included in an eternally regenerative design in a universe that has no waste? We are apparently here as local information gatherers. Being given the capacity to understand principles so we can develop instruments to go into great macrocosms and microcosms, to deal in a universe which is 99 percent nonsensorily contactable, we are able then to deal with what is common to all lives in all history: problems, problems, problems. We are here for problem solving, not to get over into some universe with no problem of any kind. No, the better you are at problem solving, the worse problems you're going to get. We are here for that. That is our function."

Bucky then came to the final challenge of his sermon—illustrated with a maple seed from the Cathedral close, which he threw sailing into the congregation.

"Humanity is now in a very critical and what I feel is a final examination. When nature has a very important function like humans, this human mind of local information gathering and problem solving, she doesn't leave it all in this one investment, she makes many alternate starts. This morning when I came in from outside, I picked up this little seed from a maple tree. Now, a tree's function is to impound the sun's radiation, to convert it by photosynthesis, and to bring energy and its circulation into the whole ecology, because you and I can't take the sun's energy through our skins. But with this seed the tree is able to get the new little trees out from under its shadow so they can get the sun. So she makes these beautiful flying machines. And I say she has many alternate plantings in our universe besides human beings. So we're into a kind of final examination right now, now that it has become clear. The information is here. We do have the capability to make it. It does not have to be you or me ever again.

"We're really going back to the first two laws we were ever taught today: Love thy God, this is the truth. And love thy neighbor as thyself. Finally you can afford to love thy neighbor as thyself. It does not have to be you or me ever again.

"I would like personally to end this way. I've witnessed one miracle after another through my whole life, extraordinary things happening. I see that God tries very hard and apparently is intent to make us a success if it is possible. So if we don't make it, it is because of each individual. You can't leave it to your politicians to represent you; you can't leave it to your ministers to pray for you. It's going to be how each individual reacts in relation to the truth."

The huge congregation stood up and burst into thunderous applause. As I thanked Bucky, I asked everyone to close their eyes as I read once again, at Bucky's request, his latest version (dated May 8, 1978) of "Ever Rethinking the Lord's Prayer."

Epiphany III

Bucky also kept his word about the solar bioshelter project for the Cathedral, and we met together twice early in the fall of 1978. In November, Bucky and Shoji Sadao came to my office bearing the completed model of wood and clear plastic showing banks of photovoltaic solar collectors along the entire 600 feet of roof. But the crowning glory was an immense tower rising 600 feet above the central dome with a geodesic globe 75 feet in diameter near its apex. People would enter the tower by elevators at its four bases and rise to the geodesic globe above. Bucky wanted the upper half of the glass globe, etched with the constellations of the night sky, to rotate according to the season so that the New York beholder's view would always be aligned with the constellation directly overhead. The lower half of the geodesic globe was to function as a solar bioshelter of living green plants and small trees, making the Cathedral's great tower indeed the midpoint compass between the stars of heaven above and the green earth below!

Epiphany IV

The task now confronting me was deeply intriguing. Bucky had indeed presented a concrete response to my question about greening the Cathedral for the solar age. But was his monumental solar Eiffel Tower a precise design to translate into an architectural program? Or was his solar bioshelter scheme an overall direction to pursue when the Cathedral resumed building? Cathedral construction had stopped with Pearl Harbor and had remained dormant for almost forty years. But since 1976 the Fabric Committee had been busy designing a new apprenticeship program to train young people from Harlem and the Upper West Side in the venerable trade of stonemasonry, in line with the Cathedral's commitment to community development. Could Bucky's most advanced, cutting-edge technology and the most ancient cathedral craft of stonecutting work in sync?

We decided that the two disciplines could and must combine—tradition and innovation together. So in late December of 1978, two important events took place in response to the challenge of Bucky's 2-by-4-foot model prominently displayed on my desk. The first event was a brief think-tank meeting of my five-man "environmental-architectural mafia,"[1] in which we agreed to an

---

[1] René Dubos (the "dean of environmentalists" and great microbiologist at Rockefeller University who developed gramicidin, the first antibiotic); John Todd (the founder of the New Alchemy Institute in Woods Hole, Massachusetts, and an old colleague of Bucky's with two geodesic bioshelters on his experimental solar farm); Tom Barry (the ecologian, geologian, and theologian, who had been with Bucky and me at D'Arros); and Ben Weese (architectural consultant to the Cathedral and trusted friend from our years in Chicago before coming to the Cathedral).

expanded two-day think tank focused on Bucky's proposal, to which we would invite for their critique three or four architects and engineers whose expertise was in solar technology and bioshelter design. Prince Chahram agreed to cover the expenses of the meeting in early 1979 as an anniversary follow-up to the D'Arros meeting with Bucky.

The second event in late December of 1978 was the widely publicized press conference announcing resumption of Cathedral construction after forty years. The first goal was to complete the half-built southwest tower at the front of the Cathedral—a task that Bucky approved and had included in his model—by training young men and women from the neighborhood to be stonecutters and masons.

Epiphany V

The two-day think tank on Bucky's solar bioshelter model took place in snowy February 1979 and was passionate and very opinionated. Bucky's solar tower scheme was deemed by our guests brilliant, inspirational, and wildly impractical, considering the still low energy yield of the best solar voltaics—even with a full acre of them side by side on the Cathedral roof—and the huge energy demands of elevators and the rotating geodesic hemisphere at the top of the tower. "Well, Bucky may say the tower will export energy to all of New York, but we see it as a huge sinkhole, a colossal energy drain!" But Bucky's notion of a solar bioshelter, on the other hand, enthusiastically carried the day and inspired the future project for a green Cathedral. My mafia again convened and proposed that we commission a small, invited group of the most innovative "solar" architects to submit schemes for a solar bioshelter as an integral part of the Cathedral. Prince Chahram was intensely interested in the development of what he called the "Bucky Fuller Cathedral Project" and once again graciously covered the expense of what is now known as the Solar Bioshelter Competition of 1979.[2]

And guess who won? David Sellers—brilliantly making the unbuilt south transept into Bucky's solar bioshelter with exterior limestone Trombe walls encased in glass and a glass roof over the attic vault: a vast solar greenhouse in the sky with the Gothic outline of Cram's original transept.

[2] Seven architects were involved: Bucky (the foundation and inspiration), Keith Critchlow, Sean Wellesley-Miller, Sim Van der Rhyn; David Sellers, Ben Weese, and Malcolm Wells. Of the six new contestants, three were students of Bucky's, and all six profoundly inspired by him. David Sellers, a teacher at the Yale Architectural School, heard of the competition through the grapevine and called up indignant—why had he not been invited? I told him we had a budget for five persons with a deadline for entries today. "I'm not interested in the money, but if you get my proposal in three days, would you look at it?" John Todd, René Dubos, and I were to judge the entries (none of us practicing architects), and we agreed to let Sellers submit his plan.

Nineteen seventy-nine also brought the Iranian revolution and a very different new life in London for Prince Chahram. But in 1980 he arranged for a second meeting of the D'Arros group in England. Bucky was not able to attend, but his indomitable spirit was with us. Tom Berry, René Dubos, John Todd, and I represented the Cathedral bioshelter mafia. In April of 1979 the one hundred Episcopal cathedral deans held their annual conference at 'St. John's and attended lectures about solar bioshelters and geodesic domes by John Todd and René Dubos. Bucky Fuller (an Episcopalian) became a member of an expanded Cathedral family.

Then on June 21, the summer solstice, the new Cathedral stoneyard was dedicated and the first five apprentices began our third building period!

Epiphany VI

Bucky died on July 1, 1983, and thirty-six hours later, on July 3, his wife, Anne Hewlett Fuller, died. On Tuesday, September 27, 1983, the great public memorial service took place at noon at the Cathedral for both Bucky and Anne. It was an extraordinary epiphany: Bucky's granddaughter, Alexandra Snyder, read the first lesson from Ecclesiasticus, "Let us now sing the praises of the heroes of our nation's history . . ."; Arthur Penn read excerpts from Bucky's *No More Second-hand God;* and Paula Robison played the flute solo sarabande from Bach's A Minor Partita. Harris L. Wofford, Jr., gave the eulogy:

Listen to what he wrote a few years ago to a friend who sent word his wife had died. Bucky wrote back: "Long ago we had irrefutable scientific evidence that whatever life may be it is not physical. I assume all humans to be immortal. Being now on the eve of seventy-nine years of age I have written to old friends only to receive letters from their wives or husbands telling me of their sometime-past death. I have always replied that receipt of their letter did not alter my sense of the absolute livingness of the friends to whom I had written, which livingness was in no way altered by the news of their being no longer temporally available. I go right on seeing them. There they are."

Or this story from his friend of fifty years Clare Boothe Luce:

When he stopped to visit her in Hawaii she decided to have dinner with the man she considered the brightest on the island, a leading admiral of the navy. She told the admiral that he was going to meet a true genius. When the two men met, Bucky asked the admiral what the navy was doing to harness windpower. The islands were ideal for great windmills, said Bucky. The admiral demurred. Bucky persisted. Frank Lloyd Wright called Bucky "a man with more absolute integrity than any man I've ever known." All of us who knew Bucky would also agree he was the man with more absolute stubbornness than anyone we've ever known. So that night in Hawaii he argued on about windpower. When the exasperated admiral left he said to Mrs. Luce, "You invited me to talk with the genius of the world, and I spent the evening arguing with a stubborn old man hipped on windmills." Three years later the admiral called Mrs. Luce with a confession. "You remember how nutty I thought your friend Fuller was," the admiral said. "Well, I want you to tell him I've just accepted chairmanship of the navy's urgent task force on windpower. Bucky was right."

Or this quote from Bucky himself after the tragedy of his first daughter's death in 1927 and after disasters in his career when he resolved "to live a new kind of thinking."

When in 1927 I began to consider what the little individual could do on behalf of his fellow man that governments and corporations could not do, it became evident that the individual was the only one that could find time to think in a cosmically adequate manner.

Wofford concluded his eulogy with these words:

The Constitution of the World is crying out for recognition of an office every human being has a duty to play—the office of Citizen of the World. Bucky was a true Citizen of the World. In being it, he showed us the special kind of fun— the delight there can be—in doing our duty as world citizens. By his life, he demonstrated how much good we "little individuals" could do if we, too, played the World Game in our lives, and took the time to try to think "in a cosmically adequate manner."

The wonderful service ended with a prayer offered by Ellen Burstyn:

Almighty and everlasting God, who didst make the universe with all its marvelous order, its atoms, worlds, and galaxies, and the infinite complexity of living creatures; Grant that, as we probe the mysteries of thy creation and the laws behind all structure, we may, even as thy servant Buckminster, come to know thee more truly, and as certainly show forth the mind of our Maker; through Jesus Christ our Lord. Amen.

and with Bucky's favorite hymn from his navy days:

Eternal Father, strong to save,
Whose arm hath bound the restless wave,
Who bidd'st the mighty ocean deep
Its own appointed limits keep:
O hear us when we cry to thee
For those in peril on the sea.

Epiphany VII

In 1991 the Cathedral received an amazing bequest from the French family of René Dubos, specifically designated to continue plans for the Solar Bioshelter inspired by Bucky and so enthusiastically championed by Dr. Dubos. Twelve years had passed since the Solar Bioshelter Competition of 1979. René and Bucky had been dead for over eight years, and with New York's financial crisis of 1987, progress on both the stoneyard and the Solar Bioshelter Project had slowed to a crawl.

So the Dubos bequest was a gigantic shot in the arm. The newly configured "mafia" determined that the most propitious use of the gift would be a worldwide competition for the solar bioshelter to put Bucky's inspired legacy in "prime time." First, sixty-five of the world's most renowned architects from seventeen countries were invited to compete by submitting statements of intention. Next a distinguished jury[3] would select six finalists,[4] four to come to New York for a two-day briefing and then submit a major project design with an accompanying model.

There was an amazing convergence of energies in the unanimous selection of Santiago Calatrava as winner. He transformed the entire attic space of the Cathedral (600 feet long, 300 feet wide at the transepts, and 45 feet high) into an immense cruciform solar greenhouse, a powerful green cross atop the Cathedral for the airplanes passing overhead. Equally important, he returned to Bucky's initial 1978 scheme with a 500-foot solar tower at the intersection of east-west and north-south axes, the compass between the starry heavens and the green earth. Bucky triumphant at last!

The competition was immensely visible, with coverage in the every architectural magazine and metropolitan newspaper across the nation. Herbert Muchamp did a splendid photographic review for the *New York Times,* and the Municipal Art Society mounted a gallery exhibit of all six finalists' models. MOMA gave Santiago Calatrava a one-man show in 1992 with the twelve-

---

[3] James Polshek, FAIA, Dean of the Columbia School of Architecture, chaired the jury of Philip Johnson, FAIA; Maya Lin, designer of the Vietnam Memorial; David Childs, FAIA, managing partner of SOM; and Professor Kenneth Frampton, all architects, in company with Lily Auchincloss, chair of the Architectural Department of MOMA and a Cathedral trustee; the Rev. Robert Parks, Cathedral trustee, Rector of Trinity Church, Wall Street, and chair of the Fabric Committee; John Todd; and myself. John W. Barton, a young architect from Santa Fe, served as administrator of the competition.

[4] The six finalists were Tadao Ando of Japan, Santiago Calatrava of Paris, and Antoine Predock, Holt-Hinshaw-Pfau-Jones, Keenen-Riley, and David Sellers (who had also won the 1979 competition), all from the United States.

foot model of the Cathedral and its solar bioshelter and tower as the center-piece. The model has since traveled worldwide to London, Madrid, Berlin, Tokyo, and Istanbul.

But will Bucky's inspiration ever be built? Here again Bucky's wisdom must prevail: it is a decision for the universe—and in the universe's time frame, not ours.

Epiphany VIII

In 1995, the centennial of Bucky's birth, Professor Haresh Lalvani, a former doctoral student of Bucky's and now professor of architecture at Pratt Insti-tute and design scientist in residence at the Cathedral, organized a mammoth celebration of Bucky's life's work and influence worldwide. Haresh, of course, wanted the exhibition to be a comprehensive retrospective of Bucky's per-sonal achievement from the Dymaxion house and car to the famous geode-sics, buckyballs, and World Game. But of even greater tribute he wanted to illustrate the robust innovations in design science and morphology of Bucky's followers, and in other seminal explorations largely unknown to the architec-tural and design community. So in addition to the multitude of Bucky's fa-mous models and photographs were models and design schemes of eighty-six architects and scientists and designers—Bucky's children—from all over the world. The two Cathedral Solar Bioshelter models—Bucky's and Cala-trava's—proudly occupied the first bay in the nave, and the immense space of the Cathedral was bejeweled with literally hundreds of the models and dis-plays that have changed our vision and understanding of the structure of the universe. The formal title of the event was "Contemporary Developments in Design Science: The Buckminster Fuller Centennial Exhibition."

The grand opening on November 9 was a truly fantastic celebration orga-nized by three distinguished co-chairs,[5] an executive committee of fifteen old friends and colleagues, and a working committee of thirty-six, all under the honorary chair Lily Auchincloss. My task was to be circus ringmaster, herding the vast crowd away from the dazzling exhibition so we could start the show, and then introducing one stunning act after another. Ed Applewhite, Bucky's associate of sixty years, beginning with the Dymaxion house at Beech Aircraft in Wichita and continuing today as Bucky's literary executor, opened the evening. Bucky's daughter Allegra Fuller Snyder followed with her wonderful recollections. She closed with a poem from Bucky's father-in-law, James Mon-roe Hewlett—"the man," Allegra said, "who taught Bucky to believe in him-self."

[5] Bucky's granddaughter Alexandra Snyder May, Priscilla Morgan, and Shoji Sadao.

Hewlett was also an architect and designer, and in later years director of the American Academy in Rome. Bucky loved these four lines and often read them to Allegra:

The flying fish has flown again, with rainbow dripping wing
He doesn't do it often and he cannot do it long
But if his flight reveals to us some rarely tried for things
'Twill prove a fitting subject for our song.

Next spoke M. C. Richards—poet, potter, painter, teacher, friend—who was with Bucky at Black Mountain College in 1947. She told of Bucky's delight when his Dymaxion dome fell down—because, he said, "it gave him an opportunity to see what was wrong so he could fix it." The next summer it stayed up. Then M.C. described the Satie festival, which John Cage organized and in which Bucky played the lead role of the Baron Medusa, with Elaine de Kooning as his daughter and Merce Cunningham as the mechanical monkey who did seven dances. John Cage played the Satie score, Arthur Penn directed, and Bucky's favorite line was "O I do love coffee, especially when it's good." M.C. concluded with an excerpt from *No More Secondhand God*.

Lim Chong Keat spoke next—a world architect, born in Singapore, with offices in Kuala Lumpur, Penang, and Singapore. He had become great friends with Bucky in the mid-1970s when the two founded the Campuan Conference on Global Issues. Lim was at the Cathedral for Bucky's sermon in June 1978 and snapped the immortal photograph of Bucky in his red doctor's gown in the pulpit.

Lim concluded his remarks with this "mental mouthful" from Bucky:

Human integrity is the uncompromising courage of self-determining whether or not to take initiatives, support or cooperate with others, in accord with all the truth and nothing but the truth, as it is conceived by the divine mind, always available in each individual. Whether humanity is to continue and comprehensibly prosper on Spaceship Earth depends entirely on the integrity of the human individuals and not on the political and economic systems. The Cosmic Question has been asked: Are humans worthwhile to universe invention?

For the principal address of the evening, we had invited William McDonough, who had met Bucky in 1972 as a twenty-one-year-old student at Dartmouth, was now dean of the Architectural School at the University of Virginia, and in 1992 would be commissioned by the city of Hanover to write the Hanover Principles for the World's Fair of 2000 on the theme "Humanity, Nature, and Technology." The core of Bill's talk was a challenge for the second industrial revolution. He called the retroactive design of the first industrial revolution brute force: if brute force does not work you are not using enough of it.

Let's design another industrial revolution that produces nothing hazardous, that measures productivity by how many people are working, that measures progress by how many buildings have no pipes, and prosperity by how much of our capital we can invest and how much of our solar income we can accrue. What Bucky called "Real Wealth." I would like to mention a few projects to show the relationship to Bucky:

For Herman Miller, we have just designed a factory to recycle furniture forever. It's a zero-emissions site full of birds and frogs—all we did was say: "no pipes, we're going to keep our water."

In Chattanooga, we're declaring zero-emissions zoning in the "Love Canal of the South." You can do whatever you want, just don't poison anybody. The people signing up for this are people like Du Pont.

We are redesigning the carpet industry so that the carpet will be leased instead of purchased. It will go back to the company when you're finished with it. Why would you want to own a carpet?

Lastly, a little "flag of the next industrial revolution," a fabric designed from Steelcase Corporation. We made the product from ramie, a perennial nettle in the Philippines, and wool from happy sheep in New Zealand. We said we're going to design this as an organic nutrient to become compost. We need to put the filters in our heads, not at the end of water pipes or on smokestacks. Here's the filter: no persistent toxins, no mutagens, no carcinogens, no biocumulative substances, no heavy metals, no endocrine disrupters. Seventy companies slammed the door and formed a consortium to block the project. We went to see the chairman of Ciba-Geigy in Basel; we explained it to him; he said, "I'm letting you in." In thirty days we had analyzed 8,000 chemicals in the fabric industry, and with that filter we had to eliminate 7,962. We were left with thirty-eight. The entire fabric line was done with only those chemicals and meets every standard known for high-quality furnishing. This is the beginning of the Next Industrial Revolution. When the factory did its first run with this product, the Swiss government came to test the effluent and thought their equipment was broken. There was nothing there. It turned out that the fabric itself was further filtering the water. When a factory would rather use its effluent than Swiss drinking water, you've hit zero emissions.

I could go on forever. Bucky could go on forever. I guess that's the point. So, in an anticipatory design kind of way, let me say: "Here's to the Earth that belongs to the Living, free to pursue and prosper, with life, liberty, happiness and free from a remote tyranny of ancestors. And the ancestors are us." Thank you, Dr. Fuller.

Interspersed between the addresses we had Bucky's favorite music by his favorite artists Paula Robison and William Wadsworth: Mozart's Andante in C Major, Massenet's "Meditation" from *Thaïs,* and the sarabande from Bach's Partita in A Minor for solo flute, which Paula had also played at Bucky's memorial.

The incredible evening ended with Irene Worth. Walter Kerr wrote of her in the *New York Times* when she opened in Ibsen's *Hedda Gabler,* "Miss Worth is just possibly the best actress in the world." She read two letters of late 1982, the first written November 19:

Dear Buckminster Fuller:
My last grandmother died recently, as expected, but I didn't expect to remember so much I forgot to talk with her about, and I'm a bit blue when I think of her. She was born about the time you were, and since your death soon could not be called unexpected, I want to send you a note of appreciation.

Insofar as I represent the generation twice removed from yours (younger in human terms, older in Universe time), I want to thank you on behalf of all of us, for a life well done. There's no question that you will be better known in fifty years than you are now, because every year your ideas, language, and general optimism appear to become more saturated in the collective consciousness.

You have known the stagecoach, yet remain on the cutting edge of the electronic age. You are so wonderfully arranged that I can be assured that you will read this letter. How you come to God, I don't understand, but God bless you,
Love,
Joe [Joseph Wheelwright]

Bucky's reply is dated one month later, December 13, 1982:

Dear Joe:
There have been two generations of Wheelwright friends in the first half of my life. They were all admirable individuals. I interpret your spontaneous writing to me in so daringly tender a way as a message of comprehension and accord from all those whom I love who have died and in my youth were so often dismayed and alarmed by all the mistakes I had to make to become shocked into discovering what I had to learn. This was to thoughtfully discover on my own, in both the biggest and most exquisite ways, what needed to be done, and could now be done for the first time in history to make the world work for everybody. It was a task that required a special-case individual to initiate, an individual who had come to the point of suicide and was inspired to commit his total experience inventory only to the advantage of all others but self.

There's lots more to be done, which it seems to me I have to do before I die, as a follow-through in accomplishing all that with which I committed myself to cope. I never pray God to do anything, because God would not be God if the eternal, absolute intellectual competence needed my suggestions.

I feel sure that God holds you in grace and will continue to do so. I thank God for an additional Wheelwright friend.

Faithfully yours,
Buckminster Fuller

Epiphany IX

Bucky's great centennial exhibition in New York did not stop there. The following June 1996 in Istanbul the penultimate United Nations summit conference on major world problems was scheduled to open.[6] Habitat II '96 was to focus on the built environment, Bucky's life work. Therefore a significant effort in 1995 was directed to bringing the Bucky centennial to its most global venue. The secretary-general of the United Nations, Boutros Boutros-Ghali, was very supportive of the project, as was the Hon. Wally N'Dow of Gambia, the deputy secretary-general for the Habitat Summit. The logistics, expense, and red tape of shipping such a large and complicated exhibition were formidable, but the universe clearly wanted it to happen. And it did!

The stated goal of Habitat II was "Adequate and Sustainable Shelter for All in Harmony with Nature," so we revised our title to "Building for the Future: The Buckminster Fuller Centennial Exhibition." Haresh Lalvani wrote these words in the prospectus :

> Geometry is the great equalizer of cultures. The fundamental principles of space and structure provide the harmonious bridge between developed and developing economics, a symbol of hope and unity as well as a pragmatic scientific tool for providing shelter using less materials, yet having greater performance.

> We saw the occasion of Buckminster Fuller's birth centennial as our opportunity to review worldwide innovations in design science. This is a robust frontier of exploration between design in nature and the art and science of architecture .

The Istanbul site for the exhibition was very beautiful and very distinguished: the Museum of Painting and Sculpture housed in the venerable Dolmabachce Palace on the Bosphorus. The exhibition opened on June 3 and continued through the summer of 1996. Architects, engineers, diplomats,

---

[6] Previous UN summits: 1990 New York, "Children"; 1992 Rio, "Environment"; 1993 Vienna, "Human Rights"; 1994 Cairo, "Population"; 1995 Copenhagen, "Social Development"; 1995 Beijing, "Women."

scientists, housing administrators, community activists, and heads of state from 185 countries saw the work, the genius, and the legacy of Bucky.

Harris L. Wofford, Jr., had stated the proper context for Bucky in his eulogy at the memorial service: "The Constitution of the World is crying out for recognition of an office every human being has a duty to play—the office of Citizen of the World. . . . By his life, he demonstrated how much good we 'little individuals' could do if we, too, played the World Game . . . and took the time to try to think 'in a cosmically adequate manner.'"

The Habitat II Summit of the United Nations in Istanbul provided the accurate and inevitable global context for Bucky's life and work as a true Citizen of the World.

Introduction to Omnidirectional Halo

from *No More Secondhand God*

The useful but infrequently used word *epistemology* means *science of the thought processes*. A total epistemological reorientation and, to the best of my knowledge, a unique philosophical reconceptioning, regarding the regenerative constellar logic of the structuring of the universe (both as a new cosmology and as a new cosmogony), seem to have followed gradually upon my hypothetically initiated querying regarding the possibility of formulating more comprehensive and symmetrical statements regarding dawningly apparent natural laws. I intuited in irrepressible degree that such a potential formulation might be accruing and harvestable in all of our acceleratingly reconsidered and progressively integrated world-around, all-history experience as now only diffusely inventoried at the middle decades of the twentieth century.

Out of multi-overlaid experience patternings there sometimes emerges an awareness of what we may call a *coincidence pattern*—a localized thickening of points. These emergent patterns of frequency congruences and concentrations display a unique configuration-integrity which has up to now been so dilute in any one experience as to be only invisibly common to many differentiated or special experiences, e.g., a pack of one hundred 4-inch by 5-inch file cards each riddled with hundreds of different sized small holes. Each card appears to be chaotically patterned with holes. However, when the cards are stacked with edges aligned three holes in each card are vertically aligned; all others are obscured by blank spaces on one card or another. A triangular pattern relationship of the light coming through three tubes in the stack of cards is now lucidly conceptual. To such persistently emergent, uniquely mutual, coincidence-patterning relationships as the same triangle array of holes in each and every card we may apply the term "pattern generalization" as used in a mathematical sense, in contradistinction to the word "generalization" as used in the literary sense. The latter often means a too-ambitious subject range which consequently permits only superficial considerations of any specific case data.

When the uniquely emergent generalized patternings become describable by us in *mentally regenerative conceptual terms,* as completely divorced from any one of the specific sensorial conditions of any of the special experiences out of which they emerged, yet apparently, as seen in retrospect to have been persistent in every special case, then we may *tentatively assume* such unique mutual pattern content to be a *generalized conceptual principle,* as for instance the conception of *tension* as opposed to *compression* independent of textures, smells, colors, sound, or size of any one tension-dominated experience.

It is in just such an epistemological process that we discover that size is not a generalized conceptual principle. Whether referring to the size of an object in respect to other objects or their *sizes* of any one object's subdivision, *size* emerges exclusively as a *frequency* concept uniquely differentiating-out each "specialized case." Generalized shape conceptioning is independent of size. A triangle is a triangle independent of size.

When a *second order* of pattern distillation as a *generalized conceptual principle* emerges, but this time exclusively from the emergently induced *co-ordinate consideration* of a *plurality of generalized conceptual principles* themselves, each independent of any special case sensoriality, and in such a regeneratively recognizable manner of patterning as to provide a means of mathematical accounting and therefrom a tentative forecasting capability, not only of generalized developments but also of special forward experiences in the terms of specific sensorial conditions, and those calculated forecast conditions materialize, and the forecasting capability is subsequently verified by recurrent experimental demonstrations under controlled generalized conditions, then we may tentatively assume that we have discovered at least a clause of "natural law." For example, we tentatively assume that *radiation* is *generalized compression* and that gravity is *generalized tension* and that tension and compression are inseparable, precessionally complementary functions of universal structure.

Newly recognized *generalizable principles* seem emergent in unprecedentedly accelerating accumulation as reported from the instrumentally extended range, velocity, and exactitude of special case experiences in the most recent moments of history's scientific venturing. The many where local probings have been meticulously organized and reported regarding measurable relationships and rates of changing relationships throughout the vast macrocosmic and exquisite microcosmic angle and frequency universe events both infra and ultra to man's direct tunability yet instrumentally tunable and transformably readable within *regeneratively informative tolerance* despite inherently limited observational exactitude.

Out of cumulative patterning overlays there emerges what seem to be generalized principles apparently governing all associative and disassociative transformings and their resultant regeneratively persistent hierarchy of constellar configurations. These hierarchies of constellar configurations disclose in turn a hierarchy of dynamically symmetrical constellation phases and their respective maxima-minima, asymmetric and complementary, accommodative transformabilities which are apparently permitted within an omnirational, omnidirectional, omniequieconomic, energy-accounting, co-ordinate system of universe. This omnirational, arithmetical-geometrical accountability is of such sublime simplicity in contrast to the awkward "mathematics" of all known yesterdays as to have occasioned an almost universal incredibility and nonconsideration of its potential significance though it has been in disclosure for one quarter of a century.

This co-ordinate system may be described as an *isotropic vector* system; that is, a generalized Avogadro system in which the energy conditions and relative quanta ratios are everywhere the same yet multi-differentiable in local patterning aspects, which aspects are interchangeably emergent without altering the comprehensive energy equilibrium or its unitary totality as implicit in the *Law of Conservation of Energy* by which it is assumed that energy may be neither created nor lost.

The discovered co-ordinate system is apparently governed by generalized laws, some of whose mathematical equatability I have been allowed not only to discern (as far as I know for the first time by anyone) but also to codify and translate into unique structural realizations. This codification governs the total co-ordinate abundance ratios of the unique pattern aspect relationships of uniquely irreducible co-operative function aspects of locally nonsimultaneous events and their equilibrious pattern totality.

Discovery of the primary and corollary laws of constantly co-ordinate relative abundance of pattern function-aspects of totality as an omnirational regularity governing all local patternings of universe as a minimum-maximum family of complexedly complementary yet uniquely identifiable conceptual function-patterning relationships followed upon intuitive formulations of the seemingly most comprehensive self-querying question I was capable of propounding to myself regarding possible detectable pattern significances accruing to progressive life experience integrations and overlays.

That most comprehensive question was, "What do you mean by the word 'universe'?" "If you cannot answer, you had best abandon use of the word 'universe' for it will have no meaning." My intuitively adopted rules for self-questioning and answering were that the answer must be made exclusively from man's experience patterns. I learned many years later that the Nobel physicist Percival Bridgeman had identified this same rule adopted by Einstein as "operational procedure," subsequently a much-abused phrase. My answer (or discard of the word "universe" as a communication tool) was operationally inherent: "Universe is the aggregate of all consciously apprehended and communicated (to self or relayed to others) experience of man." If my finite answer holds against all specific experience challenges as being comprehensively anticipatory and adequate, the *universe is finite,* and all its components *definable.* Each life as we know it is *definitive,* i.e., consists of a plurality of terminable, ergo definite, experiences, beginning with each awakening and terminating with each surrender to sleep (no man can prove upon awakening that he is the man who he thinks went earlier to sleep, or that aught else which he thinks he recollects is other than a convincing dream). The intermittent beginnings and endings of conscious experience constitute an aggregate of definitive experiences—and the aggregate is therefore finite.

In the recent moments of historical experience, men as scientists adopted the law of conservation of energy: as predicated upon the sum total experience of physicists which recalled no contradiction to this hypothesis. They thus accomplished a finite packaging of all physical behaviors of physical universe as predicated also upon the hypothesis that all physical phenomena are entirely energetic.

By embracing all the energetic phenomena of total experience, the scientists secured a synergetic advantage for all energy accounting and prospecting. "Synergy" means "behavior of whole systems unpredicted by the behavior of any of its components or by any sub-array of its components." Corollary to synergy is *the law of the whole system.* Systems are definite as they return upon themselves in a plurality of directions, ergo have concave inwardness and convex outwardness, ergo inherently subdivide universe into mutually exclusive definitive macro and micro entities. *The law of whole system* states that, given the sum of whole system pattern conception its component behaviors may be differentially discovered and predictably described as required by the already evidenced behavior functions implicit in the a priori-definitive experience and conceptioning of any given experience-verified system. Thus by the law of whole system as corollary of synergy, the component behaviors of systems may be predictably differentiated as primary and secondary componental sub-divisions of whole system and then progressively isolated and locally reconsidered for further dichotomy.

Adopting synergetic advantage, science hypothesized that the physical portion of universe is energetic and finite. Under this hypothesis Einstein wrote his equation of physical universe as $E=MC^2$. This said that the total of local system energy is the product of all concentric local systems of energy's self-interfered, shunt-holding patterns (M for mass) as multiplied by the entirely noninterfered local omnidirectional velocity of surface growth of an omnidirectional, outward-bound spherical wave of radiant energy (in terms of second power of radial wave module frequency growth rate).

In Einstein's formula mass constitutes all the patterns of precessionally self-interfered and concentrically shunted, ergo locally articulated and locally and periodically regenerative holding patterns of energy. This is also to say that M equals all the locally complex, concentric, self-associative, unique holding patterns of all energy, and C^2 equals all the eccentrically disassociative individual patternings of all energy (C being the radial or linear speed of radiant energy, which is approximately 186,000 mps.).

But physical science lacked the experience which might have persuaded it to hypothesize what all universe is. Physical science therefore restricted its comprehensive accounting strategy to the special case of definitive isolations within the *physical* portion of universe. This left the remainder of all experiences, no matter how earnestly and meticulously reconsidered, outside the *definitive* portion of comprehended experiences of universe, i.e., the physicists said all that is not physically encompassed as $E=mc^2$ is metaphysical.

However, by my definition of universe, all that was relegated to metaphysical nebulosity is now embraced by finite universe along with the physically energetic wherefore all the hitherto "inexact sciences" may become rigorously defined, enjoying equatable treatability at optimum degree of determinability.

I have found a general law of total synergetical structuring, which we may call "The Law of Structure." This law discloses that "universe" of total man experience may not be simultaneously recollected and reconsidered, but may be subdivided into a plurality of locally tunable event foci or "points" of which a minimum of four positive and four negative points are required as a "considerable set"; that is, as a first finite subdivision of finite universe. (This fourness coincides with basic quanta strategy.) All experience is reduced to nonsimultaneously "considerable sets" and holds irrelevant to consideration all those experiences which are either too large and therefore too infrequent, or too minuscule and therefore too frequent, to be tunably considerable as pertaining to the residual constellation of approximately congruent recollections of experiences. A "considerable set" inherently subdivides all the rest of irrelevant experiences of universe into macro-cosmic and micro-cosmic sets immediately outside or immediately within the considered set of experience foci.

There are two inherent twilight zones of *"tantalizingly almost-relevant recollections"* spontaneously fed back in contiguous frequency bands—the macro-twilight and the micro-twilight.

It is a corollary of this first subdivision of universe that a *considerable set* is a locally definitive system of universe returning upon its considerability in all circumferential directions and therefore has an inherent *withinness* and *withoutness,* which two latter differentiable *functions* inherently subdivide all universe into the two unique extremes of macro and micro frequencies.

The "Law of Structure" says that "local structure is a set of frequency associable (spontaneously tunable) recollectable experience relationships, having a regenerative constellar patterning as the precessional resultants of concentrically shunted periodic self-interferences, or coincidences of its systematic plurality of definitive vectorial frequency, *wave length and angle* interrelationships."

The precessionally regenerative concentricity of structure is antientropic, and evolutes toward optimally economic local compressibility and symmetry. (See *The Dymaxion World of Buckminster Fuller,* by Robert Marks, for Fuller's Law of omnioptimally economic, omnitriangulated-point-system, symmetry relationships and relative abundance of frequency modulated multiplicative subdivision of unitary local systems; i.e., M (mass) means: All the universe's self-interfering complexes having concentrically self-precessing, local-focal-holding patterns resulting in locally regenerative constellar associabilities as positive-outside-in structures. C^2 (radiation) means: All the universe's nonself-interfering complexes having eccentrically inter-precessing, omnidirectionally diffusing patterns resulting in comprehensively degenerative negative limits of dissociabilities as negative (inside-out) de-structures.

In the chapters on Energetic-Synergetic Geometry I identify *second powering* with the point population of any one radiant (eccentric) or gravitational (concentric) wave systems circumferential arrays of any given radius stated in terms of frequency of modular subdivisions of the circumferential arrays radially read systems' concentricity layering; *third powering* with the total point population of all the successive wave layers of the system; *fourth powering* with the interpointal domain volumes; *fifth* and *sixth powering* as products of multiplication by frequency doublings and treblings, etc. The *Doppler effect* or wave reception frequency-modulation caused by motions of *the observer* and *the observed* are concentric wave system fourth and fifth powering accelerations.

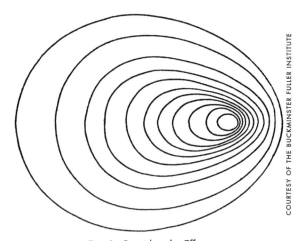

Fig. 1. Omnidoppler Effect

The Doppler effect is usually conceived of as an approximately "linear" experience. "You," the observer, stand beside a railway track (which is a "linear" model); a swift train approaches with whistle valve held open (at constant frequency pitch as heard "on board" by the engineer "blowing" the whistle). The whistle sound comes to you at approximately 700 mph, but the train is speeding toward you at 100 mph. The train's motion reduces the interval between the successive wave emissions, which in effect decreases the wavelength which gives it "higher" pitch as heard at your remote and "approached" hearing position. After the train goes by, the train runs away from each successive wave emission, thus increasing the interval between wave "crests" and therefore lengthening the waves, which apparently "lowers" the pitch as you hear it but not as others elsewhere may hear it. This is pure *observational* "relativity." But the real picture of the Doppler effect is not linear; it is omnidirectional.

The Doppler effect may also be *explained* in *omnidirectional,* experience-patterning conceptionality which is more informative than the familiar linear conceptioning of the railroad train and "you, at the crossing." "You" were flying in an air transport which exploded, and because of the sudden change in pressure differential between your in'ards and your out'ards at high altitude you personally have just been "exploded" into many separate parts receding from one another at high velocity. A series of secondary explosions follows from somewhere in the center of the galaxy of exploding debris, as one item after another of the late airplane's explosive cargo is reached by progressive local conflagration-heat concentrations. The sound waves of the successive explosions speed after your receding parts amongst which are your two ear diaphragms as yet "stringily" interconnected with your exploding brain cells, which "hear" the explosion's sound waves first at low pitch. But as your parts explode from one another at a decelerating rate because of air friction, etc., the waves of remote explosion sounds "shorten" and pitches go "up." Now consider many separate, nonsimultaneous, secondary explosions of your various exploding parts all of varying intensities of energetic content and in varying degrees of remoteness and realize that the decelerations and accelerations of Doppler effects will render some of the explosive reverberations infra and some ultra to your tuning range limits of hearing, so that the sum total of *heard* events provides very different total conceptioning as heard from various points in the whole galaxy of exploding events whose separate components would tend to new grouping concentrations.

The Halo discovered in the next chapter is that of an omnidirectional, complex, high-frequency, Doppler-effected *hypothetical-zone experience* in an omnidirectional universal maelstrom of nonsimultaneous near and far explosions and their interaccelerating refractive wave frequency patternings. Several of these fundamental concepts are also repeated several times in Halo, being reintroduced in various complex associations each of which provides unique discoveries.

In order to generate a spontaneous comprehension of the significance of the thoughts expressed in "Halo," which now follows, the latter essay will open with a swiftly martialed digest of the epistemological concepts of this introduction.

R. Buckminster Fuller was interviewed by countless news media personalities. Of all the pieces written about him, Bucky considered Calvin Tomkins's "Profile" in *The New Yorker* the most accurate. And Tomkins considered his interviews with Bucky Fuller among his most exhilarating experiences as a writer.

Tomkins wrote, "Fuller is one of this century's most astonishing works of art, and he has been devoted to exploring how nature really works. Radically innovative prototypes for mass-produced housing and a unique three-wheeled automobile and the nondistorting flat map of the world are practical applications of the generalized principles inherent in nature."

A noted author and longtime staff writer for *The New Yorker,* Tomkins contributes an account of Bucky on his beloved Bear Island, where Fullers have been going (and still go) since before Bucky was born.

Bear Island

Calvin Tomkins

In the summer of 1966 I spent four days with Bucky Fuller on Bear Island, his favorite place in the world—he had gone there for at least a short visit virtually every summer of his life. Much of his thinking, as he once wrote, had grown "out of boyhood experiences on a small island eleven miles off the mainland, in Penobscot Bay of the state of Maine," and during my visit Bucky gave me, as he was wont to do, a fairly comprehensive overview of his ideas on a wide-ranging variety of subjects, from the resolution of C. P. Snow's famous "two cultures" dichotomy to the right way to swim in frigid Maine water. (You go in and out quickly several times, warming up in between dips, and after a while the pain becomes slightly less agonizing.) Coming out of the water with him after one late-afternoon swim, I asked a question that set him off on a rousing, original, and lengthy (what else?) explanation of how the Maori tribesmen in the South Pacific had discovered the art of celestial navigation a long time before anyone in Europe did, and how their descendants had sailed around the world, interacting with other early cultures and using their precious knowledge to become high priests and shamans, preserving their great secret for many centuries, until the Ionian Greeks overwhelmed the Minoan culture at Knossos, broke the ancient Maori mathematical codes, and used them to establish the foundations of Greek science. While he talked, the sun slipped below the horizon. Suddenly noticing that my teeth were chattering—he himself wasn't the least bit cold—Bucky led the way back to the main house at a fast trot, pausing only to pick up and show me a few stones that were nearly perfect tetrahedrons—his candidate for nature's essential building block. Decades later, I still remember nearly everything he told me that afternoon. In addition to all the other things Bucky was—inventor, mathematician, architect, engineer, cartographer, philosopher, poet, cosmogonist, comprehensive social thinker—he was by far the greatest teacher I've ever come across, as well as the most endearing.

Time after time during my stay on Bear Island, he came back to the educational revolution he saw coming. His book *Education Automation,* which had originated as a lecture in 1961, was getting attention in academic circles then. Bucky was convinced that as work became progressively obsolete, education, which had been largely devoted in the past to "misinforming" the young, would change radically in almost every way, becoming a major—perhaps the major—world industry. He thought this would take place within ten years, and of course he was wrong there. The failures and broken promises of our educational system still darken the prospects of American society, but Bucky's little book, curiously enough, is as exciting to read today as it was in 1961. Whether you agree with his prescriptions or not, almost everything he says makes perfect sense, and the vigor and originality of his argument—for example, the attack on specialization as a form of slavery—is hard to resist. Nobody else in this century has thought about education with anything approaching Bucky's all-out daring, and if anyone wants to find or stimulate fresh thinking on the subject, there's still no better place to start.

Just before he came to Bear Island that summer, Bucky had been in Athens, where he took part in a symposium sponsored by Constantinos Doxiadis, the city planner. At the end of it, he told me, "Doxiadis gave a big dinner party, at which he asked me to make a speech. And when it got to be time for the speech, he got up from the table and said he was not going to introduce me— he was going to leave that to a member of my own generation. And up to the platform stepped his daughter, who graduated two years ago from Swarthmore. Lovely girl. Well, I sort of liked that." Of course he did. Like many first-rate teachers, Bucky had the gift of inhabiting both his own and his students' generations. Reading him today, most students will find that he still does.

Emergent Humanity: Its Environment and Education

from *R. Buckminster Fuller on Education*

An educational revolution is upon us.

One of the most important events of this peaceful but profound revolution is our dawning discovery that the child is born comprehensively competent and coordinate, capable of treating with large quantities of data and families of variables right from the start.

Every well-born child is originally geniused, but is swiftly degeniused by unwitting humans and/or physically unfavorable environmental factors. "Bright" children are those less traumatized. Of course, some children have special inbred aptitudes and others, more crossbred, are more comprehensively coordinated.

But the new life is inherently comprehensive in its apprehending, comprehending and coordinating capabilities. The child is interested in Universe, and asks universal questions.

This propensity of the child toward comprehensivity, given a properly patterned environment, is attested in such works as Benjamin Bloom's *Stability and Change in Human Characteristics,*[1] and Wilder Penfield and Lamar Roberts's *Speech and Brain Mechanisms.*[2]

Through electro-probing of the human brain, we are beginning to understand something of its energy patterns and information processing. We apparently start life with a given total-brain-cell capacity, component areas of which are progressively employed in a series of events initiated in the individual's brain by chromosomic

From George L. Stevens and R. C. Orem, *The Case for Early Reading* (St. Louis: W. H. Green, 1965).

[1] Benjamin S. Bloom, *Stability and Change in Human Characteristics* (New York: Wiley, 1964).

[2] Wilder Penfield and Lamar Roberts, *Speech and Brain Mechanisms* (Princeton: Princeton University Press, 1959). See also Wilder Penfield, "The Uncommitted Cortex," *Atlantic Monthly,* July 1964.

"alarm clocks." Put your finger in the palm of a newborn baby's hand and the baby will close its tiny hand deftly around your finger, for its tactile apprehending organism is operative in suberb coordination. Soon the "alarm clock" calls the hearing function into operation, and on its own unique schedule the baby will also see.

In a stimulating environment, the brain's chromosomic alarm clocks and "ticker tape" instructions inaugurate use of the child's vast inventory of intercoordinate capabilities and faculties. Children are not in fact taught and cannot be taught by others to inaugurate these capabilities. They teach themselves—if given the chance—at the right time. This provision of environmental experience conducive to the child's intellectual development has been termed the "problem of the match" by J. McV. Hunt, in his *Intelligence and Experience,*[3] he also speaks of "motivation inherent in information processing."

Bloom finds that environment has its greatet influence on a human characteristic—such as intelligence—during the period of time in which the characteristic is undergoing its greatest rate of growth or change. Thus, by age four, 50 percent of the child's total capacity to develop its I.Q. is realized.

If not properly attended to and given the chance to function, despite the brain's alarm-clock inauguration of progressive potentialities in the first four years, the brain mechanisms can be frustrated and can shut off the valves of specific capacities and capabilities to learn, then or later on, in the specific areas. The capabilities need not necessarily be employed to an important degree immediately after being triggered into inception, but must upon inception be put in use and kept in use as active tools in the human coordinating capability, else they will squelch themselves, "shut themselves off," not necessarily irreparably, but usually so.

Piaget has said: "The more children have seen and heard, the more they want to see and hear." I add: "The more children have coordinated, the more they want to coordinate."

By age eight, 80 percent of the child's total capability to self-improve I.Q. in learning how to learn is activated. By age thirteen, 92 percent of this capability is self-started into usability and by seventeen, the final 8 percent of the total capacity to coordinate and apprehend, to comprehend and teleologically employ input data, has become operative.

Traditionally, the great bulk of government educational funds has been applied *after* the critical birth-to-eight period during which 80 percent of the child's educational capacity is being established. In the light of recent re-

[3] J. McV. Hunt, *Intelligence and Experience* (New York: Ronald, 1961). See also his "Motivation Inherent in Information Processing and Action," in *Motivation and Social Interaction,* ed. J. D. Harvey (New York: Ronald).

search findings, our input of personnel, funds, and energy into education must be reversed.[4] A powerful case can be made for inverting the educational structure—for paying parents or other people responsible for the most important, formative years more than college professors—in due ratio to their greater responsibility. O. K. Moore, of "talking typewriter" fame, has made a similar suggestion.[5]

Let us focus our efforts to help the new life on the critical first thirteen years, when approximately 92 percent of brain function is progressively and automatically "turned-on," "tuned-in," "tuned-out," or "shut-off," in direct response to the positives or negatives of the individual's environmental experiences and potentials, keeping in mind that by age four 50 percent of brain function is realized, which must be properly set in use and kept in use.

Not only intelligence is developed during these formative years, but also the basic characteristics determining much of the individual's personality and behavior as well.

Will the older, adult life demonstrate that it really wants more life by designing an environment to foster the new child-life adequately—to nourish the unfolding flowers of the "cortical gardens"? As I predicted in the *Saturday Review:*

> In the next decade, society is going to be preoccupied with the child because through the behavioral sciences and electrical exploration of the brain we find that, given the right environment and thoughtful answers to his questions, *children have everything they need educationally right from birth.* We have thought erroneously of education as the mature wisdom and over-brimming knowledge of the grownups injected by the discipline pump into the otherwise "empty" child's head. Sometimes, parents say "don't" because they want to protect the child from getting into trouble. At other times when they fail to say "no" the child gets into trouble. The child, frustrated, stops exploring. It is possible to design environments within which the child will be neither frustrated nor hurt, yet free to develop spontaneously and fully without trespassing on others. I have learned to undertake reform of the environment and not to try to reform humanity. If *we design the environment properly,* it will permit children and adults to develop safely and to behave logically.[6]

[4] Operation Head Start, Title I of ESEA and recent related programs appear to represent a more realistic utilization of resources. See "The Big Federal Move into Education," *Time,* April 30, 1965.

[5] O. K. Moore, *Autotelic Responsive Environments and Exceptional Children* (Hamden, Conn.: Responsive Environments, Inc., 1963). See also Maya Pines, "How Three-year-olds Teach Themselves to Read—and Love It," *Harper's,* May 1963, p. 61.

[6] R. Buckminster Fuller, "What I Have Learned: How Little I Know," *Saturday Review,* November 12, 1966, p. 70.

The work of Bloom, Erikson,[7] and others reveals that this environment must promote *trust, autonomy,* and *initiative.* The human newborn remains helpless longer than the young of any other species. It is *in trust* to the adult's competent care, and should experience no breach of basic trust.

The child needs to have an area that is really its own—just as individuals of other species need a minimum regenerative territory. The child's room should be its "autonomy area"—the new life's learning lab complete with the expendables needed for testing tension, cohesion, etc., by tensing and tearing techniques.

This leads into *initiative*—the third element of critically controlling importance during the first four years. Psychologists have long told us that children need to *touch* to get basic information. But they also need to conduct all manner of experiments, as with gravity and inertia when they knock objects off a table. Likewise, children require experiences which indicate the coherence of things. After tearing newspapers apart and finding they give poor tensional support they will want to explore silk and other materials. As W. Gray Walter has observed, "What the nervous system receives from the sense organs is information about differences—about ratios between stimuli."[8]

Children, first taking apart and then putting together, learn to coordinate spontaneously. They learn about the way Universe works.

Children mustn't be stopped thoughtlessly as they go through their basic explorations. If parents break up that exploratory initiative by too many "don'ts" or punishments, or by having things in the child's environment that are dangerous and by which the child gets hurt too frequently, spontaneity will be stifled, probably permanently. Fortunately, a few determined and reinspired individuals whose spontaneous employment of innate capabilities has been curtailed or abandoned due to childhood frustrations, manage to later "find" themselves, but these cases are rare.

In experimental work at Southern Illinois University, we have learned that the maturing student, like the younger learner, wants privacy—a special place. We have developed a little, individual, private room-booth with a windowed door which "belongs" to each student in our project. When the students first enter they find in their private "room" all kinds of desirable items: a telephone directly and privately connected to the teacher; a good dictionary; wall charts of the periodic table of the elements; a world globe; a wall-mounted chart of the electromagnetic spectrum; a private typewriter, and other items conducive to thought and study. It becomes an obviously realized

[7] Erik Erikson, "Identity and the Life Cycle," *Psychological Issues,* Monograph 1 (New York: International Universities Press, 1959).

[8] W. Gray Walter, *The Living Brain* (New York: Norton, 1953), p. 135.

privilege to be allowed to go into this private study, where their reflexes become progressively conditioned, by association with that environment, and the students give themselves spontaneously to study, calculation, and writing. They find themselves producing. Their minds really begin to work.

As Southern Illinois University's President Delyte Morris—a true leader—has pointed out: "The assumption here is that 'dropouts' indicate inadequacy of the educational system and not of the human individual."

In my *Ideas and Integrities*[9] I have said that education, in the sense of man's being *educente* (led out from) the monological fixations of ignorance, involves also being led into, *intro-ducente*, (introduced to) the new awareness of the dynamic fluidity of the infinite persistence of complex-yet-systematic interaction of universal principles.

I consider the primary concern of education as exploring to discover not only more about Universe and its history but also about what Universe is trying to do, about why human beings are part of it, and how they may best function in universal evolution. We are finding ways to help children coordinate their spontaneous comprehension of the whole instead of becoming specialists without losing any of the advantages gained by yesterday's exclusive specialization.

Our present global civilization requires an educational approach embracing at the outset the most comprehensive review of fundamental "generalized" principles. As these are progressively mastered, the approach should continue through their subdivision and application to separate localized cases. Having established this order from "the whole to the particular" we need to take all of the advantages afforded us by the latest communications developments through which the complex patternings and behavior of Universe may be brought within reach and made part of humanity's working everyday experience.

Today, the vastness, complexity and detail of our knowledge requires restructuring into assimilable wholes, to be imparted even at the most elementary levels in terms of whole systems. We can no longer think in terms of single static entities—one thing, situation, or problem—but only in terms of dynamic changing processes and series of events that interact complexly.

Despite their venerated status, a large part, if not all of our educational institutions and their disciplines are obsolete. Virtually everything we thought we understood concerning education is fast becoming useless or worse. For example, because experiment invalidates most of the axioms of mathematics such as the existence of solids, continuous surfaces, straight lines, etc.,

[9] R. Buckminster Fuller, *Ideas and Integrities* (Englewood Cliffs, N.J.: Prentice-Hall, 1963), p. 231.

much of the mathematical curriculum sanctioned by mathematical educators, adopted by school boards, and taught in all elementary schools is false, irrelevant, discouraging, and debilitating to the children's brain functioning.

We are going to develop an environment in which the new generation is so protected from the lovingly administered nonsense of grownups that it can develop naturally just in time to save humanity from self-annihilation.

Half a century ago, in 1917, I found myself thinking that nature didn't have separate departments of physics, chemistry, biology, and mathematics requiring meetings of department heads in order to decide how to make bubbles and roses!

I decided nature had only one department and only one arithmetical, angle-and-frequency, modulating-and-coordinating system. I am quite confident that I have discovered an importantly large area of the arithmetical, geometrical, topological, crystallographic, and energetically vectorial coordinate system employed by nature itself. It is a triangular and tetrahedronal system. It uses 60-degree coordination instead of 90-degree coordination. It permits kindergarten modeling of the fourth and fifth arithmetical powers, i.e., fourth and fifth dimensional aggregations of points and spheres, etc., in an entirely rational coordinate system. I have explored the fundamental logic of the structural mathematics strategies of nature which always employ the six sets of degrees of freedoms and most economical actions.

The omnirational coordinate system which I have named *synergetics* is not an invention. It is purely discovery. With the complete and simple modelability of synergetics it will be possible for children at home with closed-circuit TV documentaries coming to them, and making their own models, to do valid nuclear physics formulations at kindergarten age. With this fundamental structuring experience, and sensing through models, children will discover with experiments why water does what it does. They will really understand what a triangle is and what it can do and does.

I agree with Jerome Bruner, whose report of the 1959 Woods Hole Conference advanced the hypothesis that "any subject can be taught effectively in some intellectually honest form to any child at any stage of development."[10]

As Margaret Mead has indicated in her classic, *Coming of Age in Samoa,* we must direct our educational efforts to preparing children for coping effectively with the choices and changes which are confronting them. We should design a *Curriculum of Change,* not merely a *changing curriculum.*"[11]

Obviously, one of the reasons why scientific education has seemed too dif-

[10] Jerome Bruner, *The Process of Education* (New York: Vintage, 1960), p. 33.

[11] Margaret Mead, *Coming of Age in Samoa,* 14th printing (New York: Mentor, 1962), pp. 144–45. See also Alfred North Whitehead, *The Aims of Education and Other Essays* (New York: Mentor, 1951), p. 28.

ficult for many is the fact that much of its mathematics is founded upon experimentally unprovable myths which must greatly offend the intuitive sensitivity of the lucidly thinking new life.

When we combine our knowledge that the period from birth to four is the crucial "school" opportunity, with the discovery that entirely new mathematical simplicities are at hand, we must realize that educational theory is entering a period of complete revolution. Excepting the mathematical-physicists, the revolution about to take place in mathematics education may be amongst history's most violent academic *reforms*. You will not have to wait long to discover that I am right.

It is very clear to me that when children stand up, breathing and coordinating exquisitely complex patterns by themselves, get their own balance and start drinking in the patterns of cosmos and earth, they are spontaneously interested in coordinating the total information—the total stimulation. Children crave to understand—to comprehend. That is why they ask their myriad questions.

New tools will make it easier for the young to discover experimentally what really is going on in nature so that they will not have to continue taking so much nonsense on experimentally unverified axiomatic faith.

Computers, suddenly making human beings obsolete as specialists, force them back into *comprehensivity* functioning, which they were born spontaneously to demonstrate.

Computers as learning tools can take over much of the "educational metabolics," freeing us to really put our brains and wisdom to work. A recent report by the President's Science Advisory Committee recommends that the government underwrite a program to give every college student in America access to a computer by 1971. I suggest that we give every preschooler access first!

One device I have invented to provide total information integration is the "Geoscope" or miniature Earth. After sixteen years of experimentation and development, I can describe it as a 200-foot diameter sphere with 10 million electric light bulbs—each with controllable light intensity—evenly covering the entire surface and hooked up to a computer to provide, in effect, an omnidirectional spherical television tube which when seen at a distance, will have as good resolution as a fine mesh halftone print. The Geoscope, accurately picturing the whole earth, will be used to communicate phenomena presently not communicable, and therefore not comprehended by the human eye-brain relay. For example, we could show all the population data for the world for the last 300 years, identifying every 1,000 human beings by a red light located at the geographical centers occupied by each 1,000 human beings. You would then be able in one minute to develop the picture of the world's population growth and geographical spread trends of recent centuries. You would see the glowing red mass spreading northwestward around the globe like a great fire. You would be able to run that data for another second or two to carry you through three or four more decades of population growth. While the edge of the data would be unreliable, the gravity and momentum centers of population would be quite reliable. Or all the cloud cover and weather information around Earth can be shown and accelerated to predict the coming weather everywhere.

If we were to flash a red light for each 1,000 "reading problems" in U.S. urban school systems, our metropolitan areas, including the nation's capital, would flicker in distress.

According to a March 1967 *Washington Post,* "At least one out of three public school students in Washington is reading two grades or more below where he should be. . . . Many are reading five, six, or seven grades behind."[12]

The question I would ask of the "reading readiness" advocates is: "When are these children going to be ready to read?"

In *The Case for Early Reading,* the authors have assembled considerable and convincing evidence that preschool children [Why not call them "school-at-home" children?] want to and will learn to read at home *given the opportunity.*

Their argument that the "before-age-six" period is the *naturally* optimum time for language learning—reading included—is increasingly supported by recent research disclosures in diverse disciplines which, coupled with the historical evidence cited, merits the closest consideration.

Of current relevance is the March 1967 *NEA Journal* report of the Denver study of 4,000 school children designed to determine whether beginning reading could be taught effectively in the kindergarten.

The experimental kindergarten group who were given special reading instruction twenty minutes a day and an adjusted program in the first and later grades showed the greatest initial and long-range gains in both reading comprehension and vocabulary. They also read faster than any of the other groups at the end of the third grade. Early reading instruction was shown, in short, to have "a positive, measurable, continuing effect."[13]

Stevens and Orem would agree that kindergarten as the time for introducing reading instruction[14] is preferable to first grade.

But, they argue, *prekindergarten children* must be provided an environment

---

[12] Susan Filson, "Reading Levels in D.C. Schools Shock New Teachers," *Washington Post,* March 16, 1967.

[13] J. Brzeinski, M. L. Harrison, and P. McKee: "Should Johnny Read in Kindergarten?" *NEA Journal,* March 1967, pp. 23–25.

[14] By *instruction* they mean informal, interesting, inductive, inner-paced learning. *Auto-education* is Montessori's term.

which epigenetically enhances early reading, for they have the prime potential for language learning—a potential which, they say, will prove to be of revolutionary significance for educational strategists.

I am confident that these authors are going to win their "case," for it is increasingly evident that the child's neurophysiology is on their side.

What we must plan now, on an even more comprehensive scale than the important Denver project, are imaginative and innovative studies to determine the optimum school-at-home combinations of these elements: (1) home environment, with enlightened parents; (2) TV, computers, and other tools and technology conveying the most cogent content; and (3) young child's motivation and sensitive period for symbol systems mastery—and communication-computational competencies.

Another device which I have invented to encourage comprehensive thinking is my Dymaxion map, originally published in *Life,* March 22, 1943, and the first projection system to be granted a U.S. patent as a cartographic innovation. Dr. Robert Marks has described it as "the first in the history of cartography to show the whole surface of the earth in a single view with approximately imperceptible distortion of the relative shapes and sizes of the land and sea masses."[15] This map, which I now call my Sky-ocean World Map, is an aid in effectively conceiving the totality of world (and Universe) events.

Because three-fourths of Earth's surface is covered by water, I have developed a "fluid geography" approach to the study of geography, to correct the "landlubber" bias prevalent in our schools.

I have also been engaged over the years in writing a comprehensive maritime reconstruction of history—a saga of the world's sailormen "seeding" civilization by ship.

When I talk about changing obsolete, ineffective, and debilitating school patterns, the established reflexive conditioning of our brains tends to expect a lag of another 100 years to bring about that change. But the rate at which information is being disseminated and integrated into our current decision-making regarding the trending topics I am discussing indicates that the changes are going to happen quite rapidly. My advice to educators who are thinking about what they may dare undertake is, "Don't hesitate to undertake the most logical solutions. *Take the biggest steps right away and you will be just on time!"*

Individuals are going to study at home in their elementary, high school, and college years. Not until their graduate-work days begin will they take residence on campus. Inasmuch as the period of greatest educational capability

---

[15] Robert Marks, *The Dymaxion World of Buckminster Fuller* (Carbondale, Ill.: Southern Illinois University Press, 1960), p. 49.

development is before age four the home is the primary schoolhouse—and kindergarten is the high school.

As John McHale, Executive Director of the World Resources Inventory, has noted: "It is now literally and technically possible to have the equivalent of the school (or even college) actually in the home dwelling. This may very well be the indicated direction for educational and training development in the emerging countries. It is not really a new concept. The home/family dwelling as the prime educational environ and its re-integration as a fully advantaged unit are new."[16]

One may already "tune in" on knowledge through the radio, TV, and telephone. With more sophisticated systems, now available, one may individually select, and follow through, complex sequences and instructional programs.

Advances in educational technology have now made available a number of measurably efficient, self-instructional programmed materials, which are swiftly developing into presequenced "packaged learning" devices employing video tape, film, book texts, plus the remote computer linkages via libraries and control centers.

As I have forecast in *Education Automation,*[17] we are faced with a future in which education will be the number one great world industry, within which will flourish an educational machine technology providing tools such as the individually selected and articulated two-way TV and an intercontinentally networked, documentaries call-up system, operative over any home two-way TV set.

In my 1938 *Nine Chains to the Moon,*[18] I outlined a number of predictions including, "Broadcast Education: the main system of general educational instruction to go on the air and screen." Frank Lloyd Wright, reviewing the book for *Saturday Review,* agreed: "Dead right. The sooner the better."[19]

With two-way TV we will develop selecting dials for the children which will not be primarily an alphabetical but rather a visual species and chronological category selecting device with secondary alphabetical subdivisions, enabling children to call up any kind of information they want about any subject and get their latest authoritative TV documentary. The answers to their ques-

[16] John McHale, *The Ten-Year Program* (Carbondale, Ill.: Southern Illinois University, World Resources Inventory), p. 79.

[17] R. Buckminster Fuller, *Education Automation* (Carbondale, Ill.: Southern Illinois University Press, 1962).

[18] R. Buckminster Fuller, *Nine Chains to the Moon* (Philadelphia: Lippincott, 1938). Copies of the original printing (5,000) of this book are now a collector's item.—Ed. Note.

[19] Frank Lloyd Wright, "Ideas for the Future," *Saturday Review,* September 17, 1938, p. 14.

tions and probings will be the best information that is available up to that minute in history.

The "lecture routine" which most teachers look forward to with as little enthusiasm as their students, will give way at all levels to the professionally best possible filmed documentaries by master teachers with hi-fi recording. Southern Illinois University's film production division at Carbondale, in collaboration with independent producer Francis Thompson, is completing a moving-picture series on my *comprehensive anticipatory, design-science explorations.*

While I confined my discourse to those unique aspects of my comprehensive thinking which provide pioneering interpretation of humanity's total experience as distinguished from the formally accepted and taught academic concepts, it required fifty-one hours to exhaust my inventory of unique experience interpretations and their derived patterns of generalized significance conceptionings. The fifty-one hours represented my net surviving inventory of unique thoughts which have been greatly modified and amplified by my progressive world-around university experiences.

(Over the years I have accepted more than 350 separate, unsolicited appointments or invitations to visit or revisit as a "Visiting Professor" or Lecturer at nearly 200 different universities or colleges in over thirty countries.)

Children must be allowed to discipline their own minds under the most favorable conditions—in their own special private environment. We'd better consider mass producing "one pupil schools"; that is, little well-equipped capsule rooms to be sent to all the homes; or we can design special private study rooms for homes. There are many alternatives but the traditional schoolroom is not one of them. I was invited by Einstein to meet with him and talk about one of my earlier books. I can say with authority that Einstein, when he wanted to study math and physics didn't sit in the middle of a schoolroom "desk prison." That is probably the poorest place he could have gone to study. As do any logical humans when they want to truly *study,* he went into seclusion—in his private study or laboratory.

Much of what goes on in our schools is strictly related to social experience, which, within limits, is fine. But we will be adding much more in the very near future by taking advantage of children's ability to show us what they need. When individuals are really thinking, they are tremendously isolated. They may manage to isolate themselves in an airline terminal, but it is despite the environment rather than because of it. The place to study is certainly not in a schoolroom.

The red schoolhouse—little or big—is on the way out. New educational media are making it possible to bring the most important kinds of experiences right into the home. With television reaching children in the privacy of their homes everywhere, we should bring education—school—to where the children are. This is a surprise concept—the school by television always and only in the home—if possible in a special room in the home. Ralph Waldo Emerson was right—"The household is a school of power."

TV is the number one potential emancipator from ignorance and economic disadvantage of the entire human family's residual poverty-stricken 60 percent. Even in the world's slums, TV antennas bristle. There is thus a wireless hook-up directly to the parents and children who watch their televisions avidly. Whatever comes over TV to the children and parents is the essence of education, for better or worse. What is now needed are educational TV advancements of high order.

Photographs I have taken of my grandchildren (without their awareness) as they looked at TV illustrate the fabulous concentration of the child.

Give children logical, lucid information when they want and need it, and watch them "latch on." Dr. Maria Montessori designed her "method" to tap the child's powers to attend and absorb.[20]

While great knowledge and ingenuity are being put into research on the channels through which education is conveyed, relatively little consideration has been given to *what* is conveyed in the communicated content. The magnitude of the task demands a most rigorous examination of "what" knowledge is to be imparted and in which *order, amount,* and *forms* it is to be conveyed.

So the home is the school, education is the upcoming major world industry, and TV is the great educational medium.

All of this has been made possible by *industrialization* which I define as the extracorporeal, organic, metabolic regeneration of humanity. Industrialization consists of tools, which, in turn, are externalizations of originally integral functions of humans. I divide all tools into the two main classes—*craft* tools and *industrial* tools. Craft tools consist of all those that can be invented and produced by one person starting and operating alone nakedly in the wilderness.

Industrial tools are those which cannot be produced by one person, such as the steamship *Queen Mary,* the giant dynamo, a concrete highway, New York City, or even the modern forged alloy-steel carpenter's hammer, with electro-insulated plastic handles, whose alloyed components and manufacturing operations involve thousands of people and the unique resources of several countries of Earth.

Words are the first industrial tools, for inherently they involve a plurality of people and are also inherently prior to relayed communication and integration of the respective experiences of a plurality of individuals. This is remi-

[20] R. C. Orem, ed., *Montessori for the Disadvantaged* (New York: Putnam's, 1967).

niscent of the scriptural account, "In the beginning was the word," which we may modify to read, "In the beginning of Industrialization was the word." Crafts are limited to a single person and involve only very local resources and very limited fragments of Earth and time, while industrialization, through the relayed experience of all people—permitted through the individualization of the spoken and written word—involves *all experiences of all people everywhere in history.*

As I stressed in my keynote address to the 1966 Music Educators' National Conference,[21] the *speech pattern of the parents* exerts a critically important formative influence during the child's early years.

If the parents take the trouble to speak clearly, to use their language effectively, to choose appropriate words, the children are inspired to do likewise. If the parents' tones of voice are hopeful, thoughtful, tolerant, and harmonious, the children are inspired to think and speak likewise. If the parents are not parroting somebody else, but are quite clearly trying to express themselves, nothing encourages more the intuitions of the young life to commit itself not only to further exploration but to deal competently in coordinating its innate faculties. However, if the parents indicate that they are not really trying, or relapse into slang cliches, slurred mouthings, blasphemy, anger, fear, or intolerance, indicating an inferiority complex which assumes an inability of self to attain understanding by others, then the children become discouraged about their own capability to understand or to be understood.

If the proper books are on the family shelves, if there are things around the house which clearly show the children that the parents are really trying to educate themselves, then the children's confidence in family is excited and the children too try to engender the parents' confidence in their—the children's—capabilities.

The child's verbal ecological patterning is a fascinating process. My granddaughter Alexandra was born in New York. She was brought by her parents from the hospital to their apartment in Riverdale, just across from the northern end of Manhattan, which is quite a high point of land directly in the path of the take-off pattern for both of New York City's major airports, La Guardia and (now) Kennedy. The planes were going over frequently, sometimes every few seconds. There was the familiar roar and, on such a high promontory, it was a fundamental event to a new life.

[21] R. Buckminster Fuller, "The Music of the New Life: Thoughts on Creativity, Sensorial Reality, and Comprehensiveness" (Keynote address presented at the National Conference on the Uses of Educational Media in the Teaching of Music, under joint auspices of the U.S. Office of Education and the Music Education National Conference, Washington, D.C., December 10, 1964). Published in the *Music Educators Journal,* part 1 (April–May 1966), part 2 (June–July 1966).

The interesting result was that my granddaughter's first word was not *Mummie* or *Daddy,* but *air*—short for airplane.

How we see the world depends largely on what we are told at the outset of life before we unconsciously or subconsciously lock together our spontaneous brain reflexings.

Children can most easily learn to see things correctly only if they are spoken to intelligently right from the beginning.

As I was quoted in the *New Yorker* "Profile":

> I've made tests with children—you have to get them right away, before they take in too many myths. I've made a paper model of a man and glued him down with his feet to a globe of the world, and put a light at one side, and shown them how the man's shadow lengthens as the globe turns, until finally he's completely in the shadow. If you show that to children, they never see it any other way, and they can really understand how the earth revolves the sun out of sight.[22]

We are learning to test experimentally the axioms given to us as "educational" springboards, and are finding that most of the "springboards" do not spring, if they exist at all!

There is growing awareness that we have been overproducing rigorously disciplined, game-playing, scientific specialists who, through hard work and suppressed imagination, earn their academic union cards, only to have their specialized field become obsolete or by-passed swiftly by evolutionary events of altered techniques and exploratory strategies.

Biological and anthropological studies reveal that overspecialization leads to extinction. We need the philosopher-scientist-artist—the comprehensivist, not merely more deluxe-quality technician-mechanics.

Artists are now extraordinarily important to human society. By keeping their innate endowment of capabilities intact, artists have kept the integrity of childhood alive until we reached the bridge between the arts and sciences. Their greatest faculty is the ability of the imagination to formulate conceptually. Suddenly, we realize how important this conceptual capability is.

Spontaneously, painters, dancers, sculptors, poets, musicians, and other artists, ask me to speak to them; or they look at my starkly scientific structures, devices, and mathematical exploration models and express satisfaction, comprehension, and enthusiasm. The miracle is that the artists are human beings whose comprehensivity was not pruned down by the well-meaning but ignorant educational customs of society.

[22] Quoted in Calvin Tomkins, "Profile," *The New Yorker,* January 8, 1966, pp. 35–36.

Artists are really much nearer to the truth than have been many of the scientists.

In a beautiful demonstration, Gyorgy Kepes, of M.I.T., took uniform-size black-and-white photographs of nonrepresentational paintings by many artists. He mixed them all together with the same size of black-and-white photographs taken by scientists of all kinds of phenomena through microscopes and telescopes. He and students classified the mixed pictures by pattern types. They put round-white-glob types together—wavy-gray-line-diagonals, little circle types, etc. together. When so classified and hung, one could not distinguish between the artist's works and scientific photographs taken through instruments. What was most interesting was that if you looked on the backs of the pictures you could get the dates and the identities. Frequently the artist had conceived of the pattern or parts in infra- or ultra-visible realms. The conceptual capability of the artists' intuitive formulation of the evolving new by subconscious coordinations are tremendously important. Science has begun to take a new view of the artist.

Philip Morrison, Head of Cornell's department of nuclear physics, talks about what he terms "right-hand" and "left-hand" sciences. Right-hand science deals in all the proven scientific formulas and experiments, while left-hand science deals in all of the as-yet unknown or unproven, that is, with all it is going to take intellectually, intuitively, speculatively, imaginatively, and even mystically, by inspired persistence, to open up the as-yet unknown.

We have been governmentally underwriting only the right-hand science, making it bigger and sharper. How could Congress justify appropriations of billions for dreams?

Pride, fear, economic and social insecurity, and the general reluctance of humanity to let go of nonsense in order vastly to reorganize are basic to the problem of education.

We adults must learn about our universe and how to modify the environment in order to permit life to operate and articulate the innate capabilities of humanity, the range and richness of which we are only beginning to apprehend. Innate cerebral and metaphysical capabilities have been frustrated by negative factors of the environment—not the least of which are the people in it who surround every individual. Today, the young people really want to know about things, they want to get closer to the truth, and my job is to do all I can to help them.

What I describe as *positive design-science reformations of the environment* must now be undertaken with the intent of permitting our innate faculties and facilities to be realized with subconscious coordinations of our organic process. *Reform of the environment* undertaken to defrustrate our innate capabilities, whether the frustration be caused by the inadequacies of the physical environment or by the debilitating reflexes of other humans, will *permit* humanity's original, innate capabilities to become successful. *Politics and conventionalized education have sought erroneously to mold or reform humanity, i.e., the collective individual.*

I have thought long and hard about architectural education and its potential for promoting environmental reform. I envision an utterly revised education of the architect, enabling successful students to operate on their own initiative in dealing with both comprehensively and in effective depth in mathematics, chemistry, physics, biology, geology, industrial tooling, network systems, economics, law, business administration, medicine, astronautics, computers, general systems theory, patents, and the whole gamut of heretofore highly specialized subjects.

This "comprehensivity curriculum" will prepare the graduating architects to gain the design initiative, performing thereafter not as economic slaves of technically illiterate clients and patron despots but as comprehensivists, integrating and developing the significance of all the information won by all the respective disciplines of the specialized sciences and humanities, converting this information into technical advantages for world society in completely tooled-up and well-organized comprehensive anticipatory *livingry systems.*

When President Eisenhower was first confronted by the strategic data on atomic warfare he said, "There is no alternative to peace," without defining the latter or indicating how it could be secured. Professor John Platt, Chicago University physicist and biophysicist, in a thorough survey of the overall shapes of a family of trend curves which comprehensively embrace science, technology, and humanity in Universe, said in 1964, "The World has become too dangerous for anything less than Utopia," but did not suggest how it might be attained. Jerome Wiesner, head of the department of nuclear physics at M.I.T. and past science adviser to Presidents Kennedy and Johnson, wrote in a recent issue of the *Scientific American* that "the clearly predictable course of the arms race is a steady downward spiral into oblivion."

So far, the only known and feasible means of arresting that spiral, by elimination of the cause of war, is the program of the World Students Design Decade. This ten-year plan of world architectural students is divided into five evolutionary stages of two years each. Phase One, "World Literacy Regarding World Problems," was on exhibit in the Tuileries Garden in Paris, France, for the first ten days of July 1965 (under the auspices of the International Union of Architects' Eighth World Congress). Emphasizing the central function of education and communication in overall planning, it dramatized the need for an informed world society to cope with the global nature of our problems.

It confronted the world with those basic facts leading the students to the research conclusion that human survival apparently depends upon an immediate, consciously coordinated, world-around, computerized, research marshalling and inception of the theoretically required additional inventions and industrial network integrations for the swiftest attainment and maintenance of physical success of all humanity.

Phase Two, "Prime Movers and Prime Metals," focusing upon the design of more efficient energy and metals utilization, was exhibited with the "Tribune Libre" section of the Ninth World U.I.A. Congress in Prague, Czechoslovakia, June–July 1967.

There is a new dedication on the part of the world's young. Students are corresponding with each other all over the globe. These young people are about to seize the initiative, to help us make humanity a success on Earth.

What I call "The Third Parent"—TV—is bringing babies half-hourly world news as well as much grown-up-authored, discrediting drivel. The students in revolt on the university campus are the first generation of TV-reared babies. They insist on social justice the world around. They sense that imminent change is inexorable.

On Southern Illinois University's Carbondale campus, we are setting up a great computer program to include the many variables now known to be operative in world-around industrial economics. In the machine's memory bank, we will store all the basic data such as the "where" and "how much" of each class of physical resources; location of the world's people; trendings and important needs of humanity, etc.

Next, we will set up a computer feeding game, called "How to Make the World Work." We will bring people from all over the world to start playing the game relatively soon. There will be competitive teams from all around Earth, testing their theories on how to make the world work. If a team resorts to political pressures to accelerate their advantages and are not able to wait for the going gestation rates to validate their theory, they are apt to be in trouble. When you get into politics, you are very liable to get into war. War is the ultimate tool of politics. If war develops, the side inducing it loses the game.

Essence of "success in making the world work" will be to make *all* people able to become world citizens free to enjoy the whole Earth, going wherever they want at any time, able to take care of all the needs of all their forward days without any interference with any person and never at the cost of another person's equal freedom and advantage.

I think that the communication task of reporting on the computerized playing of the game "How to Make the World Work" will become extremely popular all around Earth. We're going to be playing the game soon at S.I.U., and you'll be hearing more about it!

So, 1967 is the year of *The Case for Early Reading*.

Fortunately, the authors did not wait for "official permission" to use their initiative in exploring innovative approaches to answering the question: "When and under what circumstances should reading instruction begin?"

No one licensed the inventors of the airplane, telephone, electric light, and radio to go to work. It took only five men to invent these world transforming developments. The license comes only from the blue sky of the inventor's intellect.

The individual intellect disciplinedly paces the human individual, who disciplinedly paces science. Science disciplinedly paces technology by expanding the limits of technical, advantage-generating knowledge. Technology paces industry by progressively increasing the range and velocity inventory of technical capabilities. Industry in turn paces economics by continually altering and accelerating the total complex of environment controlling capabilities of man. Economics paces the everyday evolution acceleration of man's affairs. The everyday patterning evolution poses progressively accelerating problems regarding the understanding of the new relative significance of our extraordinarily changing and improving degrees of relative advantage in controlling our physical survival and harmonic satisfaction.

In 1927, I gave up forever society's general economic dictum that every individual who wants to survive must earn a living, substituting instead a search for the tasks that needed to be done that no one else was doing or attempting to do, which if accomplished, would physically and economically advantage society and eliminate pain.

By disciplining my faculties, I was able, as an individual, to develop technical and scientific capability to invent the physical innovations and their service industry logistics.

Seventeen of my prime inventions in a wide range of categories have been granted a total of 145 patents in fifty-six countries around the world; incidentally, these patents in late years have produced millions of dollars of revenue. There are over 200 licensees, a number of which are industrial "blue chip" corporations, operating under these patents.

I've never had an "expert" who ever comprehended in significant degree the importance of any new development on which I was working. While this is deplorable, it's also understandable in the big corporations because research and engineering heads, confronted with something from the "outside," become very defensive, believing that acquiescence or approval would imply an admission that they are not alert themselves.[23]

1967 is also the year of Canada's World's Fair, EXPO 67; the United States Pavilion is a 250-foot diameter Geodesic Skybreak Bubble.[24] I invented the geodesic dome in 1947, and today can count over 6,000 of my structures in

[23] See "Creativity, Innovation, and the Condition of Man," a dialogue between R. Buckminster Fuller and Stanley Foster Reed, *Employment Service Review,* Washington, D.C. March–April 1967.

[24] David Jacobs, "An Expo Named Buckminster Fuller," *New York Times Magazine,* April 23, 1967, pp. 32–33.

fifty countries, ranging from play domes to the 384-foot diameter dome which rose in Baton Rouge, Louisiana, as the largest clear span enclosure in history.

A U.S. government citation describes "air deliverable geodesic structures" as the "first basic improvement in mobile environmental control in 2,600 years." I can assure you I have never waited for any Bureau of Breakthroughs to grant me my permit to ponder, produce, or prognosticate.

Forty years ago, after my pioneering studies had revealed the low technical advance in everyday dwelling facilities as compared with transport and communication developments, I invented my 1927 Dymaxion[25] House to function as part of my concept of an air-deliverable, mass-producible world-around new human life protecting and nurturing, scientific dwelling service industry to transfer high scientific capability from a weaponry to livingry focus.

Children, born truthful, learn deception and falsehood from their elders' prohibition of truth. Much of this prohibition arises from a great, largely unconscious, parental selfishness born of drudgery and dissatisfaction (visibly rampant in the slums). The housing of children during their upbringing is the fundamental function of the home. If we solve the problem of the home, we can erase much of this unenlightenment.

The same year (1927), I published my conviction that two billion new era premium technology-dwelling devices would be needed before 2000 A.D., requiring a whole new world-encompassing service industry. I predicted it would take twenty-five years to establish that new industry. In 1952, right on theoretical schedule, the Ford Motor Company purchased the first of my large geodesic domes, which are the prototypes of the new era premium technology structures.

I anticipate that full scale industrialization of the livingry service industry will be realized by 1977, or just 60 years after the model "T" inaugurated the major world mass-production industry. Henry Ford, Sr., pioneered the long-range, world-around historical development of the application of the tools-to-make-tools system of mass production to larger end-product tools, with his motorized road vehicle in 1907.

My structures, as reported in engineering and scientific publications, can cover very large clear span spaces more economically than by any other rectilinear or other shaped systems—for example, 1,000-fold more economically weightwise than accomplished by the dome of St. Peter's in Rome, or thirty times more efficiently than by reinforced concrete.

The New York Herald Tribune in February 1962 reported that Dr. Horne of Cambridge University had announced, at a world conference of molecular biologists, the discovery of the generalized principles governing the protein shells of the viruses. "All these virus structures had proved to be geodesic

[25] Dymaxion: The maximum gain of advantage from the minimal energy output.

spheres of various frequencies. The scientists reported that not only were the viruses geodesic structures, which latter had been discovered earlier by Buckminster Fuller, but also that the mathematics which apparently controls nature's formulations of the viruses had also been discovered (1933) and published by Fuller a number of years earlier (1944)."[26]

Although the words "genius" and "creativity" have been employed to explain my being "well known," I am convinced that the only reason I am known at all is because I set about deliberately in 1927 to be a comprehensivist in an era of almost exclusive trending and formal disciplining toward specialization. Inasmuch as everyone else was becoming a specialist, I didn't have any competition whatsoever. I was such an antithetical standout that whatever I did became prominently obvious, therefore, "well known."

Luckily, as a special student at the United States Naval Academy in 1917, I had been exposed to a comprehensive educational strategy fundamentally different from the then-prevailing ivy-league model. The potential I have since developed, every physically normal child also has at birth.

What, if anything, is hopeful about my record is that I am an average human and therefore whatever I have been able to accomplish also can be accomplished, and probably better, by average humanity, each successive generation of which has less to unlearn.

Humanity is beginning to transform from being utterly helpless and only subconsciously coordinate with important evolutionary events. We have gotten ourselves into much trouble, but at the critical transformation stage we are reaching a point where we are beginning to make some measurements—beginning to know a little something. We are probably coming to the first period of direct consciously assumed responsibility of humanity in Universe.

Human beings sort, classify, and order in direct opposition to entropy—which is the law of increase of the random element—increase of disorder. Human beings sort and classify internally and subconsciously as well as externally and consciously, driven by intellectual curiosity and brain. Human beings seem to be the most comprehensive antientropy function of Universe.

Human beings, as designed, are obviously intended to be a success just as the hydrogen atom is designed to be a success. The fabulous ignorance of humanity and our long-wrongly conditioned reflexes have continually allowed the new life to be impaired, albeit lovingly and unwittingly.

Our most important task is to become as comprehensive as possible by intellectual conviction and "self-debiasing," not through ignorant yielding, but through a progressively informed displacement of invalid assumptions and dogma by discovery of the valid data. In this development, the young will lead

---

[26] R. Buckminster Fuller, "Conceptuality of Fundamental Structures," in *Structure in Art and in Science,* ed. Gyorgy Kepes (New York: George Braziller, 1965), p. 76.

the old in swiftly increasing degree. The child is the trim tab of the future. Let us respect in our children the profound contribution trying to emerge. The Bible was right: ". . . and a little child shall lead them."

As one who has spent his own lifetime comprehensively *putting things together* in an age of specialized *taking apart*—as a poet—I close with these lines—

And with Industrialization a uniformly beautiful
world race emerges
as does the fine chiseled head
from the rough marble block
certifying the god-like untrammeled beauty
of a perfect human process
implicit in the dynamic designing
genius of the mind. . . .[27]

[27] R. Buckminster Fuller, *Untitled Epic Poem on the History of Industrialization* (Highlands, N.C.: Jonathan Williams, Publisher, 1962), p. 116.

Of the myriad projects Bucky Fuller felt to be important, one of his "What am I try-ing to do?" and "How to make the world work" projects developed into the imple-mentation of World Game. Conceived originally at Southern Illinois University as the "Inventory of World Resources Human Trends and Needs," Bucky envisioned and expected World Game to be played by major world individuals and teams.

One of Fuller's most active students in World Game was Medard Gabel. With Bucky's Dymaxion Airocean World Map, the World Game Institute has been turning data into knowledge. The important work of Bucky Fuller continues through this insti-tute, which is headquartered in Philadelphia with Medard Gabel as executive direc-tor. The World Game motto is "Fostering responsible change and citizenship in a global society." The organization's "strategy 18" is a list of points that the world needs to address in order to solve the major systemic problems confronting humanity.

Buckminster Fuller and the Game of the World

Medard Gabel

The Encyclopedia of the Future, a 1,115-page two-volume work first published in 1996, took over five years to produce and featured contributions from over four hundred different experts. It covered topics ranging from Abortion to Zion and is considered to be an authoritative source on all matters concern-ing the rapidly growing field of "futurology," futurism, or futures studies. Featured in its appendix is a survey of professional futurists who were asked, "Who was the most influential futurist in the history of the world?"

Buckminster Fuller is listed first—ahead of such luminaries as H. G. Wells, Isaac Newton, Arnold J. Toynbee, and Leonardo da Vinci.

Why? Why was such a distinction bestowed on the inventor of a house that never made it past the three-foot-model stage, a car that killed its test driver and never went into mass production, a mass-produced bathroom ensemble that never made it to the masses, a structure for enclosing large spaces that was best known for how it leaked, and a bunch of social theories and policies that have been called everything from iconoclastic to bombastic? Why indeed.

The answer lies in Fuller's grand perspective, his bold synthesis of technology and human values, and his integration of these into a tool for humanity to use in solving its planetary problems. As important as his inventions were in their own right (and as a more balanced presentation than the one above would soon disclose), they pale in comparison to their impact on the world's imagination of what is possible. When Fuller proposed a housing service industry in the 1920s that would mass-produce "housing units" and air-deliver them via giant dirigible to any place in the world—and those same housing units would be hung from a central tower that contained all the services needed for the house to be autonomous—he was not just fifty to one hundred years ahead of his time, he was lighting a bonfire in the collective imagination of the world (and a firebomb in the straw house of the architectural profession). What Fuller's original autonomous house did was present a way not only of building a revolutionary house in a revolutionary way, he presented a way of looking at building, housing, shelter, and architecture in a way that swept them all away in a grand vision of housing as a basic human need that all humans have (not just the client in traditional architectural circles) and which was a global, not local or personal problem that only the rich could afford to address while the rest of humanity had to make do. Fuller's contribution went further: his methodology for addressing the housing problem was generalizable. You could, as did he, apply it to transportation, energy, education, pollution, accounting, governance, and a wealth of additional social problems.

The core of this approach was a concern with the whole: the whole Earth, the entire history of the planet, all of humanity—both those living now and those yet to be born. His approach, as he would later codify it, was:

- *Comprehensive,* starting from the whole system and working back to the special case, dealing with all facets of a problem, including the larger system the problem was a part of;
- *Anticipatory,* in that it sought to recognize the threats coming down the pike before they arrived full-blown on an unsuspecting or ill-prepared society, as well as to deal with the way things were going to be when the solution was going to be implemented, not the way things were in the present;
- A *design strategy,* in contradiction to a political, or let's-pass-a-law-and-change-human-behavior, approach, that sought to change the larger system of which the specific problem was a part;
- A *science-based methodology* that used the latest advances of science to benefit humanity.

His "Comprehensive Anticipatory Design Science" was at least as much a perspective on the problems of the world as it was a methodology for tackling those problems. When applied to contemporary problems, whether those of Fuller's day or of the twenty-first century, it leads to strikingly fresh insights and solutions. It was also the perspective that led to the World Game.

In the 1960s, Buckminster Fuller proposed a "great logistics game" and "world peace game" (later shortened to simply "World Game") that was intended to be a tool that would facilitate a comprehensive, anticipatory design science approach to the problems of the world. The use of "world" in the title obviously reflects Fuller's global perspective and his contention that we now need a systems approach that deals with the world as a whole, and not a piecemeal approach that tackles our problems in what he called a "local focus hocus pocus" manner. The entire world is now the relevant unit of analysis, not the city, state, or nation. We are, in Fuller's words, aboard Spaceship Earth, and the illogic of two hundred nation-state admirals all trying to steer the spaceship in different directions is made clear through the metaphor—as well in Fuller's more caustic assessment of nation-states as "blood clots" in the world's global metabolism.

The logic for the use of the word "game" in the title is even more instructive. It says a lot about Fuller's approach to governance and social problem solving.

Obviously intending it as a very serious tool, Fuller chose to call his vision a "game" because he wanted it seen as something that was accessible to everyone, not just the elite few in the power structure who thought they were running the show. In this sense, it was one of Fuller's more profoundly subversive visions. Fuller wanted a tool that would be accessible to everyone, whose findings would be widely disseminated to the masses through a free press, and that would, through this groundswell of public vetting and acceptance of solutions to society's problems, ultimately force the political process to move in the direction that the values, imagination, and problem-solving skills of those playing the democratically open World Game dictated. It was a view of the political process that some might think naive, if they only saw the world for what it was when Fuller was proposing his idea (the 1960s)—minus personal computers and the Internet. It was not merely a matter of leveling the playing field; the good-old-boy political process was to be subverted out of existence by a process that would bring Thomas Jefferson into the twentieth century.

In order to have this kind of power, the game needed to have the kind of information and tools for manipulating that information that empowers. It needed a comprehensive database that would provide the players of the World Game with better data than their politically elected or appointed counterparts had. They needed an inventory of the world's vital statistics—

where everything was and in what quantities and qualities, from minerals, to manufactured goods and services, to humans and their unmet needs as well as capabilities. They also needed an information source that would monitor the current state of the world, bringing vital news into the "game room" live. None of this existed when Fuller began talking about the game. And then something funny happened on the way to the twenty-first century: CNN, personal computers, CD-ROMs, the Internet and the World Wide Web, supercomputer power on personal computers. Reams of data about the world, its resources, its problems, and potential solutions started to bubble to the surface and transform the world and the way we communicate, do business, research, and govern.

The World Game that Fuller envisioned was to be a place where individuals or teams of people came and competed, or cooperated, to

"make the world work,
for 100% of humanity,
in the shortest possible time,
through spontaneous cooperation,
without ecological offense
or the disadvantage of anyone."

The goal of making the world work for 100 percent of humanity reflects Fuller's global perspective as well as his values. We are not here just to make ourselves rich, famous, or top consumer of the day or decade, or just for the 3 percent living in our part of the world, we're here for all of humanity. The phrase "spontaneous cooperation" is instructive in light of the previous discussion of the choice of the word "game" as part of the title for this activity. Fuller's statement of intent does not read "make the world work for 100% of humanity through a central government," or "through enforced coercion by a strong military"; he wanted it to work through a cooperation that would arise from a fundamental transparency of society and its needs. If everyone knows what the situation is and has a clear vision of what should be and what needs to be done, we cooperate to get it done—as we do as a society in times of emergency.

In Fuller's vision of a world peace game, participants would come to play from around the world, irrespective of their political ideology or local concerns. One model of how it could work had players focused on a problem, like food availability or hunger, for a certain time period, say a week. The team or individual that demonstrated how, using current technology and known resources, hunger could be eliminated in ten years would "win." The team that could show how it could be done in a shorter time, or by using less resources, or costing less, or accomplishing more than one thing at a time, such as providing clean water as well as eliminating malnutrition, would win round two. Round three would be won by an effort that was even "better." The next week the focus would shift to energy, or health, or education. Eventually the focus would return to food. These efforts, as pointed out above, were not intended as academic exercises. Each new strategy that incrementally improved the method for solving a problem was one step closer to implementation in Fuller's view. The strategies for solving a given problem would become ever more compelling as they demonstrated how all humanity "won"—demonstrated that the World Game was not the zero-sum I-win/you-lose variety of game, but the total-wealth-increasing kind.

Today the World Game is being developed by the World Game Institute, a nonprofit, nonpartisan, independent research and education organization based in Philadelphia, Pennsylvania. It is keeping Fuller's vision alive and vital as it takes advantage of new advances in technology and data availability. One of its products is the World Game Workshop, an interactive, experiential workshop that takes place on top of the world's largest and most accurate map of the whole world. Participants are placed in charge of the world and lead it into the twenty-first century. NetWorld Game is an Internet-based simulation of the global economy in which people from around the world assume leadership for the world for twenty years and address the problems of their region and the whole world. Some of the problems are local in origin, others are global. Both these active, information-intensive games involve people in solving the critical problems facing global humanity. Participants learn a huge amount about the world, its resources, problems, options for development, and what they can do to engage in solving the problems that their values deem important. They do this, as the word "game" implies, while having some fun in the process. Both these tools help people recognize, define, and solve global problems and local problems in a global context. They tap the creative imaginations of groups as varied as high school students and corporate executives. There is even now a Junior World Game for younger players.

In addition to these engaged-learning and problem-solving processes, the World Game Institute has also developed some research tools. One of these is the Global Data Manager CD-ROM. This is a vast collection of socioeconomic and environmental statistics for every country in the world, in time series from the 1960s on. It is one of the world's largest collections, if not *the* largest collection, of world resources data, integrating over one hundred separate sources into one powerful database. It is the statistical foundation of the World Game Workshop and the NetWorld Game.

In addition, the institute conducts research into global problems and capabilities. One such effort is its What the World Wants Project, illustrated by the accompanying chart that points out that the world has the resources, technology, and financial wherewithal to meet the basic human needs of 100 percent of humanity and clean up the environment as well. It illustrates the institute's continuing documentation of the Fullerian intuition that the world could be made to work for 100 percent of humanity—with an important addition: how it could be done.

WHAT THE WORLD WANTS

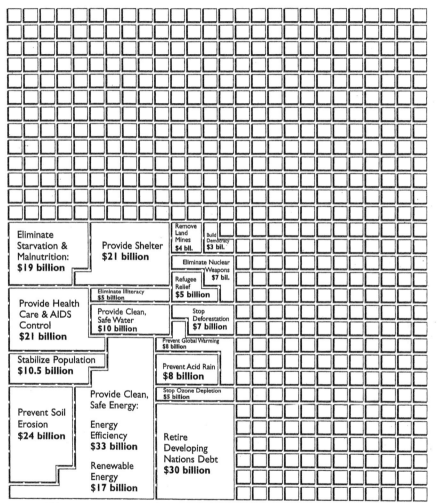

Eliminate Starvation & Malnutrition: **$19 billion**

Provide Shelter **$21 billion**

Remove Land Mines **$4 bil.**

Build Democracy **$3 bil.**

Eliminate Nuclear Weapons **$7 bil.**

Refugee Relief **$5 billion**

Provide Health Care & AIDS Control **$21 billion**

Eliminate Illiteracy **$5 billion**

Provide Clean, Safe Water **$10 billion**

Stop Deforestation **$7 billion**

Prevent Global Warming **$8 billion**

Stabilize Population **$10.5 billion**

Prevent Acid Rain **$8 billion**

Stop Ozone Depletion **$5 billion**

Prevent Soil Erosion **$24 billion**

Provide Clean, Safe Energy:

Energy Efficiency **$33 billion**

Renewable Energy **$17 billion**

Retire Developing Nations Debt **$30 billion**

AND HOW TO PAY FOR IT

Above are annual costs of various global programs for solving the major human need and environmental problems facing humanity. Each program is the amount needed to accomplish the goal for all in need in the world. Their combined total cost is approximately 30% of the world's total annual military expenditures.

Total Chart represents Annual World Military Expenditures: **$780 billion**

 = $1 billion

= Amount that was needed to eradicate Smallpox from the world (accomplished 1978): $300 million

World Game Institute, 3215 Race Street, Philadelphia, Pennsylvania 19104 • Phone: 215-387-0220 • Fax: 215-387-3009
E-mail: wgi@worldgame.org • World Wide Web: http://www.worldgame.org   © 1999 World Game Institute

The World Game—How to Make the World Work

from *Utopia or Oblivion*

To start with, here is an educational bombshell: Take from all of today's industrial nations all their industrial machinery and all their energy-distributing networks, and leave them all their ideologies, all their political leaders, and all their political organizations and I can tell you that within six months two billion people will die of starvation, having gone through great pain and deprivation along the way.

However, if we leave the industrial machinery and their energy-distribution networks and leave them also all the people who have routine jobs operating the industrial machinery and distributing its products, and we take away from all the industrial countries all their ideologies and all the politicians and political machine workers, people would keep right on eating. Possibly getting on a little better than before.

The fact is that now—for the first time in the history of man for the last ten years, all the political theories and all the concepts of political functions—in any other than secondary roles as housekeeping organizations—are completely obsolete. All of them were developed on the you-or-me basis. This whole realization that mankind can and may be comprehensively successful is startling.

In pursuance of this theme and under auspices to be announced later we are going to undertake at Southern Illinois University, in the next five years, a very extraordinary computerized program to be known as "How to Make the World Work."

Here on Southern Illinois' campus we are going to set up a great computer program. We are going to introduce the many variables now known to be operative in economics. We will store all the basic data in the machine's memory bank; where and how much of each class of the physical resources; where are the people, what are the trendings—all kinds of trendings of world man?

Next we are going to set up a computer feeding game, called

"How Do We Make the World Work?" We will start playing relatively soon. We will bring people from all over the world to play it. There will be competitive teams from all around earth to test their theories on how to make the world work. If a team resorts to political pressures to accelerate their advantages and is not able to wait for the going gestation rates to validate their theory they are apt to be in trouble. When you get into politics you are very liable to get into war. War is the ultimate tool of politics. If war develops the side inducing it loses the game.

Essence of the world's working will be to make every man able to become a world citizen and able to enjoy the whole earth, going wherever he wants at any time, able to take care of all the needs of all his forward days without any interference with any other man and never at the cost of another man's equal freedom and advantage. I think that the communication problem—of "How to Make the World Work"—will become extremely popular the world around.

The game will be played by competing individuals and teams. The comprehensive logistical information upon which it is based is your Southern Illinois University-supported Inventory of World Resources Human Trends and Needs. It is also based upon the data and grand world strategies already evolved in the Design Science Decade being conducted, under our leadership here at Southern Illinois University, by world-around university students who, forsaking the political expedient of attempting to reform man, are committed to reforming the environment in such a manner as to "up" the performance per each unit of invested world resources until so much more is accomplished with so much less that an even higher standard of living will be effected for 100% of humanity than is now realized by the 40% of humanity who may now be classified as economically and physically successful.

"The game" will be hooked up with the now swiftly increasing major universities information network. This network's information bank will soon be augmented by the world-around satellite-scanned live inventory of vital data. Spy satellites are now inadvertently telephotoing the whereabouts and number of beef cattle around the surface of the entire earth. The exact condition of all the world's crops is now simultaneously and totally scanned and inventoried. The interrelationship of the comprehensively scanned weather and the growing food supply of the entire earth are becoming manifest.

In playing "the game" the computer will remember all the plays made by previous players and will be able to remind each successive player of the ill fate of any poor move he might contemplate making. But the ever-changing inventory might make possible today that which would not work yesterday. Therefore the successful stratagems of the live game will vary from day to day. The game will not become stereotyped.

If a player resorts to political means for the realization of his strategy, he may be forced ultimately to use the war-waging equipment with which all national political systems maintain their sovereign power. If a player fires a gun—that is, if he resorts to warfare, large or small—he loses and must fall out of the game.

The general-systems-theory controls of the game will be predicated upon employing within a closed system the world's continually updated total resource information in closely specified network complexes designed to facilitate attainment, at the earliest possible date, by every human being of complete enjoyment of the total planet earth, through the individual's optional traveling, tarrying, or dwelling here and there. This world-around freedom of living, work, study, and enjoyment must be accomplished without any one individual interfering with another and without any individual being physically or economically advantaged at the cost of another.

Whichever player or team first attains total success for humanity wins the first round of the gaming. There are alternate ways of attaining success. The one who attains it in the shortest time wins the second round. Those who better the record at a later date win rounds 3, 4, and so on.

All the foregoing objectives must be accomplished not only for those who now live but for all coming generations of humanity. How to make humanity a continuing success at the earliest possible moment will be the objective. The game will also be dynamic. The players will be forced to improve the program—failure to improve also results in retrogression of conditions. Conditions cannot be pegged to accomplishment. They must also grow either worse or better. This puts time at a premium in playing the game.

Major world individuals and teams will be asked to play the game. The game cannot help but become major world news. As it will be played from a high balcony overlooking a football-field-sized Dymaxion Airocean World Map with electrically illumined data transformations, the game will be visibly developed and may be live-televised the world over by a multi-Telstar relay system.

The world's increasing confidence in electronic instrumentation in general—due to the demonstrated reliability of its gyrocompasses, and its "blind" instrument landings of airplanes at night in thick fog, and confidence in opinion-proof computers in particular, will make the "world game" playing of fundamental and spontaneous interest to all of humanity.

Ultimately its most successful winning techniques will become well known around the world and as the game's solutions gain world favor they will be spontaneously resorted to as political emergencies accelerate.

Nothing in the game can solve the problem of two men falling in love with the same girl, or falling in love with the same shade under one specific tree. Some are going to have to take the shade of another equally inviting tree. Some may end up bachelors. Some may punch each other's noses. For every problem solved a plurality of new problems arise to take their place. But the problems need not be those of physical and economic survival. They can be perplexing and absorbing in entirely metaphysical directions such as those which confront the philosophers, the artists, poets, and scientists.

The game must, however, find ways in which to provide many beautiful shade trees for each—that is to say a physical and economic abundance adequate for all. There will, of course, have to be matchings of times and desires, requiring many initial wait-listings. As time goes on, however, and world-around information becomes available, the peaks and valleys of men's total time can be ever-improvedly smoothed out. Comprehensive coordination of bookings, resource, and accommodation information will soon bring about a 24-hour, world-around viewpoint of society which will operate and think transcendentally to local "seasons" and weathers of rooted botanical life. Humanity will become emancipated from its mental fixation on the seven-day-week frame of reference. I myself now have many winters and summers per year as I cross the equator from northern to southern hemisphere and back several times annually. I have now circled the earth so many times that I think of it and literally sense it in my sight as a sphere. I often jump in eight- and nine-hour time-zone air strides. As a consequence my metabolic coordination has become independent of local time fixations.

It is my intention to initiate on several occasions in a number of places anticipatory discussion of the necessary and desirable parameters to establish for playing the world game. I intend to nominate as participants both in these preliminary discussions and in formal play only those who are outstandingly capable of discussing these parameters. The participants must also be those well known for their lack of bias as well as for their forward-looking competence and practical experience.

In July 1969, when mankind was to take its epochal step and send its first man to the moon, the world's most powerful television network chose Walter Cronkite and Arthur C. Clarke to sit in the broadcast chairs. America's senior newsman, Walter Cronkite, regarded as one of the world's most trusted newscasters, would report this historic event. Cohost Arthur C. Clarke was selected for his years as a cosmic realist and futurist and for his scientific writings. In the 1950s, Clarke had developed, from concept to mathematical reality, synchronized orbit communications satellites, which we now call COMSAT. We are all familiar with *2001: A Space Odyssey,* the imaginative 1968 film that Clarke wrote with director Stanley Kubrick. Based on a short story by Clarke, *2001* made a complete breakthrough into space-age moving picture and television concepts.

Clarke and Fuller first met at a university in the fifties, and Bucky visited him in Sri Lanka in 1978. Later, Bucky presented Clarke with a drawing of the "Earthian location" of the Fountains of Paradise, the title of a novel by Clarke. Clarke describes Bucky's response to the novel in his contribution piece.

The Fountains of Paradise

Sir Arthur C. Clarke, C.B.E.

Although we met only a few times, Bucky had a considerable influence on me—as on thousands of others. Our first encounter was at a university symposium in the late 1950s, which he effortlessly dominated with the aid of a monstrous instrument that served as both hearing aid and megaphone.

In 1978 I had the great privilege of flying him around Sri Lanka and showing him the locales I had used in my novel *The Fountains of Paradise.* A decade later this resulted in one of the most amazing coincidences in my life—one so extraordinary that I sometimes wonder if I'm in an advanced stage of terminal solipsism. But judge for yourself.

The theme of *Fountains* was the building of a "Space Elevator" reaching from Earth up to the geostationary orbit 22,000 miles above, where a satellite can hover apparently motionless over the same spot on the equator. If we could construct such an orbital tower, escaping from Earth would require less than a thousand dollars' worth of electricity per passenger—not hundreds of tons of rocket fuel.

When I wrote *Fountains* in 1977, only one material was known with sufficient strength to build what its Russian inventor, Yuri Artsutanov, called a "cosmic funicular." Unfortunately it is not available in the megaton quantities required: it is the crystalline form of carbon, better known as diamond.

However, in 1985 the necessary building material was discovered, and it turned out to be yet another form of carbon, C_{60}—immediately named "Buckminsterfullerene" because of its exact resemblance to Bucky's famous geodesic domes. How he would have enjoyed this discovery! Alas, he missed it by only two years.

Now for the amazing coincidence. In 1979 I made a recording (Caedmon TC 1606) of *Fountains II* on a 12-inch LP (remember them?). And who do you think was kind enough, despite his countless commitments, to write the sleeve notes and even make a sketch of the space elevator reaching up from Earth to the stationary orbit?

Here are Bucky's own comments on my novel. When the space elevator is built, some time in the twenty-first century, it will be his greatest memorial.

"Today, Arthur lives on that jewel-shaped Indian Ocean island of Sri Lanka (Ceylon). He loves it so much, he intends never again to travel away from it.

"I am fortunate enough to have been his guest on Sri Lanka, and to have

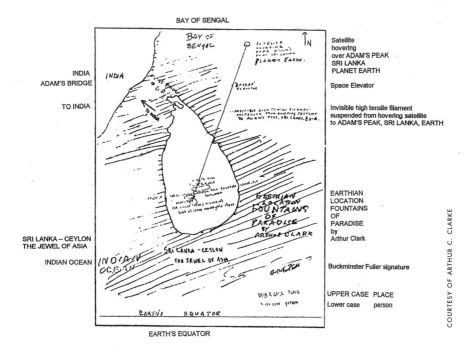

visited Taprobane (pronounced ta-PROB-a-nee) and its magic, awe-inspiring Adam's Peak (Sri Kanda), et al. We flew figure eights around Adam's Peak, the Demon Rock, and Ceylon's highest mountain, Pidurutalagala. A map of Ceylon, including only the loci featured in my story, is on the preceding page.

"I have spent some years thinking realistically about engineering undertakings somewhat akin to those Arthur Clarke incorporates into this moving story and can attest to his thorough respect of physical laws. In 1951, I designed a free floating tensegrity ring-bridge to be installed way out from and around the Earth's equator. Within this 'halo' bridge, the Earth would continue its spinning while the circular bridge would revolve at its own rate. I foresaw Earthian traffic vertically ascending to the bridge, revolving and descending at preferred Earth loci.

"I think you will be deeply and unforgettably moved by his Taprobane. I love it."

Spaceship Earth from *Operating Manual for Spaceship Earth*

Our little Spaceship Earth is only eight thousand miles in diameter, which is almost a negligible dimension in the great vastness of space. Our nearest star—our energy-supplying mother-ship, the Sun—is ninety-two million miles away, and the next nearest star is one hundred thousand times further away. It takes two and one-half years for light to get to us from the next nearest energy supply ship star. That is the kind of space-distanced pattern we are flying. Our little Spaceship Earth is right now traveling at sixty thousand miles an hour around the sun and is also spinning axially, which, at the latitude of Washington, D.C., adds approximately one thousand miles per hour to our motion. Each minute we both spin at one hundred miles and zip in orbit at one thousand miles. That is a whole lot of spin and zip. When we launch our rocketed space capsules at fifteen thousand miles an hour, that additional acceleration speed we give the rocket to attain its own orbit around our speeding Spaceship Earth is only one-fourth greater than the speed of our big planetary spaceship.

Spaceship Earth was so extraordinarily well invented and designed that to our knowledge humans have been on board it for two million years not even knowing that they were on board a ship. And our spaceship is so superbly designed as to be able to keep life regenerating on board despite the phenomenon, entropy, by which all local physical systems lose energy. So we have to obtain our biological life-regenerating energy from another spaceship—the sun.

Our sun is flying in company with us, within the vast reaches of the Galactic system, at just the right distance to give us enough radiation to keep us alive, yet not close enough to burn us up. And the whole scheme of Spaceship Earth and its live passengers is so superbly designed that the Van Allen belts, which we didn't even know we had until yesterday, filter the sun and other star radiation which as it impinges upon our spherical ramparts is so concentrated that if we went nakedly outside the Van Allen belts it would kill us. Our Spaceship Earth's designed infusion of that radiant energy of the stars is processed in such a way that you and I can carry on safely. You and I can go out and take a sunbath, but are unable to take in enough energy through our skins to keep alive. So part of the invention of the Spaceship Earth and its biological life-sustaining is that the vegetation on the land and the algae in the sea, employing photosynthesis, are designed to impound the life-regenerating energy for us to adequate amount.

But we can't eat all the vegetation. As a matter of fact, we can eat very little of it. We can't eat the bark nor wood of the trees nor the grasses. But insects can eat these, and there are many other animals and creatures that can. We get the energy relayed to us by taking the milk and meat from the animals. The animals can eat the vegetation, and there are a few of the fruits and tender vegetation petals and seeds that we can eat. We have learned to cultivate more of those botanical edibles by genetical inbreeding.

That we are endowed with such intuitive and intellectual capabilities as that of discovering the genes and the R.N.A. and D.N.A. and other fundamental principles governing the fundamental design controls of life systems as well as of nuclear energy and chemical structuring is part of the extraordinary design of the Spaceship Earth, its equipment, passengers, and internal support systems. It is therefore paradoxical but strategically explicable, as we shall see, that up to now we have been mis-using, abusing, and polluting this extraordinary chemical energy-interchanging system for successfully regenerating all life aboard our planetary spaceship.

One of the interesting things to me about our spaceship is that it is a mechanical vehicle, just as is an automobile. If you own an automobile, you realize that you must put oil and gas into it, and you must put water in the radiator and take care of the car as a whole. You begin to develop quite a little thermodynamic sense. You know that you're either going to have to keep the machine in good order or it's going to be in trouble and fail to function. We have not been seeing our Spaceship Earth as an integrally-designed machine which to be persistently successful must be comprehended and serviced in total.

Now there is one outstandingly important fact regarding Spaceship Earth, and that is that no instruction book came with it. I think it's very significant that there is no instruction book for successfully operating our ship. In view of the infinite attention to all other details displayed by our ship, it must be taken as deliberate and purposeful that an instruction book was omitted. Lack of instruction has forced us to find that there are two kinds of berries—red berries that will kill us and red berries that will nourish us. And we had to find out ways of telling which-was-which red berry before we ate it or otherwise we would die. So we were forced, because of the lack of an instruction book, to use our intellect, which is our supreme faculty, to devise scientific experimental procedures and to interpret effectively the significance of the experimental findings. Thus, because the instruction manual was missing we are learning how we safely can anticipate the consequences of an increasing number of alternative ways of extending our satisfactory survival and growth—both physical and metaphysical.

Quite clearly, all of life as designed and born is utterly helpless at the moment of birth. The human child stays helpless longer than does the young of any other species. Apparently it is part of the invention "man" that he is meant to be utterly helpless through certain anthropological phases and that, when he begins to be able to get on a little better, he is meant to discover some of the physical leverage-multiplying principles inherent in universe as well as the many nonobvious resources around him which will further compoundingly multiply his knowledge-regenerating and life-fostering advantages.

I would say that designed into this Spaceship Earth's total wealth was a big safety factor which allowed man to be very ignorant for a long time until he had amassed enough experiences from which to extract progressively the system of generalized principles governing the increases of energy managing advances over environment. The designed omission of the instruction book on how to operate and maintain Spaceship Earth and its complex life-supporting and regenerating systems has forced man to discover retrospectively just what his most important forward capabilities are. His intellect had to discover itself. Intellect in turn had to compound the facts of his experience. Comprehensive reviews of the compounded facts of experiences by intellect brought forth awareness of the generalized principles underlying all special and only superficially-sensed experiences. Objective employment of those generalized principles in rearranging the physical resources of environment seems to be leading to humanity's eventually total success and readiness to cope with far faster problems of universe.

To comprehend this total scheme we note that long ago a man went through the woods, as you may have done, and I certainly have, trying to find the shortest way through the woods in a given direction. He found trees fallen across his path. He climbed over those crisscrossed trees and suddenly found himself poised on a tree that was slowly teetering. It happened to be lying across another great tree, and the other end of the tree on which he found himself teetering lay under a third great fallen tree. As he teetered he saw the third big tree lifting. It seemed impossible to him. He went over and tried using his own muscles to lift that great tree. He couldn't budge it. Then he climbed back atop the first smaller tree, purposefully teetering it, and surely enough it again elevated the larger tree. I'm certain that the first man who found such a tree thought that it was a magic tree, and may have dragged it home and erected it as man's first totem. It was probably a long time before he learned that any stout tree would do, and thus extracted the concept of the generalized principle of leverage out of all his earlier successive special-case experiences with such accidental discoveries. Only as he learned to generalize fundamental principles of physical universe did man learn to use his intellect effectively.

Once man comprehended that any tree would serve as a lever his intellectual advantages accelerated. Man freed of special-case superstition by intellect has had his survival potentials multiplied millions fold. By virtue of the leverage principles in gears, pulleys, transistors, and so forth, it is literally possible to do more with less in a multitude of physio-chemical ways. Possibly it was this intellectual augmentation of humanity's survival and success through the metaphysical perception of generalized principles which may be objectively employed that Christ was trying to teach in the obscurely told story of the loaves and the fishes.

One of the people most admired by Bucky Fuller was his great-aunt Margaret Fuller Ossoli. She was a literary critic for the *New York Herald* and became the first woman foreign correspondent. With Ralph Waldo Emerson, she coedited the Transcendentalist quarterly *The Dial*, which was the first journal to publish Thoreau. A social reformer and a very early feminist, she wrote the influential book *Woman in the Nineteenth Century*. One summer, in the early 1980s, actress Ellen Burstyn traveled to Maine to confer with Bucky about doing a story about Great-Aunt Margaret; it would have been timely.

Bucky, who supported the feminist movement, spoke up strongly for the passage of the Equal Rights Amendment and did not hesitate to admonish those who opposed this important bill defending the rights of American women. Among the women he befriended were Barbara Marx Hubbard and Marilyn Ferguson, who had founded the Committee for the Future, to inspire effective thinking among the young and to encourage them to act with integrity. Though always apolitical, Bucky did suggest that Cleveland's Cuyahoga County Commissioner Timothy F. Hagan run for higher office and that Hubbard run for president of the United States against Richard Nixon.

Since that time, Hubbard has founded the Foundation for Conscious Evolution. In her contribution here she recounts her experiences with Fuller, particularly her Bucky-inspired presidential aspirations.

The Path of Social Evolution

Barbara Marx Hubbard

Buckminster Fuller revealed to us the potential of humanity for an ever-evolving, regenerative future. He made it clear that "our minds are designed to know the design," and that we have the resources, technology and know-how to make of this world a 100 percent physical success. The design science revolution which he called for is still emerging among us in countless innovations in all fields. I believe that we are approaching the natural connectedness and synergy among positive innovations, which together can build a "win-win" world—a world that can work for everyone. This is the crucial choice point for humanity, as he pointed out. Our "final exam" is going on a bit longer than he predicted. But we haven't failed yet, and his life and work give us the impetus to seek out the patterns of social and technological innovations that can carry us forward into the third millennium toward an immeasurable future as citizens of the local universe.

My personal experience with him gives insights into his remarkable gift of inspiring others to go beyond their normal path to reach for the highest.

The first time I met him was in the early 1970s, when the Committee for the Future was sponsoring a conference called "Mankind in the Universe" at Southern Illinois University, where Dr. Fuller was in residence. It was the time of student protests and terrified deans. I remember giving a speech one evening. An angry student stood up and violently criticized the Apollo Program for spreading the "disease of mankind" into the universe. He began a tirade against humanity. I rose up in defense of our species—for all our violence and tragedy, I said, "we are still a young species, evolving toward an unknown destiny that must not be aborted by those who criticize but do not contribute."

The next day, after this speech, Tom Turner, who was Bucky's director of special projects, called a special meeting and said that he had an announcement to make. It was a message from Dr. Fuller. The whole conference group assembled. I was on the stage. Tom turned to me and said: "Dr. Fuller believes that a woman should run for president, and he asks you to be that woman, to run on the theme of Positive Options for the Future." The whole conference sanctioned the suggestion. I would have run against Richard Nixon!

Of course, at that time, I did not do it, partially because of my family obligations.

The next time I saw him I was introducing him to a large audience. Before the event he came up to me, put his bald head upon my forehead, and held me silently for a moment. Then he said, "Darlin'"—he called many of us "darlin'"—"you're free now, aren't you. This is good."

I was astonished. There had been a shift in my personal life, and indeed, I was free to follow my own course.

At the same event, he was speaking far longer than planned. A member of the staff came out onto the stage and placed a large card on the floor in front of him. He glanced down at it and kept on talking, and then many minutes later, he picked up the placard, turned it toward the audience, and let us read it. It said: THE SESSION IS OVER.

"The session is *never* over," he said, and calmly went on speaking for another hour, way past the lunch, which was by that time cold and ruined, but nobody cared.

My next encounter with him was the most extraordinary.

I had written a manuscript based on a vivid Christ experience in 1980. I had intuited the presence of the living Christ as a field that was drawing humanity forward to do the works that he did, and even greater works. I saw that the *belief* that we would be radically transformed had in fact begun to give humanity the powers of precisely such transformation. As we understand more clearly the invisible technologies of nature—the atom, the gene, the brain—we are learning to affect matter directly. At the growing edge of evolutionary technologies such as biotechnology, nanotechnology, quantum computing, and so forth, we see the possibility of the transformation of the material world as well as of our own bodies. After this experience, I (a Jewish agnostic futurist) was guided to the New Testament. It revealed itself to be coded evolution, *coming true* through us. If we can combine our spiritual wisdom with our advancing technological abilities we will indeed "all be changed," and be able to do the work that he did, and far greater work. I wrote a sixteen-hundred-page manuscript based on the New Testament, called "The Book of Cocreation" (now published, in part, as *The Revelation*)—I sent him the manuscript.

Several weeks later, we were to speak together at the Annenberg Communication Center in Los Angeles. Word came down that Dr. Fuller wouldn't be speaking that night. We were told that he was reading something. We all thought he must not be feeling well, and I spoke alone.

However, when I had finished, someone came up to me and said that Dr. Fuller wanted to see me alone in the garden. I was led there and the door was shut. There he was, carrying the manuscript.

He told me that he knew what I had written was true, because he had had a similar experience, part of which he has told many times, but there was another part that, as far as I know, he did not mention to others.

He described to me how he had been walking down the street in Chicago when a light surrounded him and seemed to lift him off the sidewalk. Then he heard the inner words. *"Bucky, you are to be a first mini-Christ on Earth. What you attest to is true."*

Then, he said, he had turned to the New Testament, and had written a similar interpretation to mine, almost word for word. He told me he hid it, never showing it to anyone, because, given his work in the world, he could not use the words "Christ" or "God." (I had the same problem, and rarely speak of this experience, because it is either polarizing or seems unreal to so many.)

However, that evening, I had found my spiritual godfather. He put his arms around me, and touching that wonderful forehead against mine again he said gently, "Darlin', there is only God, there is nothing but God."

Subsequently Michael Toms, of New Dimensions Radio, and David Smith, television director of Xavier University (who made a beautiful video called

The Video Book of Cocreation), brought us together in a dialogue with students on television entitled "Bucky and Barbara: Our Spiritual Experience." But he never spoke the words in public.

I mention the story now because I feel it holds a profound truth. To be a "mini-Christ" was not a religious function. It meant that he deeply understood the nature of nature. He could see that the design of creation is that everyone can have life, and life more abundant. Everyone can be fed and healed. We are citizens of local universe, four and a half billion (at that time) billionaires, with wealth greater than any king of the past, if only we would understand the invisible design of nature and cooperate consciously with it.

Bucky inspired a whole generation with the thought that the world can work for everyone, because that is the way the world is designed. What is blocking it is ourselves, the "grunch of giants"—the nation-states, the organized religions, the military, the giant corporations—all institutions and organizations that have a vested interest in separation and scarcity. In that sense he was a profound revolutionary, but he saw that it was not necessary to destroy one group to put another in power. Rather what we must do is comprehend the exquisite laws of nature and work with them for an ever-regenerating future for all humanity. "The world will be saved," he told us, "by individuals of integrity freely joining."

His vision so deeply inspired me that in 1984 I finally decided to develop a new kind of political campaign. I formed a Campaign for a Positive Future to run to be selected as the vice-presidential nominee on the Democratic ticket. It was the Year of the Woman. As a futurist, I could get a hearing for my ideas.

The campaign was based on the ideas of three great founders of the positive future: Buckminster Fuller, Abraham H. Maslow, and Teilhard de Chardin. Bucky gave us the technological basis for hope. Maslow, founder of humanistic psychology, discovered the innate potential of every person for higher values and self-actualization. And Teilhard revealed spirit at work in evolution leading toward "Omega," the awakening of the planetary consciousness, or love on a mass scale.

I proposed a Peace Room in the office of the vice president that would be as sophisticated as a war room. In our war rooms we track our enemies and strategize their defeat. In our Peace Room we scan for innovations that work. We map them according to function to see the design of what is working. We connect people and projects for greater cooperation, and we communicate live from the office of the vice president the stories of our successes, breakthroughs, and models that work. "That," I said, "would help move us toward a world that works for everyone." I called for the shift from "weaponry to livingry," and the building of "new worlds on Earth, new worlds in space through new worlds in the human mind."

As I was preparing my platform I called him up. "Bucky," I said, "I want to know exactly the critical path you see that I should be for. I don't want to bother you—is there anyone else who could tell me what you see?" "No, darlin'," he said. "I want you to come and see me."

And then he died only a few days later.

I went on to have my name placed in nomination for the vice presidency of the United States, along with Geraldine Ferraro, at the 1984 Democratic convention. I proposed the Peace Room in the office of the vice president and in every capital of the world. In four years, I said at my acceptance speech, we will see the outlines of the world that is already working. This is a step toward the world that works for everyone.

Now, in Santa Barbara, California, we are committed to actually building a local Peace Room, both on the Internet and as a social process of cocreation. Allegra Fuller Snyder, Bucky's daughter, was at one of our events. She said that Bucky had begun the design science revolution, and now she saw that what is needed is *community* to act upon it. The Foundation for Conscious Evolution, which I cofounded in 1990, has moved to Santa Barbara, and it is our privilege as a community to work with many Fuller students to bring his vision one step closer to fulfillment.

Buckminster Fuller

The Lord's Prayer—First Version from *Intuition*

I feel intuitively that what is now identified as the Lord's Prayer was digested through ages from many philosophies in many lands. Also I feel intuitively that in relaying the Lord's Prayer from country to country, from language to language, from one historical period to another that many at first small, then later large alterations of meaning may well have occurred. It seems unlikely to me that the prayer's original conceivers and formulators would have included a bargaining proposal such as asking forgiveness of our trespasses or debts because we agree to forgive others. It also seems illogical to remind God of anything or to ask special dispensation for self, or to suggest that God doesn't understand various problems, or that God needs earthly salesmen for his cause. Before going to sleep, even for short naps, I always re-explore and rethink my way through the Lord's Prayer. And as I thought it through tonight, August thirteenth, 1966, I decided to inscribe it on paper.

Oh god, our father—
our furtherer
our evolutionary integrity unfolder
who are in heaven—
who art in he-even
who is in everyone
hallowed (halo-ed)
be thy name
(be thine identity)
halo-ed—

     the circumferential radiance
     the omnidirectional aura of
     our awareness of being
     ever in the presence
     of that which is greater,

144

more exquisite
and more enduring
than self—
haloed be thine identification
which is to say
the omnidirectional
vision and total awareness
is a manifest of your identity
which name or identity is
most economically stated as:
truth is your identity—
by truth we mean
the ever more
inclusive and incisive
comprehension
which never reaches
but always approaches
closer to
perfection of understanding
and our awareness is ever
a challenge by truth—
truth is embraced by and permeates
the omnidirectionally witnessible integrity
of omni-intertransforming events
which ever transpire—radiationally
which means, entropically in the physical;
and contractively—gravitationally—
which means, syntropically in the metaphysical
both of which—are characterized by either
the physical expansions
toward ever-increasing disorder,
of the entropic physical;
or the metaphysical contractions
toward ever-increasing order
of the syntropic metaphysical;
and these pulsating
contractions-expansions
altogether propagate
the wave and counterwave oscillations
of the electromagnetic spectrum's
complex integration—
of the omnienvironment's

evolutionary reality
and its concomitant
thought regeneration
which altogether constitute
what we mean
by the total *being*.

Hallowed
or haloed
"be thy name," which means
your identification to us,
Your kingdom come!—
your mastery of both
the physical and metaphysical Universe
emerges as the total reality
your will be done—
your will of orderly
consideration and mastery
of the disorderly be done
on our specialized case planet
and in our specialized case beings
and in our special-case consciousness
and in our special-case intellectual integrities
and in our special-case teleologic integrities
as it is in your generalized case he-even (heaven).

We welcome each day our daily evolution
and we forgive, post give, and give
all those who seemingly
trespass against us
for we have learned retrospectively
and repeatedly
that the seeming trespasses
are in fact the feedback of our own negatives,
realistic recognition of which
may eliminate those negatives.

For yours, dear god—
oh truthful thought
is our experience proven manifest
of your complete knowledge,
your complete understanding,

your complete love and compassion,
your complete forgiveness—
subjective and objective,
your complete inspiration and vision giving,
and your complete evolutionary volition,
capability, will, power, initiative
timing and realization—
for yours is the glory—
because you *are* the integrity
forever
and forever
amen.

Buckminster Fuller

The Lord's Prayer—Second Version from *Intuition*

This is the way I thought through THE LORD'S PRAYER on June 30, 1971, at the American Academy in Rome.   *B.F.*

Oh god
Our father
Who art in he even
Omniexperience
Is your identity.
You have given us
O'erwhelmingly manifestation
Of your complete knowledge,
Your complete comprehension,
Your complete wisdom,
Your complete concern,
Your complete competence,
Your complete effectiveness,
Your complete love and compassion,
Your complete forgiveness, giveness, and postgiveness,
Your complete inspiration giving,
Your complete evolutionary sagacity,
Your complete power, will, initiative,
Your absolute timing of all realization.
Yours, dear god,
Is the only and the complete glory!
You are the universal integrity
The eternal integrity is you.
We thank you with all our hearts,
Souls and mind—   Amen.

Bucky Fuller once starred as an actor at Black Mountain College with Merce Cunningham, Elaine de Kooning, and Arthur Penn in Erik Satie's *The Ruse of Medusa*. John Cage said Fuller's performance was outstanding. Fuller once told me that actors have "a presence to present information effectively, and they ought to think about using that talent for useful purposes."

Bucky's theater friends included director John Huston, Ellen Burstyn, and Marian Seldes. Other admirers were singer John Denver and actress Valerie Harper, perhaps best known as television's Rhoda. A recipient of four Emmies, a Golden Globe, and Harvard's Hasty Pudding Woman of the Year Award, Valerie Harper recently wrote and performed a one-person play, entitled *All Under Heaven*, based on the life of Nobel Prize–winning author Pearl S. Buck. Fuller's intersection with Madame Buck was China, for Fuller longed to visit that country. He learned from Buck that many of his earlier books had been translated into Chinese. Although Pearl Buck died of cancer in 1973, before she could return to China, in 1979 Fuller was invited to the People's Republic of China.

When I asked Valerie Harper to contribute to the book, I knew that she was the mother of a teenager and suggested she speak about Bucky in a way that might interest her daughter's generation. Fuller always talked to the young, urging them to reform the environment and not fall victim to their political or social state. We were standing in the lobby of the Century Theatre in New York City just after Valerie's performance as Pearl S. Buck, and it occurred to me that both women had a personal experience of Bucky. Oddly, we both had met Madame Buck through Bucky—there are no accidents.

Bucky: Citizen of the Universe

Valerie Harper

Although neither a scientist nor a scholar I am a devoted fan of R. Buckminster Fuller. When Thomas Zung, Fuller's architectural partner, asked me to contribute to this book, to write about my experiences of "Bucky," I was flattered. Then the agonizing began. What to write? How to write it? How do I avoid its being terrible, or worse, mediocre? Having agreed to write an essay, I had to recall the time in my life when I had written quite a few—in high school. Since multiple decades have passed, I consulted an expert, a member of the age group Bucky greatly enjoyed addressing, my sixteen-year-old daughter. Cristina diligently attempted to refresh my skills in this area, but to no avail. After tearing up several—no, many—attempts to make the most valuable contribution I could, finally I yelled at myself, "Stop trying to be significant! Forget the essay. What would you say directly to Bucky if given the opportunity?" I'd thank him. "Well, here's your chance—take it!"

DEAREST BUCKY,

Thank you for being a "see-er" who then chose to apply yourself with energy, courage, tenacity, integrity, and love to communicate your many insights to all of us who were without your vision. And for the precise, literal use of language that shakes us into thinking more authentically about the universe and what we are and who we are in it. Thank you for inspiring the movement to end hunger on the planet forever, the Hunger Project, in which I have had the honor to participate these twenty-three years. And for conceiving the World Game, now an active institute in Philadelphia dealing with issues of supply and demand.

In 1976, I was one of thousands who attended an "An Evening with Bucky" in Los Angeles. I am grateful to Werner Erhard for providing me the opportunity to experience your ideas and concern about ending world hunger.

Although the founders of the Hunger Project were Werner Erhard, John Denver, and Professor Robert Fuller of Oberlin University, certainly the inspiration and much of the thinking that went into its creation have you as their source. As a scientist, humanist, and futurist, your global knowledge and language shaped the context in which human beings could work to not just alleviate but to end hunger on earth forever.

- "Hunger and starvation are completely obsolete. It is not an issue of food but of integrity—it's not a lack of proper food distribution—it is a lack of integrity in how we treat each other."
- "A world that works for everyone—to have that world hunger must be ended."
- "Have your life make a difference—you must strive to make a difference with your life."

Whatever my preconceived notions, that evening in 1976, of what the legendary Buckminster Fuller would be like, my first thought when you came out on the stage was "He's adorable." Sweetly unassuming, in a dark business suit, a look of unhurried expectancy on a purely kind face. Your clarion blue eyes intent, alert, significantly magnified through your practical black-framed glasses. Here was a vital eighty-year-old with a very hip white crewcut and hands that looked like they could garden, fairly brimming with goodwill, wisdom, and aliveness.

And then you spoke dynamically of ideas, notions, facts, and principles and of infinite possibilities. I was mentally scampering to keep up and at the same time so gratified that I could grasp as much as I did, clearly a function of your dogged persistence to reach us, to have us understand. You presented us with a powerful question, the same one you had asked yourself over four decades: *What can the little individual do?* I often ask that question of my daughter, Cristina, and myself.

Seeing it on the paper before me, I am struck that one, if so inclined and if given half a chance, could easily misinterpret this sentence. The actor in me knows that those six words spoken angrily could express denial (no use doing anything!), or with a shrug, resignation (there's nothing I can do), or tearfully, defeat (poor helpless victim me). There are innumerable attitudes—actors call them "choices"—with which to color that "line," all in avoidance of its meaning. No lament of powerlessness this, but an insistently inspiring, energizing, compelling call to action. You extended a joyful invitation to come face to face with who we really are as human beings in relationship to and with the universe . . . and to act accordingly. "What can the little individual do?" presupposes that there is much we can do even while confronting our "littleness" and asserts the power of each little individual choice of what to do. You urge us to discover how to apply ourselves to transform the intensely discouraging conditions on our planet that Pearl S. Buck once described as "the New World *Dis*-Order."

Thank you for revealing that it is precisely what we are not doing, what we are resisting doing, that is keeping world history repeating itself treadmill-style with humanity behaving like so many rats running endlessly on the same old outdated, dangerously primitive wheel.

And for declaring in no uncertain terms, "We are all astronauts on a space vehicle called Earth and our behavior must be consistent with that knowledge." We can no longer continue the "flat-four-corners-of-the-earth-we'll-fall-off-the-edge" mentality that freezes us with fear.

Thank you for assailing "the egocentricity and selfishness of ignorantly opinionated humans, prolonging inertia by refusing to let go of yesterday's make-believe substitute worlds and come into comprehension and competence of the discovered universal principles." And for your laserlike expression of thoughts penetrating through millennia of traditionally held false beliefs to illuminate the truth . . ." pure science events represent openings of windows through the wall of ignorance and fiction to reveal the only reality—the behavior of the naked universe that always was and always will be."

Thank you for creating the "cosmic report card," in 1947, which for the first time charted the chronology in which the ninety-two elements were acquired by civilization. It is fascinating to be able to look at a profile of scientific, technological history; and it's a thrill to feel my consciousness expand as I do so. And for promoting "the application of scientific technical capabilities to produce standards of living and freedom of thoughts and actions for all without any individual being advantaged at the expense of another"—as a woman, I particularly love this one!

Thank you for exhorting us to "*rearrange* in such a way as to progressively support ever-more-lives in ever-more-ways with ever-increasing health" and for stating so clearly that being "conscious of our home, the earth, as a star in process will affect our lives and the life of our planet in every way imaginable."

Thank you for giving your very life with all your gifts—that powerful intellect, prodigious energies, great good humor, and loving spirit—in order to transform a you-or-me world into a you-and-me world.

With unending love and appreciation . . . thank you for it all, Bucky.

An Earthian,

VAL

Well, who do I think I am? How presumptuous—who am I to thank R. Buckminster Fuller? A little individual, that's who. I have chosen in this piece to quote Bucky, to entice you to wholeheartedly accept his amazing gifts. Read Bucky! Buy his books, find him on the Internet or at your local library (that's what I did). Demand that his out-of-print books be reprinted and get yourself to the magnificent Stanford archives. If you don't understand what he's saying, or get lost or disagree, *keep on reading.* He allows you to catch up—if you stay open! Let him into your mind, your heart, your experience. Listen, listen to Bucky—you may not get all the notes, but you will get his music.

Embrace Bucky's invitation, no, demand, that we give up "endarkenment" and live in a way that gives life to others. In closing here are the lyrics of John Denver's song written for Bucky on the occasion of his eighty-fifth birthday, and again, especially you teenagers like Cristina, I urge you to discover Bucky.

What One Man Can Do
by John Denver

I suppose that there are those
Who will say he had it easy,
Had it made in fact
Before he'd even begun.
They don't know
The things I know
For I was always with him.
It may sound strange
But we were more than friends.
It's hard to try and tell the truth,
When no one wants to listen
When no one seems to care
What's going on.
It's hard to stand alone
When you need someone behind you,
Your spirit and your faith,
They must be strong.
What one man can do is dream,
What one man can do is love.
What one man can do is change the world,
And make it new again.
Don't you see
What one man can do?
That's how bright his mind is,
That's how strong his love for you and me.
A friend to all the universe,
Grandfather of the future,
And everything that
I would like to be.
What one man can do is dream,
What one man can do is love,
What one man can do is change the world,
And make it young again.
Can't you see what one man can do?

Revolution in Wombland from *Earth, Inc.*

At all times nowadays, there are approximately 66 million human beings around Earth who are living comfortably inside their mothers' wombs. The country called Nigeria embraces one-fourth of the human beings of the great continent of Africa. There are 66 million Nigerians. We can say that the number of people living in Wombland is about the same as one-fourth the population of Africa. This 66 million Womblanders tops the total population of either West Germany's 58 million, the United Kingdom's 55 million, Italy's 52 million, France's 50 million, or Mexico's 47 million. Only nine of the world's so-called countries (China, India, Soviet Union, United States, Indonesia, Pakistan, Japan, and Brazil) have individual populations greater than our luxuriously living, under-nine-months-old Womblanders.

Seemingly switching our subject, but only for a moment, we note that for the last two decades scientists probing with electrodes have learned a great deal about the human brain. The brain gives off measurable energy and discrete wave patterns disclosed by the oscillograph. Specific, repetitive dreams have been identified by these wave patterns. The neurological and physiological explorers do not find it extravagant to speculate that we may learn that what humanity has thus far spoken of mystifiedly as telepathy, science will have discovered, within decades, to be ultra-ultra-high-frequency electromagnetic wave propagations.

All good science fiction develops realistically that which scientific data suggests to be imminent. It is good science fiction to suppose that a superb telepathetic communication system is interlinking all those young citizens of worldaround Wombland. We intercept one of the conversations: "How are things over there with you?" Answer: "My mother is planning to call me either Joe or Mary. She doesn't know that my call frequency is already 7567-00-3821."

Other: "My mother had better apply to those characters Watson, Crick, and Wilkins for my call numbers!" And another of their 66 million Womblanders comes in with, "I'm getting very apprehensive about having to 'go outside.' We have been hearing from some of the kids who just got out—they say we are going to be cut off from the main supply. We are going to have to shovel fuel and pour liquids into our systems. We are going to have to make our own blood. We are going to have to start pumping some kind of gas into our lungs to purify our own blood. We are going to have to make ourselves into giants fifteen times our present size. Worst of all, we are going to have to learn to lie about everything. It's going to be a lot of work, very dangerous, and very discouraging." Answer: "Why don't we strike? We are in excellent posture for a 'sit-down.'" Other: "Wow! What an idea. We will have the whole population of worldaround Wombland refuse to go out at graduation day. Our cosmic population will enter more and more human women's wombs, each refusing to graduate at nine months. More and more Earthian women will get more and more burdened. Worldaround consternation—agony. We will notify the outsiders that, until they stop lying to themselves and to each other and give up their stupid sovereignties and exclusive holier-than-thou ideologies, pollutions, and mayhem, we are going to refuse to come out. Only surgery fatal to both the mothers and ourselves could evacuate us."

Another: "Great! We had might as well do it. If we do come out we will be faced with the proliferation of Cold War's guerrillerized killing of babies for psycho-shock demoralization of worldaround innocent communities inadvertently involved in the abstruse ideological warfare waged by diametrically opposed, equally stubborn, would-be do-gooder, bureaucratic leaders and their partisans who control all of the world's means of production and killing, whose numbers (including all the politically preoccupied individuals around the Earth) represent less than one per cent of all humanity, to whose human minds and hearts the politicos and their guns give neither satisfaction nor hope. Like the women in *Lysistrata* who refused intercourse with their men until they stopped fighting, we Womblanders would win."

Until yesterday, what are now the 150 member nations of our planet's United Nations were tiny groups of humans who for two million years had been regenerating around our globe so remotely from one another that each colony, nation, or tribe was utterly unaware of one another's existence. Only through telepathy, as supposedly operative in the previous paragraphs, could those remote cells of precariously surviving human beings have been aware of one another throughout those two million years. In the last few split seconds of overall history, there emerged a dozen millennia ago from the womb of tribal remoteness a few sailors and overland explorers who began to discover the presence of other humans scattered around the mysterious world. Finding the tribes to be each unaware of either the surprising resources or the vital needs and desires of the others, they kept the whereabouts of these surprise demands and supplies secret and thus were able, through monopoly of commerce and middle-manning, to exploit to their own special advantage the vital needs, ignorance, and the wealth of life-support to be generated by expediting or slowing the physical resource interactions with humanity's available time to work the resources into higher advantage tools, environment controlling devices, and metabolically regenerative sustainers.

Throughout all the two million years up to the Twentieth Century, the total distance covered by an average man in an average lifetime disclosed to him less than one millionth of the surface of our spherical planet. So tiny was a human and so relatively large is our planet that it is not surprising that humans as yet cerebrate only in terms of a "wide, wide world—a four-cornered Earth," situated in the middle of an infinite plane, to which all the perpendiculars are parallel to one another and lead only in two directions—UP and DOWN—with sky UP there and earth DOWN here.

No matter how you may look upon the matter morally and ideologically, the assumption that humanity could or could not own a piece of land with all the earth vertically below it and all the air vertically above it is not only scientifically invalid—it is scientifically impossible. The scheme is geometrically possible only as an up-and-down make-believe flat world.

To understand the scientific impossibility of such a scheme, let us consider a cube inside of a sphere, with the cube's eight corners congruent with the surface of the sphere. Let the cube's twelve edges consist of steel structurals. A light is at the common center of the cube and sphere and casts a shadow of the twelve structural edges of the cube outwardly upon the surface of the translucent sphere. We will now see that the total spherical surface is divided symmetrically by great circle arcs into six equilateral four-edged areas. Though each of the four-sided symmetrical areas has 120-degree corners instead of 90-degree corners, each is called a spherical square. Altogether they constitute a spherical cube.

We will now suppose the spherical cube to be the planet Earth. We will suppose that war and treaties have resulted in the total Earth's being divided equally amongst six sovereign groups—each empowered by its laws to grant deeds to properties within their respective spherical square surfaces on the planet, regardless of whether covered by water or not. We will suppose that, as at present, each of the world's major sovereign nations assumes the authority to deed or lease the titles to subdivisions of each of their respective lands to corporations, subgovernments, and individuals. All the legally recognized deeds to property anywhere around our Earth date back only to sovereign claims established and maintained exclusively by military might.

Now that we have the model of a cubical subdivision of the sphere, let us color our cube's six faces, respectively, red, orange, yellow, green, blue, and violet. Let Russia sovereignly possess the red face of the cube. Consider all the perpendiculars to the red face of the internally positioned cube as being the up and down perpendiculars defining the property claims to all the land *below* the surface and all the air *above* the surface. Under these conditions, it will be seen that the red square owns all the interior of the cube which occurs perpendicularly below that red surface square. Therefore, each of the six countries would be claiming exclusive possession of the same "whole" cube, which obviously invalidates each and all of their claims to only one-sixth of the cube. This realization is mildly reminiscent of Portia's admonition to Shylock that he must be able to cut loose his pound of flesh without letting a drop of blood.

"All right," you say, "I will concede it is impossible to demonstrate the validity of the claims to the lands lying perpendicularly below my surface map without invalidating all other landowners of the world. Therefore, I will try to live on the surface of my land and just claim it and the airspace vertically—above me." "All right," we say to you, "what air are you talking about, because it just blew away." You retort testily, "I don't mean that nonsense . . . just the air geometrically above me. That is what I refer to when I say you are violating my airspace—you are violating my overhead geometry." "All right," we say to you, "which stars were you looking at when you said, '. . . that space above me'? Our Earth has been revolved away from those stars. Other stars are now above us. Not only are we revolving, but we are simultaneously orbiting around the sun, while all the planets and stars are always in swift motion, but are so far away from us and our lives so short that we are unable to perceive those motions. The distances involved are so great that the light from the next star to the sun takes four and one-half years to come to our solar system while traveling 700 million miles per hour and the distance across our galaxy is more than 300,000 light-years, while the next nearest of the millions of galaxies are multimillions of light years away from our galactic nebula. With those kinds of distances in the heavens, the amount of star motion that you and I can detect in our lifetime is humanly unrecognizable. Most of the star speeds within their galaxies are in the order of only 100,000 miles per hour, which is a negligible speed beside light's speed of 700 million miles per hour."

Because all the stars in the Universe are in motion, our planet orbits rotatingly in an ever-changing omni-circus of celestial events. There is no static geometry of omni-interrelationship of Universe events. Some of the stars you are looking at have not been there for a million years—some no longer exist. As Einstein and Planck discovered, "Universe is a scenario of non-simultaneous and only partially overlapping, transformative events." One frame in the scenario of caterpillar does not foretell the later scenario event of its transformation into butterfly. One frame of butterfly cannot tell you that the butterfly flies; only large time-sequence segments of the scenario can provide meaningful information. Cogitating on the myriads of stars apparently scattered in disorderly spherical array about the heavens, individuals often remark, as may you, "I wonder what is outside outside?"—asking for a one-frame answer, which is as unintelligent as asking, "Which word is the dictionary?" You know the order of the dictionary to be alphabetical, but its words do not read sequentially. Just hearing them read aloud, they make an only apparent, disorderly array. This is typical of the manner in which nature hides her orderliness in only apparent disorder.

Back to little space vehicle Earth and that question of property. The most that the individual could be entitled to own would be the inside of an infinitely thin blueprint of his land, because there is no geometry of space outside it and no exclusively occupiable land below. Our planet Earth is the home of all humans, but scientifically speaking it belongs only to Universe. It belongs equally to all humans. This is the natural, geometrical law. Any laws of man which contradict nature are unenforceable and specious.

Without the infinitely extended lateral plane, the words up and down are meaningless. The airman initiated the correct descriptive terms "coming IN for a landing and going OUT." It is meaningful to say "INSTAIRS and OUTSTAIRS." Say it for a week and your senses will discover and notify that you are living on a planet.

What do you mean, "astronaut"? *We are all astronauts.* Always have been— but really! Never mind your "Never-mind-that-space-stuff, let's-be-practical, lets-get-down-to-Earth" talk—brain-talk as undisturbed by knowledge as is a parrot's brain-talk by any awareness born of thought. Brain is physical— weighable; thought is metaphysical—weightless. Many creatures have brains. Man alone has mind. Parrots cannot do algebra; only mind can abstract. Brains are physical devices for storing and retrieving special case experience data. Mind alone can discover and employ the generalized scientific principles found holding true in every special case experience.

Universe has disclosed to astrophysics an elegantly orderly inventory of ninety-two regenerative chemical elements, each with its unique behaviors, all of which are essential to the success of Universe. All are in continual interexchange within the total evolutionary process of Universe. Ignorant humans aboard space vehicle Earth are now screaming, "Pollution!" There is no such phenomenon. What they call pollution is extraordinarily valuable chemistry essential to Universe and essential to man on Earth. What is happening is that the egocentricity of omni-specialized man makes him ignorant of the value with which his processing is confronting him. The yellow-brown content of fume and smog is mostly sulphur. The amount of sulphur going out of the smokestacks around the world each year is exactly the same as the amount of sulphur being taken from the Earth each year to keep the world ecology going. It would be far less expensive to catch that sulphur while concentrated in the stack, and to distribute it to the original users, than to do the original mining AND to get it out of human lungs, et cetera, when all the costs to society over a deteriorating twenty-five years are taken into account. But humanity insists on holding to this year's profits, crops, and elections. World society is lethally shortsighted.

Subconsciously reflexing to the as yet mistaken concept of an infinite plane, men have felt that they could dispose of annoyingly accruing substances with which they did not know how to deal by dispatching them outward in some cosmic direction, assumedly to be diffused innocuously in infinity. "I spit in the ocean. So what?" Humans as yet cerebrate secretly and hopefully that—inasmuch as yesterday's exhaustion of customary resources has always been followed by discovery of alternate and better resources—the great infinity is going to keep right on taking care of ignorant carelessness and waste. "So what the hell?" say the "down-to-earth" status-quoers. "Pump all the fossil fuel energy-depositing of billions of years out from the Earth's crust. Burn it up in a century. Fill all your bank accounts with ten-place figures. To hell with the great-grandchildren. Let them burn up our space vehicle Earth's oceans with hydrogen fusion. Let them do the worrying about tomorrow."

Just as biological protoplasmic cells are colonized into larger organisms, the most complex and omni-adaptable of which is the human, so too do humans colonize and inventively externalize the same organic tool functions for their mutual metabolic regeneration. We call this complex mutual tool externalization by the name industrialization, in which each of us can use the telephone or the electric light in our special, unique tasks, all of which require increasing development of world-around access to the total resources and world-around distribution of the advantages comprehensively produced in total metabolic regeneration.

The world population which, after the cell-colonizing within its controlled environment, has been emitted from the thin, protoplasmic, tissue-sheathed, human womb into planet Earth's larger biosphere-sheathed, industrial organism womb, goes on colonizing, integrating, and specializing locally as innocently and ignorantly as did the protoplasmic cells within the woman's womb, all the while mistrusting one another as they evolve their utter interdependence around Earth, as do the individual protoplasmic cells of the residents of human Wombland gather together selectively, finally to form a whole child. In due course, we will realize a one world human integrity and with each degree of physical integration a new degree of metaphysical freedom will be attained.

Earthians in their more roomy biosphere are as yet provided for, despite their utter ignorance of the infinitely exquisite reliable interactions of cosmic mechanics. Mothers don't have to invent a breast to feed the baby or invent oxygen for it to breathe. Nor do they have to tell the child how to invent its cell growth. Humans are utterly ignorant of what goes on, how, and why.

The Universe is a self-regenerating and transforming organic machine. Human womb graduates now gestating within the biosphere's world industrial organism womb are discovering and employing a few of the principles governing micro-macro cosmic mechanics, all the while ignorantly speaking of their accomplishments of the generally-disregarded obvious as "inventions" and "creations." Now humans have become suspicious of their little machines, blaming them for the continual disconnects of the inexorable evolutionary processes of cosmic gestations which—transcendental to their brain detecting—ever and again emit them into a greater, more inclusively exquisite spherical environment of automated mechanical controls that progressively decontrol humanity's thought and action capabilities—ever increasing humanity's options—emancipating it from its former almost total preoccupation with absolute survival factors.

Assuming erroneously that their day-to-day positive experiences should be rendered perpetual and their negative experiences eliminated, humans try to freeze the unfreezable evolution at specific stages. They try to make "plastic flowers" of all momentarily satisfying events and paraphernalia. In the past, they tried to do it with stone. Separated from the familiar, confronted with the unfamiliar, and reflexed only by the brain's mechanical feedback, unthinking humans—not realizing that there are no straight lines, only wavy ones, and not realizing that waves can only be propagated by positive-negative oscillating—find their straight linear strivings forever frustrated by the wave system realities of Universe. Ignorantly they speak of the evolutionary waves' regeneratively oscillating complementaries as "good" and "bad," though the scientist can find no such moral and immoral qualities in the electron or its complementary opposite, the positron.

Humanity as a whole is indeed being emitted from a two-million-year gestation within the womb of permitted ignorance, for which infantile period cosmic mechanics have been making ample provision not only to offset ignorance and waste but also to permit humanity's gradual trial-and-error experimental discovery of the relatively negligible effectiveness of its muscle—which it had at first employed not only exclusively but savagely—and the concomitant discovery of the infinite apprehending and comprehending effectiveness of the human mind, which alone can discover and employ the Universal verities—and thereby realize comprehensively the potential, progressive, nonwasteful, competent, considerate mastery of the physical environment by the metaphysical intellect.

The metaphysical integrities manifest throughout the everywhere inter-transforming Universe's omni-interacccommodative cosmic organic system apparently are from time to time emulated in meager degree by the intellect of the human passengers who are gestating within the spherical womb sheath of planet Earth's watery, gaseous, and electromagnetic biosphere.

Humanity's most recent sorties to the moon from within space vehicle Earth's womblike biosphere sheath have been tantamount to a premature, temporary surgical removal of a baby from its human mother's womb, skillfully enclosed within a scientifically controlled environment, still attached to the mother, and after successful surgery being returned into the human mother's womb to loll-out its remaining gestation days to the successful detached-action launching outwards in Universe which we ignorantly identify as "birth." Sovereign nation "landing cards" require answers to ridiculous questions: "When were you born?" "Where do you live?" Answer: "I am immortal. I check in here and there from celestial-time-to-celestial-time. Right now I am a passenger on space vehicle Earth zooming about the Sun at 60,000 miles per hour somewhere in the solar system, which is God-only-knows where in the scenario Universe. Why do you ask?"

Humanity's sorties to the moon have been accomplished only through instrumental guidance of their controlled-environment capsules and mechanical-enclosure clothing by utterly invisible electromagnetic wave phenomena referenced to instrument-aligned star bearings, with the invisible mathematical integrations accomplished by computers, uncorrupted and incorruptible by ignorantly opinionated humans. Thus has man been advantaged by the few who have thought and acted to produce the instruments, as yet relieving the vast majority of humans from the necessity of having to think and coordinate their sensings with the realities of cosmic mechanics.

Humans still think in terms of an entirely superficial game of static things—solids, surfaces, or straight lines—despite that no things—no continuums—only discontinuous, energy quanta—separate event packages—operate as remotely from one another as the stars of the Milky Way. Science has found no "things"; only events. Universe has no nouns; only verbs. Don't say self-comfortingly to yourself or to me that you have found the old way of getting along with false notions to be quite adequate and satisfactory. So was the old umbilical cord to your mother. But you can't reattach it and your mother is no longer physically present. You can't go back. You can't stay put. You can only grow and, if you comprehend what is going on, you will find it ever more satisfactory and fascinating, for that is what evolution is doing, whether you think, ignorantly, that you don't like it or do.

To each human being, environment is "all of Universe that isn't me." Our macrocosmic and microcosmic "isn't me-ness" consists entirely of widely dissynchronous frequencies of repetitions of angular changes and complex interactions of waves of different lengths and frequencies of repetition. Physics has found a Universe consisting only of frequency and angle modulations.

Our environment is a complex of frequencies and angles. Our environment is a complex of different frequencies of impingement—from within and without—upon the individual "me-nesses." We are in a womb of complex frequencies. Some of those frequencies man identifies ignorantly with such words as "sight, sound, touch, and smell." Others he calls "tornadoes, earthquakes, novae." Some he ignorantly looks upon as static *things*: houses, rocks, and human-like manikins.

Very, very slow changes humans identify as inanimate. Slow change of pattern they call animate and natural. Fast changes they call explosive, and faster events than that humans cannot sense directly. They can see the rocket blasted off at 7,000 miles per hour. They cannot see the hundred thousand times faster radar pulse moving 700 million miles per hour. Humans can sense only the position of pointers on instrument dials. What they call "radio"— electromagnetics—they learn of through scientific instrumentation. Of the total electromechanical spectrum range of the now known realities of Universe, man has the sensory equipment to tune in directly with but one millionth of the thus far discovered physical Universe events. Awareness of all the rest of the millionfold greater-than-human-sense reality can only be relayed to human ken through instruments, devised by a handful of thought-employing individuals anticipating thoughtfully the looming needs of others.

The almost totally invisible, nonsensorial, electromagnetic wombsheath of environmental evolution's reality-phase into which humanity is now being born—after two million years of ignorant, innocent gestation—is as yet almost entirely uncomprehended by humanity. Ninety-nine and nine tenths per cent of all that is now transpiring in human activity and interaction with nature is taking place within the realms of reality which are utterly invisible, inaudible, unsmellable, untouchable by human senses. But the invisible reality has its own behavioral rules which are entirely transcendental to man-made laws and evaluation limitations. The invisible reality's integrities are infinitely reliable. It can only be comprehended by metaphysical mind, guided by bearings toward something sensed is truth. Only metaphysical mind can communicate. Brain is only an information storing and retrieving instrument. Telephones cannot communicate; only the humans who use the instruments. Man is metaphysical mind. No mind—no communication—no man. Physical transactions without mind—YES. Communication—NO. Man is a self-contained, micro-communicating system. Humanity is a macro-communicating system. Universe is a serial communicating system; a scenario of only partially overlapping, nonsimultaneous, irreversible, transformative events.

As yet preoccupied only with visible, static, newspicture views of superficial surfaces of people and things—with a one-millionth fraction of reality which it has cartooned in utter falsehood—society fails to realize that several hundred thousand radio or TV communications are at all times invisibly present everywhere around our planet. They permeate every room in every building—passing right through walls and human tissue. This is to say that the stone walls and human tissue are invisible and nonexistent to the electromagnetic wave reality. We only deceived ourselves into reflexing that the walls are solid. How do you see through your solid eyeglasses? They are not full of holes. They are aggregates of atoms as remote from one another as are the stars. There's plenty of space for the waves of light to penetrate.

Several hundred thousand different wide-band radio sets can at any time be tuned in anywhere around our biosphere to as many different communications. Going right through our heads now, these programs could be tuned in by the right crystals and circuits. Crystals and circuits consist of logically structured atomic arrays. Such arrays could operate even within our brains. Tiny bats fly in the dark by locating objects ahead in their flight path by ever more minuscule radar sending and receiving, distance-to-object calculating mechanisms. Right this minute, five hundred Earth-launched satellites with sensors are reporting all phenomena situated about our planet's surface. Tune in the right wavelength and learn where every beef cattle or every cloud is located around the Earth. All that information *is* now being broadcast continually and invisibly.

For humans to have within their cerebral mechanism the proper atomic radio transceivers to carry on telepathetic communication is no more incredible than the transistors which were invented only two decades ago, and far less incredible than the containment of the bat's radar and range-finding computer within its pin-point size brain. There is nothing in the scientific data which says the following thoughts are impossible and there is much in the data which suggests that they are probable. The thoughts go as follows: The light of a candle broadcasting its radiation in all directions can be seen no farther than a mile away in clear atmosphere. When the same candle's flame is placed close in to the focus of a parabolic reflector and its rays are even further concentrated into a beam by a Fresnel lens, its light can be seen at ten miles distance. The earliest lighthouses were furnished with such reflectively concentrated beam lights of tiny oil lamps.

What we speak of as light is a limited set of frequencies of the vast electromagnetic wave ranges. All electromagnetic waves can be beamed as well as broadcast. When beamed and lensingly concentrated (as with the laser beams refracted through rubies), their energies are so concentrated as to be able to bore tunnels in mountains. The shorter the waves, the smaller the reflector and refractor may be.

We know that the human has never seen outside himself. Electromagnetic waves of light bounce off objects outside him and frequencies are picked up by the human eyes and scanningly relayed back into the brain. Because the light is so much faster than touch, smell, and hearing, men have tended to discount the billionth of a second it takes light to bounce off one's band and to get the information back into one's brain. All sensing is done by humans entirely inside the brain, with information nerve-relayed from the external contact receivers. The human brain is like a major television studio-station. Not only does the brain monitor all the incoming live, visible, audible, smellable, and touchable 3D shows, it also makes videotapes of the incoming news, continually recalls yesterday's relevant documentaries and compares them with incoming news to differentiate out the discovered new and unexpected events from the long-familiar types, and to discover the implications of the news from those previously-experienced similar events, in order swiftly to design new scenarios of further actions logically to be taken in respect to the newly-evolved challenges.

So faithful has been the 4D, omni-directional, image-ination within the human omni-sense transceiving studio-stations of human brains that the humans themselves long ago came to assume spontaneously that the information received inside the brain made it safe to presume that those events were, in fact, taking place outside and remote from the seeing human individual. The reliability of all this imagining has been so constant that he now tends to think he sees only outside himself.

The shorter the electromagnetic, air, water, sand, or rocky earthquake wavelengths, the higher their frequency. The higher the wave frequencies, the more the possibility of their interfering with other high-frequency, physical phenomena such as walls, trees, mountains. The nearer they approach the same frequencies, the less do they interfere with one another. For this reason, the very high-frequency electromagnetic waves of radio and television get badly deflected by obstacles. As a consequence, man learned to beam short-wave television programs from horizon to horizon. He developed parabolic transceiver reflectioning cups that took in and sent out waves in parallel beam-focused rays. At the transceiver relay stations on the horizons, additional energy is fed into the signals received and their projection power is boosted so that, when they arrive at final destination after many relayings, their fidelity and power are as yet exquisitely differentiated and clearly resonated.

It may well be that human eyes are just such infra-sized parabolic transceiver cups. It may be that our transceiver eyes adequately accommodate the extraordinarily low magnitude of energy propagating of the brain as electromagnetic wave pattern oscillations to be picked up by others.

Early photography required whole minutes of exposure. As film chemistry improved, exposure times decreased. Yesterday, one thousandth of a second was fast. Today's capability makes one millionth of a second a relatively slow electro-astrophotography exposure. Pictures taken in a millionth of a second today are clearer than those of yesterday which took minutes. The scanned-out picture signals travel 700 million miles per hour. The effect in terms of man's tactile, hearing, and smelling senses is instantaneous.

Speakers who appear frequently before large audiences of human beings over a period of years have learned that the eyes of the audience "talk back" so instantaneously to them that they know just what their audiences are thinking and they can converse with their audiences, even though the speaker seems to be the only one making audible words. The feedback by eye is so swift as to give him instantaneous, spontaneous reaction and appropriate thought formulation.

The parabolic reflector-beamed, ultra-ultra-high-frequency electromagnetic waves—such as can be coped with by transceivers with the infra-diameter of the human eye—are such that they would be completely interfered with by walls or other to-us-seemingly-opaque objects. However, when they are beamed outwardly to the sky in a cloudless atmosphere, no interference occurs. Ultra-shortwave radio and radar beams which are interfered with by mountains and trees can be beamed into a clear sky and bounced off the moon, to be received back on Earth in approximately one and three-fourths seconds. In a like manner, it is possible that human eyes operating as transceivers, all unbeknownst to us, may be beaming our thoughts out into the great night-sky void, not even having the sun's radiation to interfere mildly with them. Such eye-beamed thoughts sent off through the intercelestial voids might bounce off various objects at varying time periods, being reflectively re-angled to a new direction in Universe without important energy loss. A sufficient number of bouncings-off of a sufficient number of asteroids and cosmic dust could convert the beams into wide-angle sprays which diffuse their energy signals in so many angular directions as to reduce them below receptor-detection level. Eye-beamed thoughts might bounce off objects so remote as to delay their 700 million mile per hour travel back to Earth for a thousand years, ten thousand years, a hundred thousand years. It is quite possible that thoughts may be eye-beamed outwardly not only from Earth to bounce back to Earth at some later period from some celestially-mirroring object, but also that thoughts might be beamed—through non-interfering space to be accidentally received upon Earth—from other planets elsewhere in Universe. There is nothing in the data to suggest that the phenomenon we speak of as *intuitive thought* may not be just such remote cosmic transmissions. Intuitions come to us often with surprising lucidity and abruptness. Such intuitions often spotlight significant coincidences in a myriad of special case experiences which lead to discovery of generalized scientific principles heretofore eluding humanity's thought. These intuitions could be messages to the Earthian brain receiving it to "Look into so-and-so and so-and-so and you will find something significant." Intuitions could be thoughts dispatched from unbelievably long ago and from unbelievably far away.

As Holton wrote in the *American Journal of Physics* and as reported on the "Science" page of *Time* magazine, January 26, 1970:

> To fully recognize the extraordinary intellectual daring of Einstein's equations, we note the great scientist's own explanation of their origin: "There is no logical way to the discovery of these elementary laws. There is only the way of intuition."

Because humans consist of a myriad of atoms and because atoms are themselves electromagnetic frequency event phenomena—not things—it is theoretically possible that the complex frequencies of which humans are constituted, together with their angular interpositioning, could be scanningly unraveled and transmitted beamwise into the celestial void to be received some time, somewhere in Universe, having traveled at 700 million miles per hour, which is approximately 100 thousand times faster than the speed of our moon rockets a minute after blast-off. It is not theoretically impossible in terms of the total physical data that humans may have been transmitted to Earth in the past from vast distances.

Retreating from such a speculative mood, we come now to consider closer-range possibilities and probabilities. We recall that humans, who to our knowledge arrived on Earth at least two million years ago, have been regenerating aboard that small, 8,000-mile-diameter space vehicle Earth throughout all those years without even knowing that they were aboard a space vehicle. They are now emerging, however, from the womb of permitted ignorance of their early, subjective, taken-care-of phase and are now beginning to become comprehensively aware of all the matters we have discussed so far. They are beginning to understand that they are within a limited biosphere life-support system whose original excessively abundant living supply was provided only to permit humanity's initial trial-and-error discovery of its antientropic function in Universe. Humans are coming swiftly to understand they must now consciously begin to operate their space vehicle Earth with total planetary cooperation, competence, and integrity. Humans are swiftly sensing that the cushioning tolerance for their initial error has become approximately exhausted.

Each child emerging from its mother's womb is entering a larger womb of total human consciousness which is continually modified and expanded by subjective experiences and objective experiments. As each successive child is born, it comes into a cosmic consciousness in which it is confronted with less misinformation than yesterday and with more reliable information than yesterday. Each child is born into a much larger womb of more intellectually competent consciousness.

I was seven years old before I saw an automobile, though living in the ambience of a large American city. Not until I was nine was the airplane invented. As a child I thought spontaneously only in terms of walking, bicycling, horse-drawn capability. Trips on railroads and steamships were dream-provokers learned of through a few older people who traveled. My daughter was born with cloth-covered-wing biplanes in her sky and the talkie radio in her hearing. My granddaughter was born in a house with several jet transports going over every minute. She saw a thousand airplanes before she saw a bird: a thousand automobiles before a horse. To children born in 1970, trips to the moon will be as everyday an event as were trips into the big city to me when a boy. There was no radio when I was born. Television came when I was what is called "retiring age." The first Berkeley dissident students were born the year commercial television started. They have seen around the world on the hour ever since being born—they think world. The total distance covered by an average human being in a total lifetime up to the time I was born was 30,000 miles. Because of the great changes since my birth, I have now gone well over one hundred times that distance. The astronauts knock off three million miles in a week. The average airline hostess is out-mileaging my hundredfold greater mileage than all the people before me. All this has happened in my lifetime. My lifetime has been one of emerging from the womb of human-being remoteness from one another to comprehensive integration of worldaround humanity. But all the customs, all the languages, all laws, all accounting systems, viewpoints, clichés, and axioms are of the old, divided, ignorant days. The corollary of "divide and conquer" is "to be divided is to be conquered." To be specialized is to be divided. The specialization which humanity perseveres in was invented by yesterday's armed conqueror illiterates. The separation of humans into more countries made them easy to manage. Nations may unite, as at present, without success. Strife is proliferating. Not until specialization and nations are dispensed with will humanity have a chance of survival. It is to be all or none.

In my first jobs before World War I, I found all the working men to have vocabularies of no more than one hundred words, more than 50 per cent of which were profane or obscene. Because I worked with them, I know that their intellects were there, but dulled and deprived of the information of visionary conceptioning. They had no way of expressing themselves other than by inflection and shock. Conquerors invented gladiatorial wrestling, self-brutalizing games, slapstick and illusionary drama to keep their illiterate masses preoccupied when not at work. This was not changed by any scheduled system of education—it was changed by the radio. The radio broadcasting employees were hired for their vocabularies and diction. The eyes and ears of human beings were able to coordinate the words of the radio and the graphic words of the newspaper. Literacy accelerated. In a half-century, world-around man's vocabulary has been expanded to the equivalent of yesterday's scholar. Television's scientific invention and underwater and space exploration have accelerated this process of freeing humanity from its slave complex to an extraordinary degree. The young realize, as their elders do not, that humanity can do and can afford to do anything it needs to do that it knows how to do.

Those who ignorantly think of themselves as a well-to-do conservative elite are, in fact, so slave-complexed that they are shocked when the younger generation throws aside their clothes and cars of distinction and—abandoning their make-believe mansions which only are their old conquerors' castles— congregate in hundreds of thousands in shameless, innocent bands on vast beaches and meadows. It is not an unspannable generation gap that has occurred, but an emancipation of youth from yesterday's slave-complex reflexes. This has been brought about solely by the proliferation of knowledge. "The medium is the message" is the message only of yesterday's middle-class elite. It said, "Never mind the mind. It's the body that counts," or "It's the physical that can be possessed—To hell with the metaphysical. You can possess a physical brain but not the universally free mind and its thoughts. Leave that to the intellectuals. Look out for those dangerous free thinkers." Higher education was an adornment—a mark of distinction—not something to be taken seriously. The problem of man's being born into the new womb of planetary comprehension, into the new world of integrated coordination and understanding of all humanity, is one not of educating a single absolute monarch, nor of educating either a fascistic or central party elite, nor of educating only the middle class. It is a matter of educating everyone everywhere to the realities of the emerging of man from the womb of permitted ignorance into the womb of required comprehension and competence. That education will have to be brought about by the extraordinary discarding of yesterday's inadequate amusements, shallow romances and drama, and make-believe substitute worlds to cover up the inadequacies of misinformed and underinformed, physically slavish or bureaucratically dogmatic, thoughtless life.

All the foregoing observations of human misorientations constitute but a minor fraction of those which can be truthfully and cogently made today with some chance of their not only being heard but heeded. And all this brings us all to this book by Gene Youngblood—an excellent name for one of the first of the youth who have emerged from childhood and schooling and "social experience" sufficiently undamaged to be able to cope lucidly with the problem of providing world-around man with the most effective communication techniques for speaking universal language to universal man—for helping universal man to understand the great transitions, to understand the reasonableness of yesterday's only-transitional inadequacies, to understand that the oldsters are victims of yesterday's ignorance and not Machiavellian enemies of youth, to understand that any bias—one way or another—utterly vitiates competent thinking and action, to understand that 100 per cent tolerance for error of viewpoint and misbehavior of others is essential to new-era competence—and, finally, to understand that man wants to understand. Nowhere have we encountered a youth more orderly-minded regarding the most comprehensively favorable, forward functioning of humans in Scenario-Universe than in Gene Youngblood. His book *Expanded Cinema* is his own name for the forward, omni-humanity educating function of man's total communication system.

Isaac Newton, as the greatest Olympian of classical science whose influence reigned supreme until the turn of the nineteenth into the twentieth century, assumed the Universe to be *normally at rest* and abnormally in motion. Einstein realized that the experimental data regarding the Brownian Movement and the speed of light made it clear that Universe was not normally at rest, for when its energies were released in a vacuumized tunnel they traveled linearly at 186,000 miles per second. This he assumed to manifest its norm, since that is how Universe behaves normally when unfettered in a vacuum. Any seemingly motionless phenomena, he reasoned, such as seemingly solid matter, consisted of energy moving at 186,000 miles per second but in such small local orbits that their speed and the exquisitely small, self-huddling orbit made them impenetrable; ergo, apparently solid. This was the basis of his formulation of his extraordinary $E=Mc^2$, which, when fission and fusion occurred, proved his locked-up-energy formulation to be correct. The utter difference between Newton's norm of *at rest* and Einstein's norm of 186,000 miles per second provides humanity's most abrupt confrontation regarding the epochal difference of conceptioning between that in the womb of yesterday's ignorance and in the womb of new-dawning awareness, from which and into which, respectively, man is now experiencing the last phases of delivery.

Thinking in terms of 700 million miles per hour as being normal—and informed by the experiments of scientists that no energies are lost—Einstein abandoned the Newtonian thought of Universe and assumed in its place Universe to be "a scenario of nonsimultaneous and only partially overlapping transformative events." Einstein's observational formulations, however, are subjective, not objective. In the mid-1930s I suggested in a book that Einstein's work would eventually affect the everyday environment of humanity, both physically and mentally. After reading what I had written, Einstein said to me, "Young man, you amaze me. I cannot conceive of anything I have ever done as having the slightest practical application." He said that to me a year before Hahn, Strassmann, and Lisa Meitner had, on the basis of $E=mc^2$, discovered the theoretical possibility of fission. You can imagine Einstein's dismay when Hiroshima became the first "practical application."

Gene Youngblood's book is the most brilliant conceptioning of the objectively *positive use* of the Scenario-Universe principle, which must be employed by humanity to synchronize its senses and its knowledge in time to ensure the continuance of that little, three-and-one-half-billion-member team of humanity now installed by evolution aboard our little space vehicle Earth. Gene Youngblood's book represents the most important metaphysical scenario for coping with all of the ills of educational systems based only on yesterday's Newtonian-type thinking. Youngblood's *Expanded Cinema* is the beginning of the new era educational system itself. Tomorrow's youth will employ the video cassette resources to bring in the scenario documents of all of humanity's most capable thinkers and conceivers. Only through the scenario can man possibly "house clean" swiftly enough the conceptual resources of his spontaneous formulations. Tomorrow's Expanded Cinema University, as the word uni-verse—toward one—implies, will weld metaphysically together the world community of man by the flux of understanding and the spontaneously truthful integrity of the child.

Even as Bucky was constantly on the go, his wife, whom we referred to as Lady Anne because she was a consummate lady, was always in his thoughts. When Bucky had a chance to be with his family, he immersed himself in being a husband, a father, and a grandfather. Summers at Bear Island were usually set aside for family. One individual who knew the entire family well was Marian Seldes. Seldes, an actress who has won the Tony, Drama Desk, and Outer Critics Circle Awards and is the widow of Academy Award–winning playwright Garson Kanin, was a childhood friend of Bucky's daughter, Allegra. They met as children in New York City, where Marian and Allegra both attended the Dalton School. Though they are separated by geography, Marian is still one of Allegra's best friends.

Long before the world ever heard of Bucky Fuller, Marian Seldes had an opportunity to get to know her best friend's father. So, young readers everywhere, take note: Fathers of friends do sometimes share information worth noting. Here Seldes affectionately remembers her lifelong relationship with the Fuller family in her contribution.

My Best Friend's Father

Marian Seldes

In 1935, Allegra Fuller and I were in the third grade together at the Dalton School in New York City. Our friendship started the day we first met, and a ritual soon began. In the afternoons we walked from 108 East 89th Street to her home on the second floor of a small building at 105 East 88th Street. I would telephone my mother and ask if I could have dinner at the Fullers'.

"Where are you now?" she would ask.

"At Allegra's."

I was supposed to go right home after school, but my mother, though strict, was also very kind. She understood. She always let me stay at the Fullers'.

Their apartment was small, tidy, and shadowy until evening when the lights were turned on. In Allegra's room, there was a large chest packed with dresses, scarves, costume jewelry, ribbons, feathers, and pieces of fabric left over from the pretty dresses her mother made for her. We matched and mismatched, traded and discarded the finery. We invented stories to go with our outlandish costumes. We tried out lipstick colors and eye shadows and powders. The time flew by. I never wanted to leave.

The Fullers and the Seldeses came to Dalton to see their daughters in the plays and dances we performed. Because of Allegra's passion for dance I was swept into her world of the ballet. In a "Color Ballet" at school, she wore white as Light, I wore purple as Sorrow. In *Everyman,* the morality play, I played the title part and Allegra was Good Deeds. In the Christmas pageants, she was an Angel, and so, surprisingly, was I. Our halos were held on by tight ribbons of elastic that gave us headaches, but I felt I was in heaven.

Allegra was a beautiful and beloved child. There had been a tragedy in 1922 when her sister, Alexandra, four years old, died of spinal meningitis. Allegra told me, and I brought the knowledge of it with me the first day I met her parents. I could not imagine that a child could die that way. It haunted me. The Fullers never spoke of it in my presence.

In all my years of knowing them, I never heard self-pity, unkindness, sarcasm, or pettiness in anything they said to each other. The depth of their love and respect for each other reached out into the world around them.

Mrs. Fuller, the delicate, aristocratic Anne, was usually dressed in black, her hair pulled back tightly from her brow. Her voice, sweet and pitched rather high, was soft. Her laughter was almost silent, her smile fleeting. She would sit and sew and watch us act out scenes in our improvised costumes: a few stitches, a glance toward the fairy-tale princess (Allegra) and the wicked queen (me), and then back to her stitching.

Mr. Fuller, Allegra's father, was a presence in the room. His eyes looked huge and bright behind the thick lenses of his glasses. Before I ever heard his voice I was welcomed by his eyes.

He sat like a Buddha on the couch at the far end of the living room, often wearing what looked like a Japanese kimono and sandals. It set him apart from any other adult I had seen. When he wore a simple black suit, white shirt, and narrow tie, I missed the kimono. It was, I thought, his costume.

When it was time for dinner, he moved to join us. The meals were delicious, the portions small. Allegra's father would talk. Never having heard either an inventor or a genius in my first seven years of life, I was mesmerized.

On a low table was a model of the Dymaxion house and a copy of his book *Nine Chains to the Moon*. He gave me the book to hold in my hands. It is dedicated to his daughter Alexandra. He explained the Dymaxion house to me. I wanted to live in it. The *New York Times* said, "Mr. Fuller's space frame and enclosures represent the greatest advance in building since the invention of the arch." And as far as I was concerned, he had invented the moon.

Soon Mr. and Mrs. Fuller became Bucky and Anne to me. And the names

I heard at the dinner table—Morley, Noguchi, Kirstein, Cunningham, Graham—became Chris, Isamu, Lincoln, Merce, and Martha. Bucky was interested in everything and everyone. In the early 1940s when my father, Gilbert Seldes, began planning television programs for CBS, Bucky went to the studio to talk with him and to appear on one of the first TV programs ever to be aired.

In August 1940, I spent my twelfth birthday with the Fullers on Bear Island off the coast of Maine. Here Bucky was at his most content. We were a Utopian community of family and friends living and working together in the big house or small cabins. His sister Rosie—energetic, passionate about the same things as Bucky, and equally generous of spirit—helped to organize the simple daily routines. She became a lifelong friend. The sun and moon were our lights. Lying on the beach we stared at the night sky—the stars above us seemed near enough to touch. Bucky taught us the shapes and names of the constellations.

Allegra went to boarding school, then college. I went to theater school and became an actress. We saw each other as often as possible, and she kept me in touch with her parents. When I was on tour with Judith Anderson in *Medea,* in 1948, I went to one of Bucky's lectures in Chicago. He wove his spell on the audience of students; his quickly paced words poured out of his brain.

More than ten years had passed since the evenings on East 88th Street, and I was still not able to understand everything he said, but I was enthralled by his view of a universe in which we could all take part.

Years later, I attended a lecture Allegra gave about her discoveries in the world of dance. Before my eyes was an incarnation of her father—the same utter selflessness, the passion to communicate the material, not the personality. And because of this, the purity of her mind informed every spoken word. Like her father, Allegra had the qualities of an inspired missionary, a preacher of truth in art and life.

In 1955, when my mother died, Anne Fuller's letter was the most beautiful I received. Her exquisite handwriting—even on the envelope—moved me; her words comforted me.

When my father died, Bucky took me to dinner at a restaurant called Louise's on East 58th Street. He ordered rare steak and sliced tomato for both of us. He drank no liquor, had no dessert, but the food was like a feast. Louise hovered over him like a loving friend. Bucky showed me many closely typed pages of his appointments and itinerary for the next two years. Every day was accounted for and almost every hour. Dates with friends, world leaders, businessmen, and family were all noted. I was amazed—how could he accomplish all he had planned and also live his private life? I saw that it was all one. Life and work and family were all included in each day's tasks. Looking at his face that did not seem to age, I saw a happy man.

The universe I saw now included Bucky's shapes everywhere. The small Dome Restaurant in Woods Hole, Massachusetts, the 200-foot-high U.S. Pavilion at the Expo in Montreal, movie theaters, play domes for children, stage designs, maps of the world—and when the astronauts landed on the moon, I remember thinking of their flight as being made of nine invisible, integral chains.

In 1953, Bucky's first grandchild, Alexandra Snyder, was born, and I became a godmother. In 1955 on April 28, on the same day my own daughter Katharine was born, Jaime Snyder, his grandson, was born. These two marvelous people bring Bucky and Anne into my life whenever I see or think of them. And now there are great-grandchildren, and Bucky lives on in them, too.

People magazine quoted "philosopher-architect" Bucky from a statement in 1974: "I am convinced all of humanity is born with more gifts than we know. Most are born geniuses and just get de-geniused rapidly."

November 1967: Bucky drew a geodesic sphere for my twelve-year-old Katharine and they discussed the concept of "doing the most with least." She already had his drawing of a Thanksgiving turkey with instructions on how to carve it in a frame over her desk.

April 1996: PBS aired *Buckminster Fuller: Thinking Out Loud* on the American Masters television series. I prerecorded some of the narration. One critic called the program "jaunty." Bucky would have loved that.

It is August 1999. In the electronic weekly *New York Press* for the week of August 4–10, writer Alan Cabal began his article on scientists: "I don't know much about science, but I know what I like. I like Nikola Tesla, Wilhelm Reich, L. Ron Hubbard, R. D. Laing. I love Bucky Fuller. . . ." My birthday is today and I am remembering Bucky on Bear Island.

I miss him always and love him, too.

Prologue from *Tetrascroll: Goldilocks and the Three Bears, A Cosmic Fairy Tale*

One day in 1930, when our daughter Allegra was three years old, she said, "Daddy, tell me about Goldilocks and the Three Bears." The story had been read to her many times from a child's illustrated book. As I started telling it, I began to think of new and heretofore unknown details of the famous story. Allegra was delighted with the innovations. From time to time she would ask me to tell her the story again. Gradually, with new insights into their characters, both Goldy and the Three Bears became much more interesting personalities as may the personalities in the comic strips.

At that time I was studying Einstein and others whose work promised to revolutionize the frontiers of thought. One day, when Allegra asked me to tell her a Goldilocks story, I decided to try out a scientific seminar conducted by Goldy with the Three Bears as students asking pertinent questions and getting lucid answers and explanations from Goldy. This intrigued Allegra much more than facetious behavior on the part of the bears and Goldy.

After the bears and Goldy made ice cream sodas and were comfortably seated in their famous chairs—with Goldy in a new portable movie director's chair—they would start talking about this and that, which would always lead to the most scientifically and philosophically challenging subjects. But the bears and Goldy never called it science or mathematics.

This was the beginning of my spontaneous thinking-out-loud discourses such as I now give publicly.

I became convinced that through imagined expansion of the recallable inventory of fundamental experiences of the child, achieved through description of analogous experiences of others, altered only in magnitude and always similar in principle to the child's experience recalls, it would be possible to effectively induce that child's discovery of the most complex and profound phenomena.

I was also convinced that the best way to study the thoughts of the scientists I was reading was to test myself by disclosing what I understood to a child.

Tricap 1 from *Tetrascroll: Goldilocks and the Three Bears*

Here is Goldy having a sky party with her three friends, the Polar Bear family. Goldy says the sky party is a "system" because Goldy plus the Three Bears equals four entities (or star events), and it takes four events to produce a system. A system divides all the universe into six parts: all the universe outside the system (the macrocosm), all the universe inside the system (the microcosm), and the four star events *A, B, C, D,* which do the dividing.

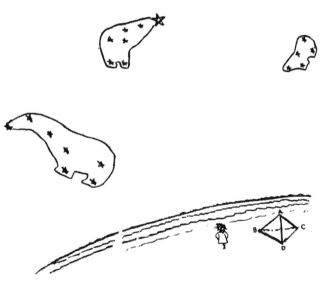

The tetrahedron's four-corner star events do not have to occur at the same time. Goldy found that light traveled six and one-half trillion miles in a year, and was fascinated when an astronomer told her that the star in the nose of the Big Bear is a live show taking place 210 light-years away-and-ago, as the American colonists are first thinking about revolting from English rule: and the pole star at Mommy Bear's nose is a live show taking place 680 light-years away-and-ago, as Dante is writing *The Inferno*: and the star at Wee Bear's front toe is a live show taking place forty-three light-years away-and-ago as Franklin Delano Roosevelt is being elected to the USA presidency for the first time, at the depth of the great 1929–39 Depression: While she, Goldy, is also a live show taking place no time away-and-ago. Altogether Goldy's four live shows constitute a scenario of nonsimultaneous but omni-interrelated events, which can and do define the four corners of a minimum system—the tetrahedron.

She now understands Einstein's concept that Universe is a scenario and not a single simultaneous structure. One picture of a caterpillar does not tell you it is going to transform into a butterfly, and it takes many frames of the cinema to inform you that the butterfly can fly.

Here is Goldy having a sky party with her three friends, the Polar Bear family. Big Bear gave Mommy the Pole Star to wear on her nose when she gave birth to

EE BEAR.

Daddy—Ursa Major, "The Big Bear"—is often called the "Big Dipper" because he drinks so much iced tea. Mommy—Ursa Minor, "The Little Bear"— is usually called the "Little Dipper" because being right at the cold North Pole she drinks much hot tea but in little cups—that way it doesn't have time to get cold, she says. Wee Bear is sometimes called Cassiopeia because he sits in the high chair Cassiopeia used when she was a baby.

Goldy says, "I have drawn Mommy Bear in reverse. I forgot when I was drawing her that if it is to be printed directly from my drawing, it requires an original mirror-image master. But I am going to leave her that way because it's well to remind everyone at the outset that we can only get from here to there by a series of errors—errors forwardly to the right, then a correcting forwardly error to the left, each time reducing error but never eliminating it. This is what generates waves; this what generates the experience life."

Goldy says the sky is a "system" because Goldy plus the Three Bears equals four entities (or star events), and it takes four events to produce a system. A system divides all the universe into six parts: all the universe outside the system (the macrocosm), all the universe inside the system (the microcosm), and the four star events *A, B, C, D,* which do the dividing.

Two star or three star event-entities have only "Betweenness" but no "insideness."

Insideness outsideness separation begins only with completion of the six interrelationship lines of the four separate entity-producing events. The four star events *A, B, C, D,* have six separate, unique and most economical inter-relationship lines *AB, AC, AD, BC, BD, CD.* These six lines and their four interconnected star-corners inadvertently produce four triangular facets of the minimum polyhedron—which four facets completely enclose the system to exclude the macrocosm and include the microcosm. A system consists at minimum of four nonsimultaneous but co-occurrent, because overlapping, yet dissimilarly beginning and enduring star entity-events of six interrelationship lines and four nothingness-window-facets plus twelve unique intercovariant vertex angles—twenty-six conceptual, topological components of a system to which must be added the multiplicative, ultravisible, macrocosmic outside-ness and infravisible, microcosmic insideness as well as the inseparably co-occurring inside concavity and outside convexity and the bipoled axis of rotation of all systems: for a total component inventory of thirty-two items.

Three thousand years ago the Greek geometers named this minimum system the tetrahedron—tetra = four, hedron = sides.

A system cannot have less than four triangular polygon "faces" (or sides or windows) or less than three triangular (polygon) faces surrounding each of the system's four event corners. The triangle is the minimum polygon face. You cannot have a polygon of less than three edges. You cannot have a location fix-point that is less than one fix-point, you cannot have an event tracing line that is less than a line, you cannot have an angle that is less than a minimum angle, and you cannot have a system of less than thirty-two uniquely differentiable and geometrically describable characteristics. All the characteristics of a system are absolute because each of its components is the minimum-limit case in its respective conceptual category, for all conceptuality, as the great mathematician Euler discovered and proved, consists at minimum of points, areas, and lines. Goldy further clarifies and simplifies Euler by saying an area is nothingness, a plurality of areas are framingly separated views of nothingness, a point is a somethingness. A line is a relationship between two somethingnesses. An enlarged, seemingly single somethingness may prove to consist of a plurality of somethingnesses between which the defined interrelationship lines fence off the nothingness into a plurality of separate views of the same nothingnesses. Points are unresolvable, untunable somethingness occurring in the twilight zone between visible and subvisible. Nothingness is the unresolvable untunableness occurring in the twilight zone between visible and supravisible experience.

Life minimally described is "awareness," which is inherently plural, for at minimum it consists of the individual system which becomes aware and the first minimum "otherness" of which it is aware, the otherness being either integrally internal or separately external to the observing system's fourteen integral, topologically componented subsystems ($4V + 4A + 6L$).

Together the observer and the observed constitute two points differentiated against an omni-environment of nothingness with one inherent line of "awareness" interrelationship running between these two points. Euler's generalized formula, which he named topology, says the number of points plus the number of areas will always equal the number of lines plus the number 2, which Goldy finds to be at minimum $2P + 1A = 1L + 2$, which minimum set of awareness aspects of life adds to four, i.e., (A) the observer, (B) the observed, (C) the line of interrelationship, and (D) the nothingness area against which the somethingness is observed.

There are no known experimentally demonstrable absolute maximum limits.

Only the minimum limit is demonstrably absolute. The minimum limit experienceable is always a system—even when it looks only like a point. A point is a system so macro remote or micro small as to appear only as an indivisible something in a specific direction relationship to the observer's integral systems arrangement of, for instance, head, toe, front, and back. "That's nifty," says Wee Bear. "It's magnificent," says Big Bear. "I call it both nifty and magnificent," says Mommy Bear.

Goldy has a tetrahedron beside her on the beach. Its four vertexes (which may also be called: locations; stars, event-fixes; points) are oriented as are Goldy and the Three Bears, with Mommy and her pole star at A, Daddy at B, Wee Bear at C, and Goldy at D.

The tetrahedron's four-corner star events do not have to occur at the same time. Goldy saw a man way down the beach pounding to drive a post into the sand. She heard each pounding a moment after she saw it occur. When she was told about light's speed having been measured in a vacuum by scientists, she understood that the light with which she saw the event, like sound, also had a limit speed. She was told that sound traveled at about 700 miles an hour and that light traveled a million times faster, which though very fast is much slower than 700 million miles in no time at all.

Multiplying 700 million by the number of hours in a year, Goldy found that light traveled six and one-half trillion miles in a year, and was fascinated when an astronomer told her that the star in the nose of the Big Bear is a live show taking place 210 light-years away-and-ago, as the American colonists are first thinking about revolting from English rule: and the pole star at Mommy Bear's nose is a live show taking place 680 light-years away-and-ago, as Dante is writing *The Inferno:* and the star at Wee Bear's front toe is a live show taking place forty-three light-years away-and-ago as Franklin Delano Roosevelt is being elected to the USA presidency for the first time, at the depth of the great 1929–39 Depression: while she, Goldy, is also a live show taking place no time away-and-ago. Altogether Goldy's four live shows constitute a scenario of nonsimultaneous but omni-interrelated events, which can and do define the four corners of a minimum system—the tetrahedron. Goldy, too, is a nonsimultaneous system. She, too, is a nonsimultaneous and only partially overlapping complex of insideness and outsideness experiences. The interrelated experiences of Goldy and the Three Bears are a scenario. She now understands Einstein's concept that Universe is a scenario and not a single simultaneous structure. One picture of a caterpillar does not tell you it is going to transform into a butterfly, and it takes many frames of the cinema to inform you that the butterfly can fly and many thousands of frames to permit a possible replicative engineering discovery of how it can fly. Because any one single "frame" or picture in the scenario filmstrip cannot disclose "what the story is all about," Goldy says to the bears, "When people look at you stars and say, 'I wonder what is outside the outside of the stars,' they are asking for a timeless, simultaneous, static-system concept where none exists. Their question is as ignorant as would be asking, 'Which word is the dictionary?'"

"You are right, Goldy," says Daddy Bear. "Minds think exploratorily, sort and compose. One thought, which is one metaphysically conceptual system, which at minimum is one tetrahedron, can interrelate any four event points or subsystems in nonsimultaneous Universe. Because of inherent nonsimultaneity all thinking is tetratuning. The (system-thought) tetrahedron can and always does include four identities: (1) the thinking individual, (2) the present otherness, (3) the past otherness, (4) the future otherness.

To which thought of Daddy Bear Mommy adds, "And the ignorant question asking of the brain occurs because brains do not think. They only play back yesterday's recordings. Brains are pretuned like bells and sound off when struck."

Tricap 2 from *Tetrascroll: Goldilocks and the Three Bears*

Using the vast, water-smoothed surface of the many-miles-long sandy beach, and walking along as she talks, Goldy keeps drawing pictures large enough for the bears to see. Goldy says to the bears, "Let's try an experiment with our tetrahedron.

By pushing successively on the tetrahedron's top vertex, Goldy keeps rolling the tetrahedron ahead of her across the beach. This succession of rolling makes a long, parallel-edged ribbon with a line zigzagging between its edges to produce a succession of adjacent triangles.

Goldy says to the bears, "We have discovered a triangularly sub-divided ribbon-printing machine—a wave-printing machine." And Daddy bear says, "That is also the sand print patterning made by our four (*A, B, C, D*) bear's feet when we are running. We can start our run with our right hind food *D* elevated. We then lunge for-wardly over the hinge line running between our two front feet *C, D,* as foot *A* goes forwardly and down while foot *B* is elevated. Because a bear's foot is itself a triangle, Goldy makes a pattern of Big Sky Bear's footprints as he walks or runs eastwardly along the beach. Goldy uses the successive triangles as the frames for the succession of illustrations of her conversation with the bears. She says the rib-bon is like a scenario filmstrip with the successive triangular pic-tures overlapping instead of being vertically separated.

Using the vast, water-smoothed surface of the many-miles-long sandy beach, and walking along as she talks, Goldy keeps drawing pictures large enough for the bears to see. The pictures illustrate each of her experimentally demonstrated explanations. She is re-counting to the bears what humans have thus far learned regarding the principles employed by Scenario Universe to accomplish its eternal regeneration.

Goldy says to the bears, "Let's try an experiment with out tetra-hedron. Let's see what happens if I use three of its four corners, *A,*

COURTESY OF THE BUCKMINSTER FULLER INSTITUTE

B, C, representing you three bears, as the tetrahedron's base on the sand—because you have been together for billions of years—and corner *D,* which is the newcomer me—Goldy—I put at the top of the tetrahedron. Now I am taking hold of my corner *D* at the top and am rolling it over around its bottom edge *BC* until corner *D* lies down in the sand, pointing in an easterly direction to my right, along the beach." This rolling leaves the triangular print *ABC* in the sand westward, to Goldy's left, where the tetrahedron had been sitting before that first rollover. Goldy now takes hold of corner *A,* which rose from the sand base to replace *D* at the tetrahedron's top. She pushes the new top corner *A* again over and downwardly to her right around edge *CD,* which means farther eastwardly along the beach until *A* hits the beach, as corner *B* rises to take the top corner place of the tetrahedron, leaving a second triangular print *BCD* in the sand with its edge *BC* congruent with *BC* of the first printed triangle *ABC.* She now pushes top corner *B* eastward, rolling it over and down around bottom edge *DA,* which brings corner *C* to the top and leaves a third successive triangular print *CDA.* She next rolls top corner *C* eastward over bottom edge *BA* to leave a fourth successive print *DAB,* making an altogether eastwardly developing ribbon of triangular prints. Goldy then pushes top corner *D,* then successively tops *A, B, C, D* again and again, as each rotates to the top to replace the one Goldy last rolled over and downward, eastwardly to her right. By pushing successively eastward the tetrahedron's successively latest top vertex, Goldy keeps rolling the tetrahedron ahead of her along the beach. This succession of rollings makes a long parallel-edged ribbon with a line zigzagging between its edges to produce a succession of adjacent triangles. One of the ribbon's edges reads successively *AC, AC, AC, AC,* while its other edge reads *BD, BD, BD, BD.* The succession of eastwardly occurring tops reads *D, A, B, C, D, A, B, C, D,* and so on.

Goldy says to the bears, "We have discovered a triangularly subdivided ribbon-printing machine—a wave-printing machine." And Daddy Bear says, "That is also the sand print patterning made by our four *(A, B, C, D)* bear's feet when we are running. We can start our run with our right hind foot *D* elevated. We then lunge forwardly over the hinge line running between our two front feet *C, D,* as foot *A* goes forwardly and down while foot *B* is elevated."

Because a bear's foot is itself a triangle, Goldy makes a pattern of Big Sky Bear's footprints as he walks or runs eastwardly along the beach. Goldy uses the successive triangles as the frames for the succession of illustrations of her conversation with the bears. She says the ribbon is like a scenario filmstrip with the successive triangular pictures overlapping instead of being vertically separated. "You may notice," says Wee Bear, "That the starry pattern of the chair Cassiopeia left for me looks like the first three triangular frames of that scenario filmstrip." "Yes," Goldy replies, "and I see that if I print these triangular frames of the scenario strip of overlapping conceptual events on a heavy paper ribbon, the strip can be spooled onto a tetrahedron. This will make a tetrahedron book that can be progressively unrolled from a tetrahedron at one end and rerolled to form another tetrahedron at the other end of the strip, with the progressively exposed strip in between telling the picture story—the scenario—of "nonsimultaneous universe," with both the four-dimensional tetrahedronal othernesses scrolls of tomorrow and yesterday as yet unrolled or already rolled back in, and therefore consciously and directionally identifiable but inscrutable. Because the tetrahedra are serving as four-dimensional scrolls, we will call our first such book *The 4D Tetrascroll.*

After graduating from Milton Academy in Massachusetts, R. Buckminster Fuller entered Harvard University, the fifth generation of Fullers to do so. But Bucky was not accepted into any of Harvard's so-called final clubs. Bucky found another interest, as he called it: being "a stage-door Johnny." His chief attraction was an actress named Marilyn Miller, who was starring in *George White's Scandals,* a musical extravaganza similar to the *Ziegfeld Follies.* One day Bucky took Marilyn and the chorus to New York City and treated the whole gang to dinner at Churchill's, thereby blowing his entire year's tuition. Since the tuition wasn't paid, Harvard dismissed Bucky. His mother's disappointment led Bucky to reapply to Harvard, but he quickly left the college a second time—Bucky himself used the word "fired"—because he was bored and felt he was wasting his mother's resources.

Almost half a century later, in 1961, Harvard appointed Bucky to the prestigious Charles Eliot Norton Professorship, a poetry chair. One of the Fuller company's attorneys, Robert D. Storey, in later years served as a member of the Harvard University Board of Overseers, and he has told me how pleased Bucky always was when he donned cap and gown on ceremonial occasions and marched through the Yard with President Derek Bok. Bucky would beam, a twinkle in his eyes, no doubt feeling redeemed in his mother's eyes for his double dismissal.

Dr. Arthur L. Loeb of the Department of Visual and Environmental Studies, located at the Carpenter Center, was Bucky's longtime associate and friend at Harvard. It was Loeb who introduced students to the patterns of Bucky's geometry and the new language Bucky called "synergetics." Through Loeb's academic support and interest in Fuller's work, Bucky lives on at Harvard. Loeb's contribution offers an analysis of some of Bucky's mathematical theories.

Continuity, Discreteness, and Resolution

Dr. Arthur L. Loeb

As a student at Milton Academy, Buckminster Fuller would be puzzled by his teacher's assertion that points have no dimension, but that placed side by side, such dimensionless points could constitute a line having no width. There, the teacher claimed, such insubstantial lines could be juxtaposed to form a plane having no thickness, and such planes could be stacked to form a solid. Young Buckminster felt that it would not be possible to assemble substance out of such ethereal objects having no finite dimensions.

Of course, Fuller was quite correct. Not only was he correct, but he put his finger on a flaw in the teaching of mathematics which has led to a great deal of muddled thinking. Fuller's skepticism at this early stage of his development prepared him for many of his trail-blazing discoveries. The traditional approach to geometry asserts that a square generated out of line segments having unit length, if it itself has unit width, will have unit area. This claim assigns a fundamental importance to the square, which is, however, not necessarily correct, because it cannot be proved or disproved experimentally. It is an arbitrary convention; one might equally well, and frequently to greater advantage, assign unit area to an equilateral triangle having unit edge length.

The fact is that length, area, and volume are as different from one another as are force and electric charge, acceleration, electric current, voltage, and magnetic field. Experiments relating them measure *changes* in variables caused by *changes* in other variables, not absolute values; they establish *proportionalities*. Proportionality constants are introduced as a result of these experiments, such as *dielectric constant, mass, resistance*, etc. Analogously, relationships between length, area, and volume involve proportionalities: area and volume are proportional to the second and third power respectively of linear dimension.[1,2,3] The proportionality constants relating length, area, and volume depend on the shape of the object being measured: square, triangle, and tetrahedron each have their own shape constant; geometrically similar shapes will have the same shape constant. Geometric formulas relating different shapes are obtained by comparing geometrical shapes and transforming them, as explained in the three references given above.

Fuller noted that, in contradistinction with his teacher's introduction of insubstantial points and lines, every point, line, and surface is made up of finite particles. A chalk line, he asserted, is made up of chalk particles, which in turn are made up of calcium, carbon, oxygen, and other ions, which in turn consist of protons, neutrons, and electrons.

Fuller espoused the view that matter is discrete. His line was a chalk line, made up of discrete particles. A circle to him was a polygon having a great

[1] A. L. Loeb, "Buckminster Fuller versus the Irrational, A Double Entendre, in Morphology and Architecture," a special issue of the *International Journal of Space Structures*, ed. Haresh Lalvani, 11, 141–154 (1996).

[2] A. L. Loeb, "Buckminster Fuller and the Relevant Pattern," in *Beyond the Cube,* ed. J. F. Gabriel (New York: John Wiley, 1997).

[3] A. L. Loeb, "Deconstruction of the Cube," in *Beyond the Cube,* ed. J. F. Gabriel.

many, but a finite number of, sides. The number π, which is irrational, may be approximated by the rational number 22/7. However, the fact that Fuller finds significance in the numbers 22 and 7 makes me uncomfortable.

In his tetrahelix, tetrahedra are located directly above and below each other after five rotations, because the dihedral angle of the tetrahedron equals essentially 72°, he believed. Fuller eschewed irrational numbers: just as a circle is really a polygon, he asserted, the dihedral angle of the regular tetrahedron, which is actually arc cos (1/3), should be rounded off to the rational one fifth of a complete revolution.

Here I believe that Fuller failed to realize that he was violating one of his own fundamental rules. Although the angles 72° and arc cos (⅓) differ in *magnitude* by only a little more than 1°, the former occurs in icosahedra, dome structures, etc., in planes perpendicular to axes of fivefold rotational symmetry, the latter between axes of threefold rotational symmetry. By ignoring these different *spatial* orientations, Fuller was in point of fact guilty of *linear* thinking. This is too bad, because in nature helical structures are significantly mismatched: if leaves were juxtaposed exactly above each other, they would intercept sunlight, hindering growth! We have pointed out (ref. 3) that most angles significant to Fuller, although not rational fractions of a complete revolution, do have rational trigonometric functions.

Whether our world is *discrete* or *continuous* has been a point of contention for thousands of years. In our century Buckminster Fuller was definitely on the side of discreteness. Fractal theory is on the side of continuity: coastal and cloud formations display geometrically similar structures at vastly different levels of scale. Of course, the clouds eventually consist of drops of water, etc., but a continuous model is significantly successful at many levels. Thus it behooves us to compare Fuller's polygonal model of the circle and other shapes with the mathematician's definition of such forms, and, if possible, to bring them into some sort of conformance.

A circle, in mathematics, is defined as the locus of *all* points equidistant from a given point. This definition as I state it here is in fact redundant, because a locus by definition already encompasses *all* points that satisfy a given condition, so that the word *all* in the definition of a circle is not necessary. However, it is the very word which distinguishes the mathematicians' definition of the circle from Fuller's! Indeed, all vertices of Fuller's polygonal circle are equidistant from a given point, its center, but there are many other points equidistant from that center. Hence Fuller's circle is not the mathematicians' circle. Neither is Fuller's straight line that of the mathematician, namely the shortest distance between two points, because for some points on a "line" having finite width the sum of the distances from the two endpoints is greater than for others, so that the points constituting Fuller's line cannot correspond to the shortest distance.

All this may be considered abstract philosophizing without substance, just like the question of how many angels can occupy the tip of a pin. However, the problem of optimizing distances is a very real one when one considers Fuller's great circles of a sphere or the shortest distance between two points on an irregular surface. Here we need the mathematicians' concept of limits, hence the concept of infinity, so hateful to those who believe in a discrete universe. Indeed, we tread on dangerous ground when we allow a variable to "go to infinity," for there is no number *infinity*: it is just larger than any preassigned value. The fact is that we need to bring Fuller's chalk circle and line, as we can visualize and draw them, into consistency with the mathematicians' idealized concept.

We recognize a drawn chalk circle as a circle precisely because we do not have X-ray vision: we cannot see the location of the ions in the chalk because our eyes can only respond to a limited range of the electromagnetic spectrum. We can see those objects whose scale conforms to the wavelength of visible light. Our eyes do not respond to radio waves or infrared radiation, whose wavelength is larger than that of visible light, nor to ultraviolet or X-rays, whose wavelength is shorter. As a matter of fact, X-rays used to examine teeth or bones have a much larger wavelength than those used to locate the positions of ions, whose scale is so much smaller. Physical structures are therefore *hierarchical:* the scale at which we observe them depends on the power of resolution of the tools with which we examine them. Fuller's chalk circle is identified as a circle precisely because with visible light we cannot distinguish its atomic components. Conversely, X-ray diffraction would not enable us to identify the object as a circle, but would tell us a great deal about the ionic structure of the chalk itself.

Fuller considers a circle as a polygon having lots of sides. We can easily distinguish it from a triangle, square, pentagon, hexagon, octagon, and even a triacontagon, which has thirty sides. However, sooner or later a polygon will turn up that has so many sides that we can no longer visually distinguish it from a circle. We can then say that *within our power of resolution* this polygon is indistinguishable from a circle. No mathematical infinities necessary here, only the finite level of resolution appropriate for the hierarchical level at which we are examining the structure. We can then refine the mathematics as long as we remain within the level of resolution appropriate for the measurements. This concept of resolution is very real in the use of computers, where the density of pixels determines whether or not a curve will be perceived on the screen as smooth. It also plays a role in wave (quantum) mechanics, where the uncertainty principle tells us that energy levels, while discrete, cannot be determined exactly over a finite time interval.

Let us examine on this basis Fuller's rounding off of the dihedral angle of the regular tetrahedron to 72°. When we bring five regular tetrahedra together, each sharing an edge with one of each of the other four tetrahedra, we will find that they do not fill the space around the common edges: a gap of over 5° can readily be observed. The roundoff is therefore not permissible within our power of resolution.

These concepts may be applied to the following well-known problem.[4] Four bugs find themselves at the vertices of a square whose sides run from east to west and from north to south. The bug at the northwest vertex faces east, the one at the northeast vertex faces south, the one at the southeast corner faces west, and the one in the southwest faces north. Each bug thus faces the bug clockwise from it along an edge of the square. The bugs have been conditioned to travel at the same speed at any moment, although they may speed up or slow down at any time, as long as each remains traveling at the same speed as every other one. Each bug will now travel to the one it faces, and the problem is whether or not they will ever meet.

As each bug travels, the bug it faces will also travel. Therefore each bug will need to adjust its direction of travel in order to keep tracking its target; as a result, the bugs will spiral into the center of the original square. But Buckminster would remonstrate that the bugs would have a finite reaction time, so that their paths would not be smooth spirals, but a series of straight line segments. Whether or not we perceive these paths as smooth spirals depends on the bugs' reaction time as well as our power of resolution.

The first question to be answered is whether the path traveled into the center has finite length. It turns out (cf. ref. 4) that as long as the path is smooth within our power of resolution, its length just equals the initial distance between each bug and its target. This would mean that each bug would easily be able to meet its target at the center of the original square. However, other factors play a role. If the bugs travel at constant speed, then as they approach the center of the square, their angular velocity will increase alarmingly. Now the size of the bugs will begin to play a role: if they are relatively large, the centrifugal forces on their inner and outer shoulders will differ sufficiently so that they will fly apart, although they may touch shoulders before this catastrophe would occur. The smaller they are, the more their angular velocity will increase, but the less the centrifugal forces will differ, so that they well might touch shoulders before critical angular velocity is reached. Buckminster Fuller will not permit us to shrink them to a point, so the outcome is still uncertain.

Suppose, however, that we would avoid catastrophe by using the fact that the bugs may accelerate or decelerate together. Let them travel at constant *angular velocity* so that they will not fly apart. The problem now will be that as

[4]A. L. Loeb, "A Look at Infinity," in *Concepts and Images,* Basel, Boston, Berlin: Birkhäuser, 1993) 122–127.

they approach the center of the square, their speed will decrease, with the result that they will never reach the center. And consider this embarrassing question: if they do reach the center, from which direction will they do so? A tangent to the spiral path at all times makes an angle of 45° with the radial vector from the bug to the center of the square, so that the bug will never actually be directed *into* that center. So can the bugs ever meet in the center? Within our power of resolution the answer could be affirmative, but woe to anyone who would let the bugs shrink to a point!

An analogous problem is the following one. An airplane leaves Boston with its automatic pilot set so that it will fly and continue to fly in a northwest direction. It will spiral toward the north pole, and there a landing strip will be readied for the plane. The question from the north pole is how the landing strip should be oriented. The answer is that any orientation is all right, because the plane will not be able to land as long as its automatic pilot remains fixed on a northwest course: the north pole will always remain to the right of the programmed trajectory, but since the plane is not a point, it will, within its power of resolution, find itself directly over the pole, and can then land as it pleases.

These examples illustrate the indeterminacies inherent in a model for continuous structure, and the pitfalls avoided by Buckminster Fuller by postulating a discrete structure. Nevertheless, mathematical abstractions such as calculus provide idealized constructions whose behavior simulates that of real structures *within the limits of their resolution*. It would appear that the concept of limited resolution would reconcile Buckminster Fuller's hierarchical model of real, discrete structures with the idealized ones of the mathematician.

Discoveries of Synergetics from *Synergetics: Explorations in the Geometry of Thinking*

250.01 Discovery

250.02 Discoveries are uniquely regenerative to the explorer and are most powerful on those rare occasions when a generalized principle is discovered. When mind discovers a generalized principle permeating whole fields of special-case experiences, the discovered relationship is awesomely and elatingly beautiful to the discoverer personally, not only because to the best of his knowledge it has been heretofore unknown, but also because of the intuitively sensed potential of its effect upon knowledge and the consequently improved advantages accruing to humanity's survival and growth struggle in Universe. The stimulation is not that of the discoverer of a diamond, which is a physical entity that may be monopolized or exploited only to the owner's advantage. It is the realization that the newly discovered principle will provide spontaneous, common-sense logic engendering universal cooperation where, in many areas, only confusion and controversy had hitherto prevailed.

250.10 Academic Grading Variables in Respect to Science Versus Humanities

250.101 Whether it was my thick eyeglasses and lack of other personable favors, or some other psychological factors, I often found myself to be the number-one antifavorite amongst my schoolteachers and pupils. When there were disturbances in the classroom, without looking up from his or her desk, the teacher would say, "One mark," or "Two marks," or "Three marks for Fuller." Each mark was a fifteen-minute penalty period to be served after the school had been let out for the others. It was a sport amongst some of my classmates to arrange, through projectiles or other inventions, to have noises occur in my vicinity.

250.11   Where the teacher's opinion of me was unfavorable, and that, in the humanities, was—in the end—all that governed the marking of papers, I often found myself receiving lower grades for reasons irrelevant to the knowledge content of my work—such as my handwriting. In science, and particularly in my mathematics, the answers were either right or wrong. Probably to prove to myself that I might not be as low-average as was indicated by the gradings I got in the humanities, I excelled in my scientific classes and consistently attained the top grades because all my answers were correct. Maybe this made me like mathematics. But my mathematics teachers in various years would say, "You seem to understand math so well, I'll show you some more if you stay in later in the afternoon." I entered Harvard with all As in mathematics, biology, and the sciences, having learned in school advanced mathematics, which at that time was usually taught only at the college level. Since math was so easy, and finding it optional rather than compulsory at Harvard, I took no more of its courses. I was not interested in getting grades but in learning in areas that I didn't know anything about. For instance, in my freshman year, I took not only the compulsory English A, but Government, Musical Composition, Art Appreciation, German Literature, and Chemistry. However, I kept thinking all the time in mathematics and made progressive discoveries, ever enlarging my mathematical vistas. My elementary schoolwork in advanced mathematics as well as in physics and biology, along with my sense of security in relating those fields, gave me great confidence that I was penetrating the unfamiliar while always employing the full gamut of rigorous formulation and treatment appropriate to testing the validity of intuitively glimpsed and tentatively assumed enlargement of the horizon of experientially demonstrable knowledge.

250.12   My spontaneous exploration of mathematics continued after I left Harvard. From 1915 to 1938—that is, for more than twenty years after my days in college—I assumed that what I had been discovering through the post-college years, and was continuing to discover by myself, was well known to mathematicians and other scientists, and was only the well-known advanced knowledge to which I would have been exposed had I stayed at Harvard and majored in those subjects. Why I did not continue at Harvard is irrelevant to academics. A subsequent special course at the U. S. Naval Academy, Annapolis, and two years of private tutelage by some of America's leading engineers of half a century ago completed my formally acknowledged "education."

250.20   My Independent Mathematical Explorations

250.21   In the twentieth year after college, I met Homer Lesourd, my old physics teacher, who most greatly inspired his students at my school, Milton

Academy, and who for half a century taught mathematics at Harvard. We discovered to our mutual surprise that I had apparently progressed far afield from any of the known physio-mathematical concepts with which he was familiar or of which he had any knowledge. Further inquiry by both of us found no contradiction of our first conclusion. That was a third of a century ago. Thereafter, from time to time but with increasing frequency, I found myself able to elucidate my continuing explorations and discoveries to other scientists, some of whom were of great distinction. I would always ask them if they were familiar with any mathematical phenomena akin to the kind of disclosures I was making, or if work was being done by others that might lead to similar disclosures. None of them was aware of any other such disclosures or exploratory work. I always asked them whether they thought my disclosures warranted my further pursuit of what was becoming an ever-increasingly larger body of elegantly integrated and coordinate field of omnirationally quantified vectorial geometry and topology. While they could not identify my discoveries with any of the scientific fields with which they were familiar, they found no error in my disclosures and thought that the overall rational quantation and their logical order of unfoldment warranted my further pursuing the search.

250.30 Remoteness of Synergetics Vocabulary

250.301   When one makes discoveries that, to the best of one's knowledge and wide inquiry, seem to be utterly new, problems arise regarding the appropriate nomenclature and description of what is being discovered as well as problems of invention relating to symbolic economy and lucidity. As a consequence, I found myself inventing an increasingly larger descriptive vocabulary, which evolved as the simplest, least ambiguous method of recounting the paraphernalia and strategies of the live scenario of all my relevant experiences.

250.31   For many years, my vocabulary was utterly foreign to the semantics of all the other sciences. I drew heavily on the dictionary for good and unambiguous terms to identify the multiplying nuances of my discoveries. In the meanwhile, the whole field of science was evolving rapidly in the new fields of quantum mechanics, electronics, and nuclear exploration, inducing a gradual evolution in scientific language. In recent years, I find my experiential mathematics vocabulary in a merging traffic pattern with the language trends of the other sciences, particularly physics. Often, however, the particular new words chosen by others would identify phenomena other than that which I identify with the same words. As the others were unaware of my offbeat work, I had to determine for myself which of the phenomena involved had most logical claim to the names involved. I always conceded to the other scientists, of course (unbeknownst to them); when they seemed to have prior or more valid claims, I would then invent or select appropriate but unused names for the phenomena I had discovered. But I held to my own claim when I found it to be eminently warranted or when the phenomena of other claimants were ill described by that term. For example, quantum mechanics came many years after I did to employ the term *spin*. The physicists assured me that their use of the word did not involve any phenomena that truly spun. *Spin* was only a convenient word for accounting certain unique energy behaviors and investments. My use of the term was to describe a direct observation of an experimentally demonstrable, inherent spinnability and unique magnitudes of rotation of an actually spinning phenomenon whose next fractional rotations were induced by the always co-occurring, generalized, a priori, environmental conditions within which the spinnable phenomenon occurred. This was a case in which I assumed that I held a better claim to the scientific term *spin*. In recent years, spin is beginning to be recognized by the physicists themselves as also inadvertently identifying a conceptually spinnable phenomenon—in fact, the same fundamental phenomenon I had identified much earlier when I first chose to use the word *spin* to describe that which was experimentally disclosed as being inherently spinnable. There appears to be an increasing convergence of scientific explorations in general, and of epistemology and semantics in particular, with my own evolutionary development.

250.32 Because physics has found no continuums, no experimental solids, no things, no real matter, I had decided half a century ago to identify mathematical behaviors of energy phenomena only as *events*. If there are no things, there are no nouns of material substance. The old semantics permitted common-sense acceptance of such a sentence as, "A man pounds the table," wherein a noun verbs a noun or a subject verbs a predicate. I found it necessary to change this form to a complex of events identified as *me,* which must be identified as a verb. The complex verb *me* observed another complex of events identified again ignorantly as a "table." I disciplined myself to communicate exclusively with *verbs*. There are no *wheres* and *whats;* only angle and frequency events described as *whens*.

250.40 The Climate of Invention

250.401 In the competitive world of money-making, discoveries are looked upon as exploitable and monopolizable claims to be operated as private properties of big business. As a consequence, the world has come to think of both discoveries and patents as monopolized property. This popular viewpoint developed during the last century, when both corporations and government supported by courts have required individuals working for them to assign to them the patent rights on any discoveries or inventions made while in their employ. Employees were to assign these rights during, and for two years after termination of, their employment, whether or not the invention had been developed at home or at work. The drafting of expert patent claims is an ever more specialized and complex art, involving expensive legal services usually beyond the reach of private individuals. When nations were remote from one another, internal country patents were effective protection. With today's omniproximities of the world's countries, only world-around patents costing hundreds of thousands of dollars are now effective, with the result that patent properties are available only to rich corporations.

250.41 So now the major portions of extant inventions belong to corporations and governments. However, invention and discovery are inherently individual functions of the minds of individual humans. Corporations are legal fabrications; they cannot invent and discover. Patents were originally conceived of as grants to inventors to help them recover the expenses of the long development of their discoveries; and they gave the inventor only a very short time to recover the expense. Because I am concerned with finding new technical ways of doing more with less, by which increasing numbers of humanity can emerge from abject poverty into states of physical advantage in respect to their environment, I have taken out many patent claims—first, to hold the credit of initiative for the inspiration received by humanity's needs and the theory of their best solution being that of the design revolution and not political revolution, and second, to try to recover the expense of development. But most importantly, I have taken the patents to avoid being stopped by others— in particular, corporations and governments—from doing what I felt needed doing.

250.50 Coincidental Nature of Discoveries

250.501 What often seems to the individual to be an invention, and seems also to be an invention to everyone he knows, time and again turns out to have been previously discovered when patent applications are filed and the search for prior patents begins. Sometimes dozens, sometimes hundreds, of patents will be found to have been issued, or applied for, covering the same idea. This simultaneity of inventing manifests a forward-rolling wave of logical exploration of which the trends are generated by the omni-integrating discoveries and the subsequent inventions of new ways to employ the discoveries at an accelerating rate, which is continually changing the metaphysical environment of exploratory and inventive stimulation.

250.51 I have learned by experience that those who think only in competitive ways assume that I will be discouraged to find that others have already discovered, invented, and patented that which I had thought to be my own unique discovery or invention. They do not understand how pleased I am to learn that the task I had thought needed doing, and of which I had no knowledge of others doing, was happily already being well attended to, for my spontaneous commitment is to the advantage of all humanity. News of such work of others frees me to operate in other seemingly unattended but needed directions of effort. And I have learned how to find out more about what is or is not being attended to. This is evolution.

250.52   When I witness the inertias and fears of humans caused by technical breakthroughs in the realms of abstract scientific discovery. I realize that their criteria of apprehension are all uninformed. I see the same patterns of my experience obtaining amongst the millions of scientists around the world silently at work in the realm of scientific abstract discovery, often operating remote from one another. Many are bound to come out with simultaneous discoveries, each one of which is liable to make the others a little more comprehensible and usable. Those who have paid-servant complexes worry about losing their jobs if their competitors' similar discoveries become known to their employers. But the work of pure science exploration is much less understood by the economically competitive-minded than is that of inventors. The great awards economic competitors give to the scientists make big news, but no great scientist ever did what he did in hope of earning rewards. The greats have ever been inspired by the a priori integrities of Universe and by the need of all humanity to move from the absolute ignorance of birth into a little greater understanding of the cosmic integrities. They esteem the esteem of those whom they esteem for similar commitment, but they don't work for it.

250.53   I recall now that when I first started making mathematical discoveries, years ago, my acquaintances would often say, "Didn't you know that Democritus made that discovery and said just what you are saying 2,000 years ago?" I replied that I was lucky that I didn't know that because I thought Democritus so competent that I would have given up all my own efforts to understand the phenomena involved through my own faculties and investment of time. Rather than feeling dismayed, I was elated to discover that, operating on my own, I was able to come out with the same conclusions of so great a mind as that of Democritus. Such events increased my confidence in the rersourcefulness and integrity of human thought purely pursued and based on personal experiences.

250.60   Proofs

250.61   I know that many of the discoveries of synergetics in the book of their accounting, which follows, may prove in time to be well-known to others. But some of them may not be known to others and thus may be added to the ever-increasing insights of the human mind. Any one individual has inherently limited knowledge of what total Universe frontiering consists of at any one moment. My list embraces what I know to be my *own* discoveries of which I have no knowledge of others having made similar discoveries earlier than my own. I claim nothing. Proofs of some of my theoretical discoveries have been made by myself and will be made by myself. Proofs may have been made by others and will be made by others. Proofs are satisfying. But many mathematical theorems provide great living advantages for humanity over long periods of time before their final mathematical proofs are discovered. The whys and wherefores of what is rated as mathematical proof have been evolved by mathematicians; they are formal and esoteric conventions between specialists.

251.00 Discoveries of Synergetics: Inventory

251.01 The ability to identify all experience in terms of only angle and frequency.

251.02 The addition of angle and frequency to Euler's inventory of crossings, areas, and lines as absolute characteristics of all pattern cognizance.

251.03 The omnirational accommodation of both linear and angular acceleration in the same mathematical coordination system.

251.04 The discovery that the pattern of operative effectiveness of the gravitational constant will always be greater than that of the radiational constant—the excess effectiveness being exquisitely minute, but always operative, wherefore the disintegrative forces of Universe are effectively canceled out and embraced by the integrative forces.

251.05 The gravitational is comprehensively embracing and circumferentially contractive—ergo, advantaged over the centrally radiational by a 6 : 1 energy advantage; i.e., a circumference chord-to-radius vectorial advantage of contraction versus expansion, certified by the finite closure of the circumference, ergo, a cumulative series versus the independent, disassociating disintegration of the radii and their separating and dividing of energy effectiveness. (This is an inverse corollary of the age-old instinct to divide and conquer.) (See Secs. 529.03, 541 and 1052.)

251.06 The gravitational-radiational constant $10F^2 + 2$.

251.07 The definition of gravity as a spherically circumferential force whose effectiveness has a constant advantage ratio of 12 to 1 over the radial inward mass-attraction.

251.10 The introduction of angular topology as the description of a structural system in terms of the sum of its surface angles.

251.11 The definition of structure as the pattern of self-stabilization of a complex of events with a minimum of six functions as three edges and three vertexes, speaking both vectorially and topologically.

251.12 The introduction of angular topology as comprised entirely of central-angle and surface-angle phenomena, with the surface angles accounting for concavity and convexity, and the thereby-derived maximum structural advantage of omni-self-triangulating systems.

251.13   As a result of the surface-angle concave-convex take-outs to provide self-closing finiteness of insideness and outsideness, central angles are generated, and they then function in respect to unique systems and differentiate between compoundings of systems.

251.14   One of the differences between atoms and chemical compounds is in the number of central-angle systems.

251.20   The discovery of the mathematically regular, three-way, great-circle, spherical-coordinate cartographic grid of an infinite frequency series of progressive modular subdivisions, with the spherical radii that are perpendicular to the enclosing spherical field remaining vertical to the corresponding planar surface points of cartographic projection; and the commensurate identification of this same great-circle triangulation capability with the icosahedron and vector equilibrium, as well as with the octahedron and the tetrahedron. (See Secs. 527.24 and 1009.98.)

251.21   The development of the spherical triangular grid bases from the spherical tetrahedron, spherical cube, spherical octahedron, and the spherical vector equilibrium and its alternate, the icosahedron, and the discovery that there are no other prime spherical triangular grids. All other spherical grids are derivatives of these.

251.22   The spherical triangular grids are always identified uniquely only with the first four prime numbers 1, 2, 3 and 5: with the tetrahedron always identifying with the prime number 1; the octahedron with 2, the face-triangulated cube with 3; and the vector equilibrium and icosahedron with the prime number 5; with the other Platonic, Archimedean, and other symmetrical polyhedra all being complex compoundings and developments of these first four prime numbers, with the numbers compounded disclosing the compounding of the original four base polyhedra.

251.23   The number of the external crossings of the three-way spherical grids always equals the prime number times the frequency of modular subdivision to the second power times two, plus the two extra crossings always assigned to the polar axis functioning to accommodate the independent spinnability of all systems.

251.24   The mathematical regularity identifies the second power of the linear dimensions of the system with the number of nonpolar crossings of the comprehensive three-way great circle gridding, in contradistinction to the previous mathematical identification of second powering exclusively with surface areas.

251.25   The synergetic discovery of the identification of the surface points of the system with second powering accommodates quantum mechanics' discrete energy packaging of photons and elucidates Einstein's equation, $E=Mc^2$, where the omnidirectional velocity of radiation to the second power—c^2— identifies the rate of the rational order growth of the discrete energy quantation. This also explains synergetics' discovery of the external point growth rate of systems. It also elucidates and identifies the second-power factoring of Newton's gravitational law. It also develops one-to-one congruence of all linear and angular accelerations, which are factorable rationally as the second power of wave frequency.

251.26   The definition of a system as the first subdivision of finite but nonunitary and nonsimultaneous conceptuality of the Universe into all the Universe outside the system, and all the Universe inside the system, with the remainder of the Universe constituting the system itself, which alone, for the conceptual moment, is conceptual.

251.27   The definition of Universe as a scenario of nonsimultaneous and only partially overlapping events, all the physical components of which are ever-transforming, and all the generalized metaphysical discoveries of which ever clarify more economically as eternally changeless.

251.28   The vector model for the magic numbers, which identifies the structural logic of the atomic isotopes in a symmetrical synergetic hierarchy.

251.30   The rational identification of number with the hierarchy of all the geometries.

251.31   The A and B Quanta Modules.

251.32   The volumetric hierarchy of Platonic and other symmetrical geometricals based on the tetrahedron and the A and B Quanta Modules as unity of coordinate mensuration.

251.33   The identification of the nucleus with the vector equilibrium.

251.34   Omnirationality: the identification of *triangling* and *tetrahedroning* with second- and third-powering factors.

251.35   Omni-60-degree coordination versus 90-degree coordination.

251.36   The identification of waves with vectors as waviform vectors; the deliberately nonstraight line.

251.37   The comprehensive, closed-system foldability of the great circles and their identification with wave phenomena.

251.38   The accommodation of odd or even numbers in the shell-generating frequencies of the vector equilibrium.

251.39   The hierarchy of the symmetrically expanding and contracting pulsations of the interpolyhedral transformations, and their respective circumferentially and radially covarying states. (Also described as the symmetrical contraction, "jitterbugging," and pumping models.)

251.40   The provision for the mathematical treatment of the domains of interferences as the domains of vertexes (crossings).

251.41   Mathematical proof of the four-color map theorem.

251.42   The introduction of the tensegrity structural system of discontinuous compression and continuous tension.

251.43   The identification of tensegrity with penumatics and hydraulics.

251.44   The discovery of the number of primes factorial that form the positives and negatives of all the complex phenomena integratively generated by all possible permutations of all the 92 regenerative chemical elements.

251.45   The disclosure of the rational fourth-, fifth-, and sixth-powering modelability of nature's coordinate transformings as referenced to the 60° equiangular, isotropic vector equilibrium.

251.46   The discovery that once a closed system is recognized as exclusively valid, the list of variables and degrees of freedom are closed and limited to six positive and six negative alternatives of action for each local transformation event in Universe.

251.47   The discovery of the formula for the rational-whole-number expression of the tetrahedral volume of both the spherical and interstitial spaces of the first- and third-power concentric shell-growth rates of nuclear closest-packed vector equilibria.

251.50   The integration of geometry and philosophy in a single conceptual system providing a common language and accounting for both the physical and metaphysical.

Black Mountain College, near Asheville, North Carolina, was conceived at a critical time in international history. Adolf Hitler, as chancellor of Germany, had ordered the closing of the Bauhaus, and America was in the depth of the Depression. In 1933 some Americans remained optimistic that the system could be reformed through education, and John Andrews Rice and his colleagues opened Black Mountain College. It was here that Merce Cunningham formed his dance company, John Cage staged his first happening, and Bucky Fuller built his first geodesic dome.

The faculty included Josef and Anni Albers, Charles Burchard, John Cage, Merce Cunningham, José de Creeft, Willem and Elaine de Kooning, Agnes de Mille, Theodore Dreier, Lyonel Feininger, Arthur Fielder, Walter Gropius, Edgar Kaufmann, Jr., Franz Kline, Richard Lippold, Robert Motherwell, Bernard Rudofsky, Ben Shahn, M. C. Richards, and R. Buckminster Fuller, to name just a few! Among the students were Ruth Asawa and her future husband, William Albert Lanier, Kenneth Snelson, Jeffrey Lindsay, Arthur Penn, Claude Stoller, Cy Twombly, Robert and Don L. Richter, Paul Taylor, and Robert Rauschenberg, again to name just a few.

Later, Ken Snelson would recall that Bucky was "absolutely hypnotizing and electrifying" in his first lecture, and Richard Lippold noted that it was "like meeting Zoroaster speaking Islamic." Elaine de Kooning remembers that when the first Bucky dome collapsed, she renamed it the Supine Dome. Bucky explained that failure is part of experimentation and that "you succeed when you stop failing." Ruth Asawa's essay recalls her student days at Black Mountain.

Black Mountain College

Ruth Asawa

It was the beginning of the June 1948 Black Mountain Summer College. It was humid and hot. You could hear the high-pitched sound of the cicadas in the heat of the day and at night the croaking of the tree frogs. Guests were beginning to arrive. Some needed a ride from the train station, some came by car. Richard and Louise Lippold and their two daughters arrived in an old black hearse.

My summer job was working in the laundry room checking out sheets and towels and preparing beds and rooms for summer guests like Elaine and Willem de Kooning, R. Buckminster Fuller, Merce Cunningham, John Cage, and Charles Burchard.

We were waiting in front of the dining hall for our evening lecturer, Buckminster Fuller. He arrived in a yellow Studebaker pulling a dome-shaped aluminum trailer. He jumped out of his car, introduced himself as Richard Buckminster Fuller, but put us at ease by adding, "Call me Bucky." We all helped him unload his trailer, which was packed full of materials for the summer project and models of all sizes. He had not yet built a dome, which was going to be his summer project.

After dinner the lecture began. At midnight Bucky was still talking to students who had signed up for his class. This was his first teaching assignment. My study, which was an army barrack, faced the meadow where the dome would be built.

I volunteered to be the college barber that summer, although I had never cut anyone's hair. (As a child I had watched my father cut our hair.) Bucky needed a short ¼-inch haircut. After the haircut he decided I needed a barber pole. He immediately went to his study and produced a red-and-yellow barber pole out of the venetian blinds. He then nailed it up on the porch of my study. Before I knew it I was in business.

At Black Mountain College each of us was given an 8-by-8-foot cubicle called a "study" in the studies building. We could paint it whatever color we chose.

For an evening's diversion the students and faculty drove a few miles to Peek's Tavern for beer and folk dancing. Bucky was fascinated by the folk dancers' dance patterns. He made diagrams of the dance steps. On the way back from Peek's in Albert's Model A, Elaine de Kooning, eating a candy bar, sang out loud, "Oh, I have eaten all the almonds out of my joy!" The townspeople surely must have thought that we were strange.

In 1948, as a twenty-two-year-old student, I did not see the significance of that summer. But looking back fifty-one years, it was a magical time: We were there to witness the first Bucky Fuller dome experiment, which didn't work on the first try. (He returned the following year with several solutions.) Merce Cunningham was thinking about forming his own dance company. John Cage was composing music for the prepared piano. Willem de Kooning and Josef Albers were exhibiting in New York to a cool reception, and Richard Lippold was working on new sculptures in wire rods.

It was that summer that helped me decide to take charge of my own life. It did not bother Fuller that I could not understand his books—"Darling, never mind." Just as it did not bother Max Dehn, mathematician and philosopher, that I slept through his first philosophy class. He said, "Let her sleep, she was up all night." When Albert asked him if he was disappointed that he had so few serious mathematics students, Dehn replied, "At my age"—seventy-seven—"I'm not interested in mathematicians."

Since the enrollment at Black Mountain was small, everyone had to do several things. Besides the job to provide fresh laundry every week, I took my weekly turn at dishwashing, setting tables, sweeping the dining-hall floor, picking up trash, and other jobs. We did our chores with some of the faculty, who took community service seriously. The faculty and students ate together. We had lively conversations. They analyzed problems and proposed solutions.

Bucky thought it was wonderful that Albert Lanier and I were getting married. I asked him to design our wedding ring. Mary Jo Slick, a student at the Institute of Design, made it from a black stone that Bucky had picked up on the beach of Lake Huron. (We married in 1949. In 1999 we celebrated our fiftieth anniversary with our six children and ten grandchildren.) His only advice to us was "The world is your oyster."

I have asked Albert to write about the dome project, since he worked with Bucky that summer, and this is how he remembers it:

"Fuller, the most optimistic, positive-thinking man I ever met, showed us a beautiful model of the frame for a dome. It must have been 3 or 4 feet in diameter. With his directing us we were going to lay out and erect the frame of a big dome maybe 44 feet or more in diameter. Very precise calculations would be made and, with these dimensions, we lay out the actual members, which would be made from the slightly curved aluminum used in the manufacture of venetian blinds. The measuring, color coding, punching of holes was very tedious and endless. Mr. Fuller did warn us that it might not work. (We did not believe him for a minute.) His definition of success was the point at which failure no longer occurred. We punched, coded, coiled the strips, and prepared for the most important performance of that performance-rich summer, the erection of the first geodesic dome. The day finally came—overcast, maybe a slight drizzle. We went to the field selected for the raising of the dome with all the parts and the few simple tools we would need for assembly. Every man, woman, and child at the college that summer gathered on a slight bluff to watch this miracle. We uncoiled the members, we bolted them together, we kept expecting the thing to rise. It didn't. It lay there in the field supine like a giant's plate of spaghetti. We students were devastated. But not Bucky. 'Next time,' he said, 'we must use stronger material or reduce the size.' And Elaine de Kooning affectionately named it the Supine Dome."

M. C. Richards translated the Erik Satie play *The Ruse of Medusa* from French. Bucky resisted playing the role of Baron of Medusa in the play. He had never acted before and was quite unhappy with the way it was going. Arthur Penn, then a student, stepped in to direct the play. He had Bucky jumping around the dining hall to loosen him up. They finally did two versions of the play. Bucky ad-libbed the ending of the second version. Elaine de Kooning played his daughter and Merce Cunningham played the mechanical monkey. Students Mary Phelan, Albert Lanier, Ray Johnson, Forrest Wright, and I made costumes for the performance; Willem de Kooning made the main set.

That summer I took Albers's "Color and Design" classes and Merce Cunningham's dance class, worked on props for the Erik Satie play, volunteered to be the college barber, watched the students working with Bucky's dome-project, and watched Bucky practicing his part as the Baron with student director Arthur Penn in the dining room.

We also tried Bucky's experiment about new sleeping patterns. He finally divided the day into four parts, six hours of work followed by thirty minutes of cat-napping, which added up to two hours of sleep per day.

After Albert and I were married in San Francisco in 1949, we saw Anne and Bucky nearly every time they came to the Bay Area or when Bucky was on his way to India to discuss India's transportation plan with President Indira Gandhi. While in San Francisco he would make time to visit a high school, a middle school, or an elementary school to teach the children how to make a great circle with twelve bobby pins and four paper plates and the DNA pattern by folding a piece of typing paper. He also gave advice on a student-built dome of wood and plastic.

When I asked him about youth boredom, he told me a story about a young Japanese protester who came to one of his lectures, Bucky told him, "Don't waste your time in the streets protesting. Learn as much as you can and become a useful citizen." This young man went on to develop many new ideas.

When I asked him about education, he said, "Create an environment so learning can take place." In 1968, a group of parents started a program in San Francisco's public schools to bring artists and children together and empower parents.

Josef Albers was a genius at bringing unlike minds together to work. Neither Albers nor Bucky was trying to produce artists. They were not interested in trends, movements, or what was popular. Their advice to us was simple: be truthful, be honest to yourself, don't be afraid of failure, study nature's way, and learn its principles. Perhaps that is a lesson from Bucky Fuller that I impart to students for the millennium.

It was the experience of that unique summer that set the course of my life as an artist, as a parent, and as an active member in our community.

What I Am Trying to Do

from *And It Came to Pass—Not to Stay*

Always excitedly acknowledging
The a priori infinite mystery
Implicitly revealed
As that which
Though relevant
Always remains
Undiscovered and unexplained
And is popularly overlooked altogether
In the momentary excitement of preoccupation
Only with that which is discovered
And especially with realizations
Of the new human advantaging significance
Of each great scientific discovery
Wherein for instance
Isaac Newton discovers
The rational *geometrical* rate of change
Characterizing the Interattractiveness
Of any two celestial bodies
While their relative distances apart
Vary only at an *arithmetical* rate
Which interattractiveness itself
Let alone its inversely varying
Second power rate of gain
Is neither manifest nor implied
By any of the constant physical characteristics
Of either of the interattracted celestial bodies
When either is considered only separately
And only in the unique terms
Of one or the other's
Integral dimensions/mass/chemistry

And independent electromagnetic properties
And though Newton's mathematically stated law
Of exponentially differing rates
Of inversely covarying
Mass-distance interattractiveness
Of independent celestial bodies
Was thereafter found by science
To be always unfailingly operative
In all macro or micro cosmic
Constellating
As well as in
The dynamic interpositioning
Of all individually remote bodies
Of all complex movements
Newton's discovered law
Never explains
Why
The interattractiveness
Of the remote-from-one-another bodies
Occurs or exists
Nor does it even suggest
What
The invisible interattractiveness
Is
Though whatever it is
Is so comprehensively embracing in importance
As apparently to guarantee
The eternal integrity
Of omni-complexedly
And everywhere ceaselessly
Intertransforming Scenario Universe
And even though we give that unexplained behavior
The name gravity
The name does not explain the mystery
For gravity
Like all scientifically generalized principles
Is inherently synergetic
And synergy means the unique behaviors manifest
Only by whole systems
Consisting at minimum
Of two independent variables

Whose unique system behaviors
Are entirely unpredicted
By any behaviors or characteristics
Of any of the system's components
When each is considered only separately
Wherefor it is ever experimentally demonstrable
That the unpredicted synergetic behaviors
Of strictly assessed system components
Constitute scientific manifest
Of the a priori mysterious
Cosmic integrity context
Within which major scientific discoveries occur
Which context itself always remains unexplained
By such discoveries as Newton's
Of the first power arithmetical
Vs. the second-power geometrical
Constantly intercovarying
Gravitational interattractiveness
Of separate bodies in Universe

And to the best
Of our experientially derived knowledge
Only humans' *minds* can discover
These nonsubstantial interrelationships
Existing only between
And not of or in
Any of the synergetically behaving system's
Plurality of independently orbiting components

For it is also experientially demonstrable
That in contradistinction to humans' minds
Humans' brains always and only
Inventory differentially
The succession of separate
Definitively sensed-in
Special case inputs and recalls
Characterizing each and every
Separate and terminalled experience
Whose separate special case data
Never contain integral physical clues
Explanatory of the synergetic behaviors

Of the omni-interaccommodative
Mathematically generalized principles
Always demonstrably governing
The complexedly overlapping episodes
Of Scenario Universe's
Nonsimultaneous
Multifrequenced and magnituded
Differentially covarienced
Ever intertransforming
Aberrationally limited complementations
And energetic transactions

And being also acutely aware
Of our own corporeal limitations
Yet ever renewably inspired
By personal rediscoveries
Of the cosmic integrity
Which only intuitively suggests
Imminent discovery of further synergetic potentials
Innate and gestating in Scenario Universe
First hints of which
Are realistically noted and sorted
Only within our subconsciously operative faculties
Which subconscious sortings it seems
Once in a while inadvertently produce
Synergetic relationship insights
News of the arrival of which
Are teleo-intuitively communicated
To our consciousness
Only as spontaneous urges
To look again in certain directions
Or to reconsider certain
Tentative concepts
And aware that those subconscious faculties
May be intuitively programmed
To search consciously
For a reliable and orderly means
Of inducing systematic conceptualizing
Of the significant import to humanity
Implicit in the synergetically gestating events
At the earliest possible system moment

Consciously apprehendable
By truth-coupled human minds
And also assuming that such conceptualizing
Can be directionally oriented
In respect to the environmental geometry
Of self's momentary circumstances
As the conceptualizing relates angularly
To self's head-to-toe observer's axis
And realizing that the dawning sense
Of significant import
Gestating in the subconscious
Which is suddenly emerging
As a discernible generalized principle
Together with spontaneous recall
Of other already proven principles
All of which may be frequency tuned
And discretely quantated
Into special case realizations of now
Thus to serve as a communicable means of entry
Into constructively competent participation by humans
In the strategic decision making
Concerning the special case
Formulative options
To be selectively and successfully realized
And employingly managed
Within the relevant set
Of generalized principles
Governing humanity's
Own evolutionary trending options
While also thereafter permitting humanity's
Responsibly followed-through
Anticipatory design science accomplishments
Of ever more effective and satisfying
Human life support artifacts
Ever increasing those artifacts' functional performances
Per each unit of resource reinvestments
As stated in physical measures
Of work or structure produced
Per each ounce of matter
Each second of time
And each erg of energy therein invested

And employing only
The unique and limited advantages
Inhering exclusively in those individuals
Who all by themselves
Take and maintain the economic initiative
By inventing artifacts
Developing their production prototypes
And proving the latters' capabilities
And safety of use
All conducted responsibly
On behalf of all humanity
As well as on behalf
Of the integrity of Universal regeneration itself
The individual initiatives being always undertaken
In the face of the formidable
Physical capital and credit advantages
Of the massive corporations, foundations
Trade and other special interest unions
And political states
All of whom seek entropically to turn
All such syntropically potential gains
Of Advantage for all
To their own exclusively special advantage

And deliberately avoiding
Political ties and tactics
As well as all negative activities
Social movements and reforms
While also concurrently endeavouring
By experiment, exploration and published data
To inductively excite
All individual Earthians'
Awareness of all humanity's
Syntropically multiplicative potentials
And techno-economically feasible options
As well as humanity's awareness
That only by use of such options
Can humanity effectively cope with
And divest itself
Of its entropically dissipative
Dilemmas and compromises

I *seek*

Through comprehensively anticipatory
Design Science
And its reductions
To physical practices
In the form of inanimate artifacts and services

To participate in nature's multi-optioned
Continuous and inexorable
Reforming of both the physical and metaphysical
Environmental events and circumstances
Instead of trying to reform
Human behaviors and opinions
Which latter is all
That history's political powers
Have ever sought to do

For I am intent
Exclusively through artifact inventions
To accomplish prototyped capabilities
Of providing ever more performance
With ever less resources
Whereby in turn
The wealth augmenting prospects for all humanity
Of such design regenerations
Will induce their spontaneous
And economically successful
Industrial proliferation
By world around
Exclusively *service*-oriented industries
As the regeneratively escalating effectiveness
Of the latters' responsible resource reinvestments
Per each unit of resources reinvested
Render comprehensively obsolete
Any and all economic necessity
To sell buy or own anything
While coincidently obsoleting as well
The economically degenerative practices
Of irresponsible one-way selling-off
Of the world's resources

All of which chain reactions will trend
To ever higher performance attainments
Of the ever improving artifact instrumented services
And thereby will swiftly
Both permit and induce
All humanity
To realize full lasting
Economic and physical success
Plus enjoyment of all Earth
Without one individual interfering with
Or being advantaged
At the expense of any other humans
Now alive or henceforth to be born
And I purpose
Through such responsibly evolving
Performance improvements
Of exclusively service oriented
Artifact instrumented industries
To accomplish universal economic success
Well being and freedom of humans
Together with a sustained abundance
For all foreseeable generations of humans to come
Of all the human life-support essentials
Thereby to eliminate all reason
For further existence
Of any and all varieties of politics
Each of whose ideologies mistakenly say
"We have the best and fairest way
Of dealing with the fundamental inadequacy
Of life support
On our under-resourced
And lethally over-populated
Planet Earth"

And with such design-science-attained
Sustainable abundance for all
Proven to be feasible
And attainable for all humanity by 1985
Will also come obsolescence
Of all the political powers'
Historically demonstrated
Ultimate recourse

To hot official
And cold guerrilla warfaring
And conscripted sacrifice
Of the lives
Of whole generations of youth
Upon the fallacious assumption
That war constitutes
The only means of proving
Which political system
Is the fittest to survive
In the misassumedly
Inherently lethal
Game of life
As we have been taught by history
That it must be played

And with such design science attainable
Elimination by obsolescence
Of all that humanity
Has learned by experience to be undesirable
Such as power corruptible
Politics and war
There will also come cessation
Of the history-long misassumed
Necessity for individuals
To *earn* their livings
That is to prove themselves to be
Extraordinarily valuable exceptions to the dictum
That humanity is meant to be a failure

And all the sovereign political states
Will also become obsolete
As will also all economic competition
For any special profit
For any state, corporation or individual
And the world will become preoccupied
With recirculating all its chemical elements
Which as they are totally recycled
At an average rate of every twenty-two years
Will be progressively reinvested
With the ever more effective
Interim gained "know how"

Thus making also obsolete
One-way dead ended wastes
And scrap monger withholdings
To escalate price structures

And the most important obsolescence making
Will be the elimination
Of humanity's most self-destructive weapon
The Lie
For Universe operates only on truth

And it is the most outstanding truth
Of this moment
That we have arrived at the threshold
Of new human functioning in Universe
Which functioning can only be effectively performed
With the prime struggle just to survive
Being completely disposed of
And with it all the debilitative fears
Attendant upon that long struggle
But midway of the threshold crossing
Comes the realization
That the only alternative to that success
Is the self-destruction of humanity

And whether we are to be
A complete success or utter failure
Is in such critical balance
That every smallest
Human test of integrity
Every smallest moment-to-moment decision
Tips the scales affirmatively or negatively
Wherefor we recognize that
It is both fear and ignorance
That delays popular comprehension
Of its historically unprecedented option
Of total human success

For fear secretly grips all wage and salary earners
As well as all political and private bureaucracies
Wherefor I realize that all humanity

Is plunging into an unprecedented state of revolution
Which if proving to be predominantly bloody and physical
Will probably terminate human occupancy of EARTH

Or if, predominantly, metaphysically coped with
As a comprehensive design revolution
Will probably insure the presence of Earthians
For millenniums to come

For the design revolution
To which I am committed
Is the metaphysical and positively objective counterpart
Of Mahatma Gandhi's
Subjectively passive resistance
And if design science wins
Each human henceforth will function
In predominantly metaphysical ways
In our cosmically designed role
As the most effective
Local Universe problem detector and solver
In the spontaneous support
Of the vast complex reciprocal scheme
Of celestial energy's
Increasingly disorderly and expanding
Local entropic stellar exportings
And their elsewhere concomitant
Increasingly orderly and contracting
Syntropic planetary importings
(As for instance here upon planet Earth)
This being the now dawningly realized
Cosmic function of Earthian humans
To solve locally evolving Universe problems
Both physical and metaphysical
Thus locally fortifying the integrity
Of eternally regenerative Universe
Whose cosmic omnicircumferential embracement
Integrates as gravity
Gravity being thereby
Always more vectorally effective
Than the sum of all local and nonsimultaneous
Radial disintegrations of radiation

Even as do a few barrel hoops
Successfully resist
The radially outward escape
Of the individual
Exclusively disintegrative
Crowded together
Individual staves of the barrel
As also myopically preoccupied
Ignorantly disintegrative humans
Are bound together upon our planet
By cosmically embracing benign laws
And necessities of Universe
Such as those guaranteeing the integrity
Of eternally regenerative Universe
Manifest in the invariable relationship
Of cosmically greater effectiveness of gravity
Than the disintegrative effectiveness
Of equally energetically endowed radiation
Which invariable superiority relationship
Is also manifest but ever more comprehensively
In the eternally greater
Cosmic-integrity-guaranteeing effectiveness
Of the always synergetic and syntropic
Metaphysical capabilities
Of comprehensively operative generalized intellect
Over the always energetic and entropic local
Physical capabilities
Which are exclusively operative
Only as local terminal special case episodes
All of which principles succinctly elucidate
The necessitous nature of Universe's grand design strategy
For installing developing and maintaining
The complex problem-solving humans
And evolving their complex ecological support
Upon our local Universe monitoring planet Earth
And it is only
By elimination of all self-deception that the highest probability
Of humanity's discovery
Of its metaphysical capabilities can occur
And do so within cosmic evolution's critical time limits
And because it is also true

That only by elimination of all this self-deception
Can this self-discovery be accomplished
Because the a priori otherness
Inherent in the fact of self's birth
And essential to the awareness
Of life itself
Wherefor comes full realization
Of the omni-interdependence
Of all humanity
Which if cultivated instead of frustrated
Will increase synergetically
To provide total physical success of all humanity
Truthful self-discovery being the initial step
Toward humanity's ultimately qualifying
For its unique local-Universe-monitoring functioning
We humans were designedly born
Naked ignorant and helpless
Endowed with the few conscious drives
Of hunger thirst procreation curiosity and fascination
To learn only by ourselves
And only through millions of years
Of often painful trial-and-error experiences
And lonely realizations
That our muscles and physical power
Are utterly subordinate
To our mind's Universe-embracing comprehensions
Wherefor in the trial-and-error struggle for survival
Lasting throughout the past
Millenniums of millenniums
Those exceptional individuals
Who have been fortunate enough
To be able to earn the right to live
Have had to do so
By proving that they had special capabilities
Which seemed functionally essential
To continuance of the successful heirs
Of those fortunate few
Who only by force of superior
Bodily size strength and cunning
Later augmented by arms
Had originally commandeered

The supposedly limited resources
And as yet have the superior weapons
With which to sustain those claims

However the spontaneous
Full life-support franchise
Which has always been accorded
By all humanity
To all new born children
Usually sustained in the past
Until the age of six
Has been progressively extended
In recent decades
First to include primary school years
After which came successively high school
And college years
Then post graduate and doctoral years
For those who wished them

And now with full life support becoming feasible
The scholarship franchise
Will soon become spontaneously extended
By revolutionary necessity
To cover all the living years
Of all humanity

And with life time fellowships for all
Will also come elimination
Of the defensive inferiority complexes
And the historical survival fears
Of all people of all circumstances
Which in the past
Have always frustrated
Humanity's urge to learn for itself
By direct experience and observation
How most effectively to comprehend phenomena
Instead of judging and guessing opinionatedly
As a consequence of credos and dogmas
Relayingly inculcated by others

And with all survival fears of all people eliminated
By universal FULL-LIFE-SPAN Fellowships

To study Universe
And seek understanding
Of the cosmic functions of humanity
Humans will swiftly develop their innate urge
To demonstrate competence
Thus qualifying themselves to participate
In producing services to all humanity
Wherefor the most spontaneous
And popularly sought for privilege of individuals
Will be to qualify for membership
On humanity's research and development teams
Or on its equally significant
Production and service teams
Just as humans now qualify
For participation in amateur athletics
Such activity having become
Completely divorced from
The concept of earning a living

And with the majority of humanity
Engaged in study and self disciplining
Direct experimental discovery
And familiarization with advantages to be gained
By understanding and use
Of the thus far discovered
Generalized scientific principles
Governing all Universe events
Will also come elimination
Of all wasteful daily travel
To the 90 percent of all jobs
Which are the non-wealth-producing
Specialized bureaucratic jobs
Which had been invented
By both private public and governmental
Power administrators
To keep people dependently obligated to them
As well as too busy to make trouble
For the power masters
Whose instincts said
"Divide to conquer
And to keep conquered
Keep divided"

This having always been the prime strategy
Of all the grand strategies
Of all history's power masters
And with such daily travel elimination
Will come vast reductions
In world energy resource consumption

The significance of which
Can be appreciated
When we recall (A) that science has standardized the quantifying
Of energy characteristics and behaviors
In terms of lifting a given weight
A given distance outwardly from Earth
In a given quantity of time
As for instance the work done
In lifting one pound one foot in one minute
I.e. in foot-pounds per minute
Or gram-centimeters per second

And (B) that it has also been discovered by geophysics
That each gallon of petroleum
Photosynthesized from Sun radiation
Into hydrocarbon molecules
Regeneratively proliferated by ecological organisms
Harvested and buryingly stored
As concentrated fossil residues
Within this Earth's crust
By wind water and gravitational power investment
Is so long drawn out a process
Requiring so much energy
That it costs nature
One million U.S.A. 1960 dollars worth of energy
As combined work pressure and heat
And chemical interexchanging quantities
Stated in scientifically defined constants
Operatively sustained
Over the requisite millenniums of time
Necessary to produce each gallon of petroleum
When that much energy
For that much time
Is priced at the same rate

At which electrical energy is charged to us today
By the public utility companies
Stated on our monthly bills
In money units for each kilowatt hour
Delivered to us of that much energy

Wherefor in Cosmic Costing
Of the regenerative affairs
Of Physical Universe
Wherein all metabolic interexchanging
Is meticulously accounted
To the last unit
Of electron-rest-mass value
Each average automobile commuter
Costs nature several million dollars each day
To go to his ecopolitical-system-invented job
For which work the economic system
Usually pays these right-to-live earners
Far less than one hundred dollars a day
While less than 10 percent of them
Produce any real wealth
Of direct human life support

Which means that a few humans
Who pooled together enough money
To buy enough machinery
To poke pipes deeply in the Earth
Are tapping Universe's progressive
Local energy accumulating
Which when attaining a critical mass magnitude
Some ten billions of years hence probably
Has been designed to inaugurate a new star
To replace others which have burned out

All of which energy accumulating
Cost Universe the one million dollars per each gallon
Which the human needs exploiters
Misassume to have cost nothing
Wherefor they take for themselves
What the money-game calls "pure" profit
Over and above the cost of their mechanical operation

And sell a million dollars cosmic-cost-gallon
For less than one cosmic-cost paper dollar
Which million-for-one is not difficult to do
And dwarfs the historical folly
Of the American Indians
Who sold the Island of Manhattan in New York
To the Europeans
For one bottle of whiskey

And with the popularly proven adequacy
Of life support for all humans
And with Universal recognition
That being born
Makes it mandatory upon world society
That each human is entitled
To healthful growthful lifelong support
Constructive and instructive travel
And access to the full inventory
Of information and knowledge and news
And all of the thus far accumulated wisdom
90 percent of humanity will spontaneously
Disengage from their specious "jobs"
And 90 percent of all the modern
Fireproof business buildings of the world
Will become vacated by the obsolete non-wealth-producing
Exclusively money or political-kudos-making businesses
Permitting those buildings' conversion
Into dwelling studio facilities
Having all contemporary conveniences
Already installed
Most of which conveniences are utterly lacking
Or are only inadequately provided
In the slum flats and squatteries
In which the majority of humanity now dwells

All of which foregoing changes
And their elimination of the survival fears
Exploitation of which always has underlain
All trade union management and political strategies
Will permit technological automation
To produce at 24 hours per day
Rather than for only eight hours

While computer automated travel facility booking
And video-automated matchings
Of all "availables" with all "wanteds"
Will be operative as a 24-hour information services continuum
Eliminating all newspapers'
Personal want and merchandise advertising
With vast savings of the newsprint forests
And with satellite sensoring
Of each and all humans' positive and negative
Individual electromagnetic field responses
To given problem solution propositions
Integratively read out
By computerized world monitors
And with world around Satellite relayed telemation
And universal cable TV feedback
And universal two-way TV
Beam and cable feedback
Humanity will come to know instantly
What all humanity's
Reactions and dispositions are
In respect to each and all problems
As they arise
And all humanity will dare to yield spontaneously
To the will of the majority
For if the majority
Makes a wrong judgment
The effect will be swiftly manifest
And the majority can immediately alter the course
Without any negative scapegoating

All of the foregoing and
Its many similar ramifications
When compounded with our now proven capability
To house and service all humanity
At higher standards than
We heretofore have ever experienced
Plus our organized knowledge
Of how to provide all humanity
With an annual energy advantage
Equivalent to the U.S.A.'s of 1972
And do so by 1985
While simultaneously phasing-out

All use of fossil fuels and atomic energy
And deriving our sustainable energy supply
Entirely from our annual energy income
Of Sun power and gravity effects
Ever operative around our planet
As wind water wave tidal
Sun methane alcohol
Eruptive and thermal sources
All of which foregoing
Design science revolution
Will phase out warfare
And release all of humanity's
Highest production technology
From killingry into livingry production
And a world around computerized
Integrated life support system
And with all the foregoing events
Humanity will be reoriented
From its one way entropic
Me-first energy wastings
To its syntropic circulatory
Synergetical you-and-we
Cosmic ecology regenerating functions

All of the foregoing will release
Our minds to perform their unique
Universe searching
Inventing capabilities
And information conserving
In support of which functioning
We humans alone
Amongst all known organisms
Were given conscious intellectual access
To the family of exclusively mathematically stateable
Metaphysical principles
Ever demonstrably governing
The cosmic integrity
Of eternal regeneration
And because the meaning of design
Is that all the parts are interconsiderately arranged
In respect to one another
And because all the generalized principles

Are omni-interaccommodative
Which is to say
That none ever contradict any others
The family of thus far scientifically discovered
Generalized principles constitutes a cosmic design
To which human mind has
The only known access
Other than that
Of the comprehensive
Absolutely mysterious
Intellectual integrity context
Of Universe itself.

In 1996, three chemists, Dr. Harold W. Kroto from Great Britain's University of Sussex and Dr. Robert F. Curl and Dr. Richard E. Smalley from Rice University in Houston, Texas, were awarded the Nobel Prize in Chemistry. They made their discovery by zapping graphite with a laser beam and mixing the resulting carbon vapors with a stream of helium. They found the molecules made up of sixty carbon atoms arranged like a soccer ball in a geodesic sphere that resembled R. Buckminster Fuller's well-known geodesic dome. At Kroto's suggestion, they graciously named their discovery, C_{60}, "buckminsterfullerene"—the "buckyball" for short. Their work opened an entirely new branch of chemistry and gave scientists a greater understanding of how nature bonds carbon atoms together.

Robert F. Curl wrote to me after notification of the trio's Nobel Prize, "I have never had the pleasure of meeting R. Buckminster Fuller in person, but no aware person could live for decades in the twentieth century without meeting him many times in his work."

Sir Harold W. Kroto, in addition to being a co-recipient of the 1996 Nobel, has received the Longstaff Medal of the Royal Society of Chemistry, the Hewlett Packard Europhysics Prize, and the Moët Hennessy/Louis Vuitton Science pour l'Art Prize. He has spoken at the Buckminster Fuller Institute in California.

Macro-, Micro-, and Nanoscale Engineering

Sir Harold W. Kroto

Meccano Magazine, in 1928, printed a letter from Mr. A. H. Finlay of Holyrood, County Down, in Ireland (Fig. 1) in which a truncated icosahedral cage constructed from Meccano pieces (Meccano is known as Erector Set in the United States) was depicted.[1] The legend suggested that the structure might be useful as a lampshade—but not much good as a football! A curious error,

---

[1]Incidentally, I lay the blame for the dearth of engineers and scientists in the West fairly and squarely on the fact that Lego has replaced Meccano as a "toy" of choice in many homes in the developed world. I use the word "toy" advisedly, because Lego is a toy, whereas Meccano is not only a toy but also an amazing introduction to the world of real engineering. It gives the child a real understanding of how nuts and bolts should be put together. They learn how to tighten a

as this shape has subsequently become the standard pattern for the construction of soccer balls. This letter also makes fascinating reading in light of the development of Buckminster Fuller's geodesic domes (Fig. 2), the discovery of C_{60}, buckminsterfullerene[2] (Fig. 3). Jonathan Hare—who while working with me extracted C_{60} (essentially simultaneously with the breakthrough of Kraetschmer, Lamb, Fostiropoulos, and Huffman[3] during some excitingly fraught weeks in September 1990—has pointed out that the world is about 100 million times larger than a soccer ball and a soccer ball is about 100 million times larger than a C_{60} molecule—i.e., the buckyball (Fig. 4). The well-known drawings of Haeckel of siliceous sea creatures, such as that of Aulonia hexagona (Fig. 5), which appear in the wonderful book *On Growth and Form* by D'Arcy Thompson,[4] led David Jones in 1967 to suggest that' carbon might be coerced into forming graphite balloons.[5] Furthermore in 1970, while watching his son playing with a soccer ball, Eiji Osawa[6] came to the equally

A MECCANO FOOTBALL?

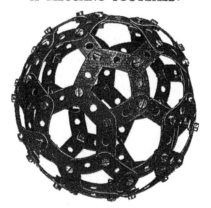

COURTESY OF
SIR HAROLD W. KROTO

Fig. 1. Page from *Meccano Magazine*, 1928

---

nut and bolt sufficiently tightly to not misthread it, and they get a feel for materials and learn how metal structures actually work. (An anecdote: A retired gentleman, who kindly came to display his amazing pantographs to young children at the 1999 BA festival in Sheffield, told me that he made a four-speed gearbox when he was eleven, and when he got his first job, at twelve, he was essentially already trained for it.)

[2]H. W. Kroto, J. R. Heath, S. C. O'Brien, R. F. Curl, R. E. Smalley, *Nature* (London) 1985, 318, 162–163.

[3]W. Kraetschmer, L. D. Lamb, K. Fostiropoulos, D. R. Huffman, *Nature* (London), 1990, 347, 354–358.

[4]D'A. W. Thompson, *On Growth and Form* (Cambridge: Cambridge University Press, 1942).

[5]D. E. H. Jones, *New Sci.* 32 (3 November 1966) 245; D. E. H. Jones, *The Inventions of Daedalus* (Oxford: Freeman, 1982) 118–119.

[6]E. Osawa, *Kagaku* (Kyoto) 1970, 25, 854–863 (in Japanese); *Chem. Abstr.* 1971, 74, 75698v.

Fig. 2. Photograph from Graphics 132 of Buckminster Fuller's geodesic dome at Montreal

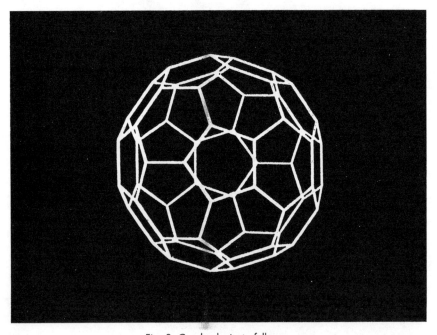

Fig. 3. C_{60}, buckminsterfullerene

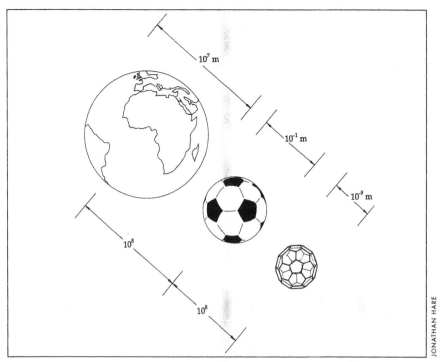

Fig. 4. Depiction of Hare's ratio—the last fundamental physical constant

Fig. 5. Drawing by Ernst Haeckel of Aulonia hexagona

imaginative conclusion that the carbon cage C_{60} molecule with the truncated soccer-ball structure that Bob Curl, Jim Heath, Yuan Liu, Sean O'Brien, Rick Smalley, and I discovered in 1985[7] would be stable—if it could be made. A year later, in 1971, Osawa, together with Yoshida,[8] wrote about this molecule in a book on aromaticity in chemistry. As the geodesic dome images such as that in Fig. 2 were an important clue to the likely structure of C_{60}, I suggested that we name the molecule after Buckminster Fuller, and after a little struggle they agreed. If we also add to the potpourri the studies by artists such as Leonardo da Vinci and Piero della Francesca (Fig. 6), we see that this amazingly charismatic framework has been a geometric force underlying the creativity of a wide range of people as well as numerous natural and physical processes.

As we enter the new millennium, it is worth throwing all these observations into a (think) tank, stirring them all up together, and ruminating (hopefully imaginatively?) about the possible implications (hopefully beneficial!—this was, after all, the pattern of the charges used to "detonate" the atomic bomb) on the fact that the icosahedral structure has now surfaced as a ubiquitous unifying geometric "engineering" principle in the nanoscale world. Of course, the C_{60} molecule probably first formed on Earth when the forest fires started—though in minute amounts. The molecule has, of course, been around since time immemorial, since carbon atoms (mass 12)—now in our bodies—were synthesized by the curious triple-alpha process in stars. In this process gravitational attraction caused three He nuclei (mass 4; $3 \times 4=12$) to fuse together in carbon stars, which then exploded as red giants blasting carbon out into space as atoms, polyyne chain molecules, carbonaceous sootlike dust, and almost certainly also C_{60} as well as other fullerenes. C_{60} is, however, probably not in high enough abundance to be definitely detectable at this time. The C atoms and molecules then wafted around in space to end up by some "Umberto Eco–type" chance on our Earth—some in our bodies, yours and mine, and unfortunately also in the bodies of some others whose atoms should, in my opinion, still be waiting about in space.

The term "engineering" tends to evoke the names of those giants of yesteryear who created the civil engineering structures that transformed our world—Isambard Kingdom Brunel and Richard Trevithick and James Watt and the Wright brothers. Today, however, the term embraces a far wider range and covers even more fascinating areas. It has expanded to cover architectural constructions such as the amazing Sagrada Familia of Antonio Gaudí, Frank Lloyd Wright's Marin County Civic Center, and Norman Foster's Hong

[7]Kroto, Heath, O'Brien, Curt, Smalley, op. cit.

[8]Z. Yoshida, E. Osawa, *Aromaticity* (Kyoto: Kagakudojin, 1971) 174–178 (in Japanese).

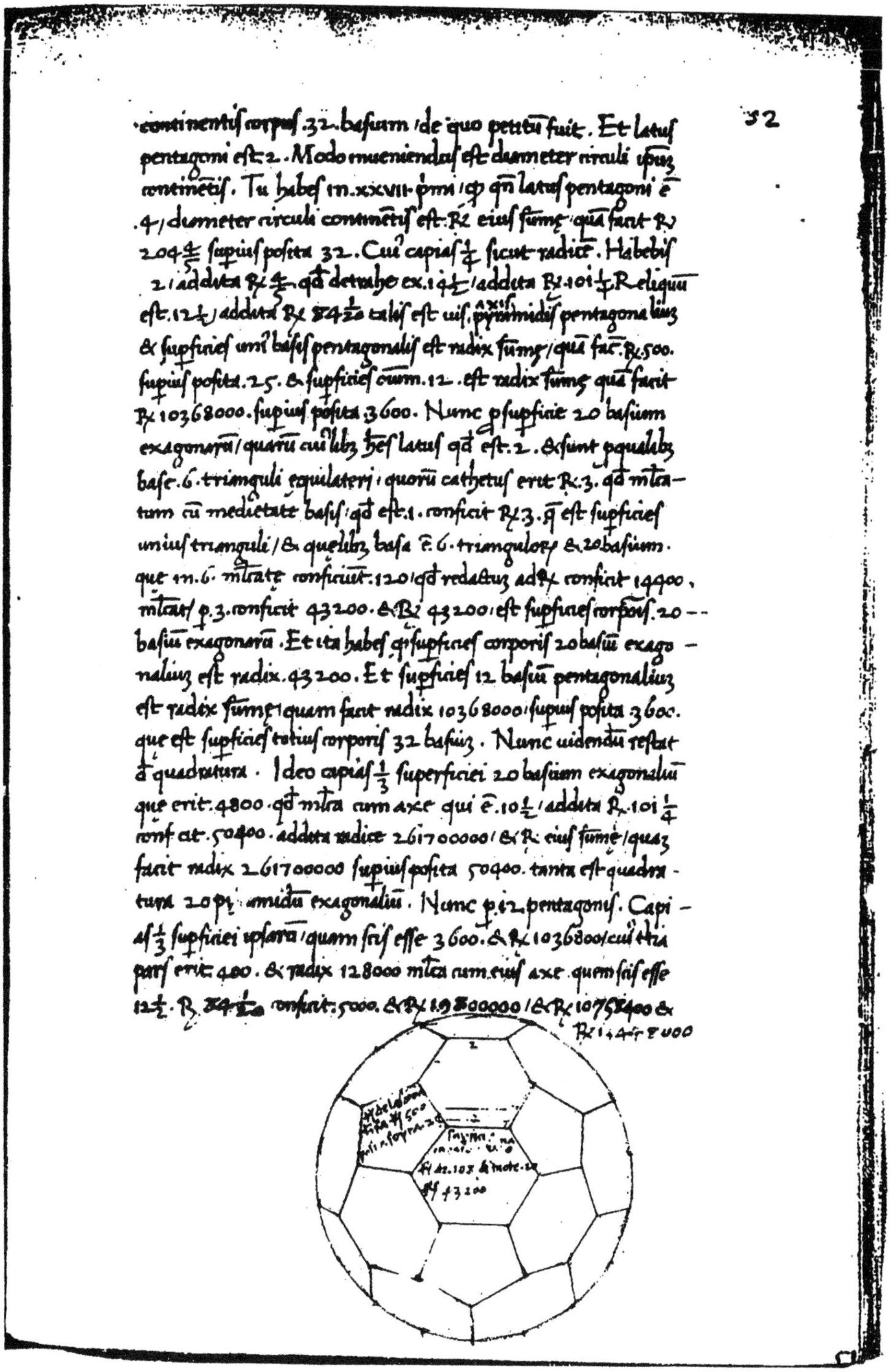

continentis corpus .32. basium /de quo petitu fuit . Et latus
pentagoni est .2 . Modo inueniendu est diameter circuli ipsu
continetis . Tu habes in .xxvii. pmi / cp qn latus pentagoni e
.4/ diameter circuli commetis est R eius sume / qua facit R
204⅘ supius posita 32 . Cui capias ½ sicut radice . Habebis
21 addita R⅘ qd detrahe ex .14½ / addita R 10i½ R Reliquu
est .12¼/ addita R 84½/ talis est vis / pyramidis pentagona lius
& supficies uni bases pentagonalis est radix sume / qua fac R 500.
supius posita .25 . & supficies ouum .12 . est radix sume qua facit
R 10368000. supius posita 3600 . Nunc psupficie 20 basium
exagonaru / quaru cui libz hes latus qd est .2 . & sunt equalibz
base .6. trianguli equilateri / quoru cathetus erit R.3. qd mta-
tum cu medietate basis qd est .1. conficit R.3. q est supficies
uniuis trianguli / & quelibz basa e.6. triangulorz & 20 basium .
que in.6. mltate conficiut .120 /qd redactus ad R conficit 14400 .
mltatu p.3. conficit 43200 . & R 43200 est supficies corporis .20 -
basiu exagonaru . Et ita habes qp supficies corporis 20 basiu exago -
naliuz est radix .43200 . Et supficies 12 basiu pentagonaliuz
est radix sume quem facit radix 10368000 supius posita 3600 .
que est supficies totius corporis 32 basiuz . Nunc uidendu restat
d quadratura . Ideo capias ⅓ superficiei 20 basium exagonaliu
que erit .4800 . qd mlta cum axe qui e .10¼ / addita R .10i¼
conficit .50400 . addita radice 2.61700000/ & R eius sume / quez
facit radix 2.61700000 supius posita 50400 . tanta est quadratura 20 pi amidu exagonaliu . Nunc p 12 pentagonis . Capias ⅓ supficiei ipsaru / quam scis esse 3600 . & R 10368000/cui tia
pars erit .400 . & radix 128000 mlta cum eius axe quem scis esse
12¼ R 84½/ conficit .5000 . & R 1.9800000 / & R 10752400 &
R 1.4 7 8000

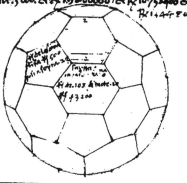

Fig. 6. Page from the work of Piero della Francesca

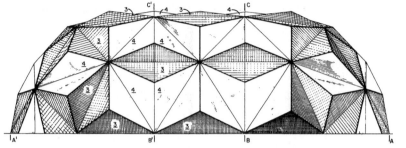

Fig. 7. A side elevational view of an elongated geodesic structure patented by Buckminster Fuller

Kong Shanghai Bank. Perhaps no other building combines the fundamentals of structural design and architecture more organically than does the geodesic dome of R. Buckminster Fuller. Perhaps it epitomizes Lloyd Wright's axiom that "form and function are one" even more purely than any of Lloyd Wright's creations. On top of all this we should not forget that, as we reach the end of the twentieth century, similar intellectual processes are involved in the creation of microchips—those tiny constructs that are revolutionizing our lives. Under the microscope they are seen to be elegant abstract mosaics, providing some of the most elegant art forms of the twentieth century.

For all its excitement, our discovery of C_{60} was most important because it showed that sheet materials on an atomic/molecular scale spontaneously form closed cages. Almost exactly five hundred years after 1492, we have come full circle and, just as Columbus explored a New "Round" World, we can now explore a new aspect of carbon chemistry only to discover a new round world of sheet materials. We should, however, have known about it much earlier—at least in the 1960s—as some clues were being uncovered by the combustion community, which had already recognized that some soots and carbon blacks consisted of graphene sheets organized in roughly concentric shells. What is more, some carbon fiber studies had revealed structures that were thought to be Swiss-role structures rather than the concentric carbon nanotubes (buckytubes) that we now recognize them to be today. These amazing new materials hark back to the patents of Buckminster Fuller[9] in which the elongated geodesic structures diagrammed (Fig. 7) are essentially macroscopic prototypes of C_{70} and the buckytubes first recognized by Sumio Iijima.[10] These structures are not just restricted to carbon, as other multiple sheet materials such as molybedenum sulphide—as shown by Tenne and coworkers[11]—also can form related structures (Fig. 8).

[9]R. B. Fuller, *Inventions—The Patented Works of Buckminster Fuller* (New York: St. Martin's Press, 1983).

[10]S. Iijima, *Nature* (London) 1991, 354, 58.

[11]R. Tenne, L. Margulis, M. Genut, G. Hodes, *Nature*, 1992, 360, 444.

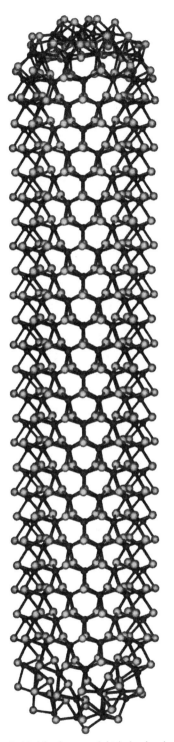

Fig. 8. Molybedenum sulphide buckytube

So now we have the tantalizing glimpse of a new material, which could revolutionize civil and electronic engineering much as steel and aluminum did in earlier times. The buckytube discovery promises a material some fifty to one hundred times stronger than steel at about one-sixth the weight, a material which could provide nanometer-diameter, low-resistance wires to interconnect the nanoscale components on the molecular computer chips that appear to be just over the horizon. However, to harness the exceptional tensile strengths as well as the electronic and magnetic properties of these molecular structures, advances in synthetic technology that rival the ingenuity breakthroughs of the past will be required. We need molecular engineering with a precision as yet well beyond us. It will require us to leap forward from the past approach in which chess-type logic was used to outmaneuver entropy and create new, ever more complex molecules. This was the board on which such wizards of synthetic chemistry as Robert Burns Woodward, Robert Robinson, and John Cornforth played and on which their heirs such as Jean Marie Lehn, K. T. Nicolau, and Fraser Stoddart are playing today.

Perhaps we should take a leaf out of the Natural World's book and attempt to create the nanoscale jigs that will allow us to hold in place the components of the nanoscale (pure carbon) Meccano components we have discovered, so that we can build truly complex, exactly specified giant molecules. To achieve its aims, "Life" has invented proteinaceous enzymes which are effectively gantries much like scaffolding we use to build the skyscrapers of today. A particularly delightful example, which exists in our bodies, is the amazing enzyme ATP synthase (Fig. 9) of Paul Boyer (UCLA) and John Walker (MRC Cambridge), who were awarded the 1997 Nobel Chemistry Prize for elucidating its structure and mode of action.[12] It seems, at least to me, to be a portent of fascinating chemistry of the future. The advance has similarities with the breakthrough made by Max Perutz, who discovered that deoxyhemoglobin (another favorite of mine) had developed the knack of readjusting its structure on oxygen uptake in order to facilitate further oxygen uptake (up to four molecules). This is really unusual, as in general one would expect oxygen uptake to get harder as successive molecules are added—as in the case of when one blows up a balloon which gets more and more difficult. Boyer and Walker found that there are tiny biological electric motors in our bodies which are essentially molecular machines consisting of a segmented pumpkin-shaped protein structure with an axial central shaft. As an asymmetric rotor in the shaft turns, driven by a flow of H ions through a disc at its base, it distorts cavities in the walls of three of the segments. The motion enables this amazing molecular machine to drive the addition of phosphate (P) to adenosine diphosphate (ADP) to form adenosine triphosphate (ATP) in the cavi-

[12]D. Voet, J. G. Voet, C. W. Pratt, *Fundamentals of Biochemistry* (New York, Wiley, 1999) 514–517.

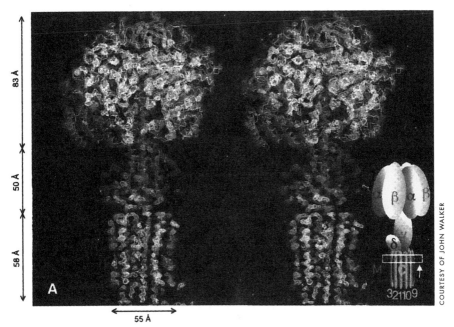

COURTESY OF JOHN WALKER

Fig. 9. The enzyme ATP synthase

ties.[13] In muscle tissue the ATP can revert back to ADP + P, and the energy stored in ATP is released, enabling us to move our arms and legs. What an amazing delight it is to discover that our muscles are powered by myriads of microscopic electric motors. We turn over at least half our body weight of ATP each day. The more we learn about nature, the more we seem to find that we are reinventing the wheel—or in this case the armature.

Perhaps a new age of chemistry—truly nanometer-scale molecular engineering—is on the horizon, an age in which we will start by modifying existing enzymes to make new molecules perfectly. New advances in genetic engineering will be required. Perhaps in the next stage we will learn how to build our own enzyme-like gantries—perhaps mainly out of carbon—and how to perfect a synthetic chemistry technology capable of rivaling the staggering efficiency of DNA replication (error rate of $1:10^9$). Any higher error rate than $1:10^9$ for DNA would be too high for us to exist for long enough! Such synthetic strategies seem a far cry from our present capabilities. Hopefully, however, now that we know what is achievable, the young geniuses of the new millennium will develop such techniques, spurred on by aesthetic visions of the novel molecular architecture which nature has created.

Acknowledgments: I wish to thank my colleagues Jonathon Hare, Humberto Terrones, and David Walton for their help with this article.

13 Voet, Voet, Pratt, op. cit.

Definition: Universe from *Synergetics 2: Further Explorations in the Geometry of Thinking*

304.00   Our definition of Universe provides for the undiscovered and for the yet-to-be discovered. Do not worry about that farthermost star which is yet to be consciously apprehended by any human being. Do not think we have not provided for those physical or chemical phenomena as yet not observed and recorded by human or mechanical sensing devices. The existence of such phenomena may not have even been postulated, but they can all be accommodated by our definition of Universe. Because we start with whole Universe we have left out nothing: There is no multiplication by amplification of, or addition to, eternally regenerative Universe; there is only multiplication by division. The farthermost star and the most unfamiliar physical phenomena are all accommodated by further arithmetical subdividing of our aggregate of overlapping experiences. Nothing could have been *left out* when you start with whole Universe. (See Secs. 522.32, 537.31, 540.03, and 1050.13.)

310.10   Odd Ball

310.11   In synergetics we find the difference of exactly one whole integer frequently manifest in our geometrical interrelationship explorations. Beyond the one additional proton and one additional electron that progressively characterize the hierarchy of the already-discovered family of 92 regenerative chemical elements and their short-lived transuranium manifestability by high-energy physics experiments, we find time and again a single integer to be associated with the positive-negative energetic pulsations in Universe. Because the energetic-synergetic relationships are usually generalized relationships independent of size, these single rational integer differentials are frequently found to characterize the limit magnitudes of asymmetric deviations from the zerophase vector equilibrium. (See Sec. *1043.*)

310.12   The minor aberrations of otherwise elegantly matching phenomena of nature, such as the microweight aberrations of the 92 regenerative chemical elements in respect to their atomic numbers, were not explained until isotopes and their neutrons were discovered a few decades ago. Such discoveries numerically elucidate the whole-integer rationalization of the unique isotopal system's structural-proclivity agglomeratings.

310.13   There is a phenomenon that we might describe as the eternal disquietude of the Odd Ball promulgating eternal reorderings, realignments, and inexorable transformings to accommodate the eternal regeneration integrity of intellectually differentiable Universe. This suggests philosophically that the individual metaphysical human viewpoint—the individual ego of the human—is indeed an essential function of the eternally regenerative integrity of complex law-governed Universe.

310.14   Possibly this mathematical Odd-Ball-oneness inherently regenerates the ever-reborn ego. Just when you think you are negative, you find you are positively so. This is the eternal wellspring of positive-negative regeneration of acceleratingly heating entropy and cooling-off syntropy, which is synergetically interoperative between the inherently terminal physical differentiating and the inherently eternal metaphysical integration.

311.10   Humans as Ultimate Complexities

•[*311.10–311.18   Complex Humans Scenario*]

311.11   Synergetics presents a picture of the multioptioned operational field of cosmic favorabilities, intertransformabilities, and complementary interaccommodations within which each human individual, his life, his world, is always one alternately elective, complex-integrity way Universe could have evolved.

311.12   No man or woman can ever prove when they wake up in the morning that they are the same person who they think they remember went to sleep. They may dream that they had other dreams, but there is now no way to prove that there are not a great many alternate "me's," and that one of them may have awakened under one set of alternate possibilities while the others may have awakened under other circumstances, and that each of them thinks that it is the only "me."

311.13   Let's assume that you have the realistically imaginative capability to invent a Universe wherein there are no substantive "things," a Universe wherein there are only events. Life is first of all awareness. No otherness: no awareness. Much otherness is difference-from-me-ness. Events are cognizable and re-cognizable only through awareness of the occurrence of interrelationship differences in sequentially observed conditions. Your Universe would have to be a Universe of ever-changing events, differentially adding here, subtracting there, multiplying, and dividing, reaching out or coming in, differentially including and integratingly refining, either locally gaining or losing, while continually and complexedly transforming at a plurality of rates and magnitudes. Within this Universe you also invent all the 92 regenerative chemical elements and their respectively unique, repulsive or associative behaviors compounding to form billions of unique substances while catalytically disassociating such substances to form all the chemical-element isotopal agglomeratings. You may go on to invent all the generalized principles of synergetic mass-interattractiveness and entropic radiation, as well as the latter's reflectivity and refractivity. And you will have to invent precession and all the unique frequencies of the chemical elements as well as each and all cyclic events. Then you start playing your game of Universe. You have all the stars and galaxies of stars entropically exporting energy and planets syntropically importing energies that, when they reach critical mass, become new stars as fragments of the exploded old stars become cores for the beginning of new planets. As you invent, you arrive at more and more intertransformative complexities and timing problems; the whole game gets more and more complicated. Quite clearly experiences must multiply, so the complexity of your invented Universe multiplies exponentially at a fourth-power rate in respect to the arithmetical progression of time. And since every event is always accomplished with six equieconomical, four-dimensionally directioned moves—the four dimensions being (1) inwardly, (2) outwardly, (3) circumferentially and equatorially around (axial spinning), and (4) polarized, involutionally-evolutionally, inside-outingly around—by means of the alternative, equimaximally economical optional move events you might find by the close of each event that you are six, five, four, three, or two radial zones outwardly in any direction from where you started—or right back where you started—and you soon come to the complex invention of thus-far-discovered billions of galaxies consisting of approximately 100 billion stars of the macrocosm and multibillions of invisible-to-naked-human-eye, microcosmic, associating and disassociating behavioral identities, altogether camouflaged beneath the blanket of make-believe "reality" of exclusively terrestrial politics and economic me-firstings, whose nonsensical preoccupation we call "today." It is doubtful that the million-light-year-distanced Andromeda nebula has any interest in Republicans or Democrats, communists or capitalists, or any other terrestrial partisanships.

311.14 So it could be that human beings, wherever they occur in Universe, may be introduced as a means of coping metaphysically with the most complex kinds of local Universe problems, so that each one of us is where the problem-solving of Universe is being transacted. If we were to think of ourselves as things—as china dolls, as kinds of china dolls that would just get smashed up or would just get worn or eroded away—that wouldn't be very good thinking. It would be much closer to actual Universe to think of ourselves as an absolutely continuous complex process. We are quite possibly the most complex of the problem-solving challenges of the invention that is eternally regenerative Scenario Universe. In this way each of us might be a department of the mind of what we might call god.

311.15   It is reasonable to suppose that there must be an overall physical-metaphysical cosmic accounting system that is always omniconsiderately integrative of all the a priori set of generalized interrelationship principles that we have found scientifically to be unfailingly operative in Universe. It may well be that each of us humans is an important function in sustaining the eternally regenerative integrity of Universe. The invention of the game of limited and terminal local awareness that we call "life" is in contradistinction to the concept of eternally total cosmic knowledge, intellect, and wisdom, whose totality of comprehensive comprehension would answeringly cancel out all questions and all problems, which would result in the eternally timeless, sublime 0=0 equation of absolute perfection.

311.16   Local life in Universe involves the invention of time. Time involves nonsimultaneity and limited information—ergo, only partial equatability. By introducing time and the myriad differential of interevent lags consequent to the exponentially multiplying, reinterpositioning distance variables inherent in the six alternative moves for each of the myriad of differentially frequenced, never-simultaneous events, the exponentially multiplying complexity of interevent lags accounts for both the micronuclear and macroastronomical progressive range of distance differentials of experience-limited, local-Universe human observing. Since each of the only-mathematically-statable, scientifically generalized laws of physical Universe constitutes a statement of truth, and since science has discovered a plurality of observable and ever redemonstrably operative truths—all of which are always omniinteraccommodative—it may be said that truth is complex. It is mathematically hypothesizable that all of the truths are potentially integratable and that the resulting integral truth constitutes the cosmic that humans intuitively sense to be in governance of Universe and speak of to one another with the inadequate sound-word god.

311.17   We can thus invent a hypothetical Universe with a limited game of individual human participation in only-locally-occurring, time-distance morphation awareness conceptualizations integratingly cognizant, as the sequential reality of human realization, of the complexedly interacting plurality of omniinteraccommodative cosmic laws that altogether enact events in pure principle, so reliably pure as to be sensorially apprehended by human brains and partially comprehended by human minds.

311.18   We may logically assume that the intellectual cosmic integrity of the timeless, sublime integral of absolute perfection must continually test its integrity of eternal regenerativity; wherefore the integral that we inadequately identify as god requires a plurality of local sensing monitors to be omni-deployed in the time-distance-differentiated, nonsimultaneously conceptual, serial Universe to continually observe the local Universe events, while also being progressively advantaged with an ever larger inventory of the discovered laws governing cosmic regeneration integrity, and thus equipped, to cope metaphysically with each of the profusion of unprecedentedly unique regenerative complexities that we speak of as problems. In effect god differentiates cosmic integrity into a time-distance-differentiated plurality of limited, local, metaphysical, intellectual experiences with which to test the capability of "god" to reintegrate and restore its timeless zerophase unity.

320.00   Scenario Universe

321.03   Humans have always tried to conform their concepts of Universe to their own human prototype. Humans stop and start; sleep and wake; are born, grow, decay, and die: so humans thought the Universe must also have a beginning and an end. Now astrophycists find this to be untrue. There could never have been a primordial chaos, simply because scientists now know that the proton and neutron always and only coexist in the most exquisite interorderliness. There could never have been a time when their integrity was not an integrity.

321.04   Universe is a scenario. Scenario Universe is the finite but nonunitarily conceptual aggregate of only partially overlapping and communicated experiences of humanity. Uni-verse is a momentarily glimpsed, special-case, systemic-episode takeout. When we start synergetically with wholes, we have to deal with the scenario within which we discover episodes—like the frog the snake is swallowing.

321.05   Time is only now. Time and size are always special-case, asymmetric episodes of *now* whose systemic aberrations are referenced to the cosmic hierarchy of primitive and symmetrical geometries through which they pulsate actively and passively but at which they never stop. The rest of Scenario Universe is shapeless: untuned-in. (See Sec. 982.62 for cosmic hierarchy and compare text at Sec. *1033.103*.)

325.10   Analogy of Rope-making and Film-strips

325.11   As seen by an individual human observer or as recorded by any humanly devised instruments, Scenario Universe is progressively reaggregated within the recorded, remembered, recalled, and progressively reconsidered information inventory of ever more macro-comprehensive-outward and micro-exquisite-inward ranges of the compositely growing individual's experiencing of life-in-time.

325.12 Scenario Universe is to any and all human observers very much like a rope-making experience—a rope that grows ever greater in total complexity but not in total diameter, and is comprised of ever more exquisitely diametered and ever stronger separate and differently lengthed fibers, a rope of which each of the myriad of progressive information events are in themselves terminal.

325.13 Each fiber enters into the scenario of rope-making by being twisted with others into a small thread of successively introduced and only partially overlapping fibers. This composited thread in turn is twisted with other threads into more complex strands. The strands are twisted with strands—always consistently clockwise or counterclockwise (never both)—until the totally twisted complex is brought together with a similarly twisted but turned-around and now oppositely directioned rope of equal complexity, whereat, when side by side, their respective tendencies to untwist interwhip them together to block one another's untwisting and produce an overall stabilized rope.

325.14 Employing the concept of individual fibers in this rope-making analogy and substituting for the word *fibers* the word *photons,* we can comprehend Einstein's curved-space assumption of the manner in which the omniremotely, entropically dispersed, individual energy increments, radiationally disassociated from former star sources at maximum remoteness from other entities, now progressively enter the gravitational neighborhood of radiationally disassociated energy increments—emanated from many sources—and become thereafter progressively reassociated with one another in forming new celestial aggregates, thereafter—as substantive matter—converging to a terminal complexity and density, thereafter once more to become radiantly dispersed.

325.15 The balancing of the gravitational and radiational exchanges is again analogous to the patterns of rope making. We simply splice together the ends of the stabilized overall ropes to produce a plurality of looped-back, cosmic rope-making in a Scenario Universe of nonsimultaneous, local, episode twistings and untwistings. While some loops are unraveling, their strayed-away strands got caught elsewhere in the new intertwinings.

325.16 The whole analogy of the rope-making and unmaking can be retransformed into the cinema concept, and the words *fibers* or *photons* being replaced by the word *atoms.* We can conceive now of all the separate atoms in the chemical compounds comprising the photo-negative celluloid ribbon of long-ago-exposed, financially exploited, stored-and-forgotten footage, the significance of whose once novel special case information has long since been incorporated in popularly accepted generalized viewpoints. The old film-strip has been chemically dissolved and its atoms disassociatively dispersed, migrated, and subsequently reassociated in a new inventory of on-the-shelf, unexposed, film-strip footage upon which may be recorded the ever-changing but progressively increasing inventory of comprehensive human experiences of tomorrow's today. In this analogy there is a plurality of fresh individual film-strips being nonsimultaneously and only overlappingly-in-time exposed for each individual observer in the Universe, as each is overlappingly intertwined into more complex information strands.

325.20 Epistemography of Scenario Universe

325.21 Synergetics always commences its considerations and explorations with finite but nonunitarily conceptual Scenario Universe, which is inherently nonsimultaneous. Scenario Universe is the totality of all humanity's consciously apprehended and communicated experiences. It is inherently prohibited for the totality of physical Universe to be quantitatively increased—ergo, in synergetics we have differentiation in the family of primitive generalized conceptual systems. Multiplication is always and only a special-case, relative time-size phenomenon—ergo, multiplication is accomplished always and only as special-case frequency of subdivision of the primitive epistemographic family of interassociations of primitive concepts.

325.22 All special-case multiplicity is considered and expressed in unique frequency, angle, vector, and topological quantation terms. The epistemography of synergetics discovers operationally, experientially, and experimentally that the most primitive of the conceptual systems to be divided or isolated from nonunitarily and nonsimultaneously conceptual Scenario Universe must inherently consist of the simplest minimum considerability none of whose components can exist independently of one another. The system components of minimum considerability are inherently recollectable only because they are experientially components of observable otherness. The words to describe inherencies of systems were invented by humans in spontaneous recognition of co-occurring observabilities. The word *part* could have been invented by humans only after having discovered a holistically considerable system.

325.23 Within the time frame of any one given observer *simultaneous* means the unitary consideration of a plurality of experientially observable, concurrently focal episodes with differently occurring births, deaths, and longevities—but they are overlappingly co-occurring with their omnidirectionally comprehensive environment. Omnicircumstance conditions are often forgotten in the recall of only the focal episodes of the considered systems' most prominent central memorabilities. Con-sideration—or the simultaneous co-reviewing of a plurality of stars—is always simultaneous. But simultaneous is not instantaneous. Apprehending and comprehending require a time lapse.

Simultaneous is episodal in Scenario Universe. Because our sight is only a light-wave frequently phenomenon, *instantaneous* cannot accommodate apprehension because instantaneity inherently lacks time span or wavelength.

326.00   Universe as Metaphysical and Physical

•[*326.00—326.50   Metaphysics Scenario*]

326.01   The support of life on our planet consists of two kinds—metaphysical and physical. Both cosmic and terrestrial energetic regeneration, organic and inorganic, are physical; while the know-what of pure science and the know-how of applied science are both metaphysical. The know-what of science's experimental evidence informs technology's know-how to employ efficiently the substantive resources and synergetic metaphysical patterns progressively found to be operative in Universe. These are essential to the maintenance of life on board our planet as well as in mounting local-Universe exploring excursions from our mother-spaceship Earth.

326.02   All that is physical is energetic. All that is metaphysical is synergetic.

326.03   All the energetic physical consists of two phases—(1) energy associative as matter, and (2) energy disassociative as radiation—with each being reconvertible into the other. All the synergetic metaphysical consists of two phases—(1) subjective information acquisition by pure science exploration, and (2) objectively employed information by applied science invention.

326.04   We can refine all the tools and energy capability of single and commonwealth into two main constituents—the physical and the metaphysical. The physical consists of specific, measurable energy quantities; the metaphysical consists of specifically demonstrable know-how capabilities. Only the metaphysical can designedly organize the physical, landscape-forming events to human advantage, and do so while also maintaining

(a) the regenerative integrity of the complex ecological-physiological support of human life aboard our planet, and
(b) the integrity of the chemical-element inventory of which our planet, its biosphere and co-orbiting hydraulic, atmospheric, ozonic radiation-shielding spheres, ionosphere, Van Allen belts, and other layerings all consist.

326.05   Only the physical is alterable; the metaphysical is unalterable. All the physical is continually intertransforming in orderly ways discoverable only by the weightless metaphysical mind. The local physical systems are everywhere energy exportive, which is humanly misinterpreted to be entropic and dissynchronously expansive only because the exported energies are electromagnetically (i.e., nonsubstantially) dispatched *only as information*—which is purely metaphysical—to be always eventually imported as information by electromagnetic reception in elsewhere-newborn, regenerative assemblages of cosmic systems. The local exportings appear to be dissynchronous only when viewed from too short a time span to permit the tuned-in occurrence of the next synchronous moment of the eons-apart frequencies often involved in celestial electromagnetics.*

326.06 We look at the stars, and they look very randomly scattered throughout the sky. But we can say that the number of direct and unique interrelationships between all the stars is always $\frac{N^2-N}{2}$. This equation demonstrates the principle of order underlying all superficially appearing disorderliness or randomness (see Sec. 227) and tells us quite clearly and simply that we are mathematically justified in assuming order to be always present despite the superficially appearing disorder. This gives us a personal sense of the order-discovering and -employing power of the weightless mind and at the same time a sense of our human anatomy's negligible magnitude in Universe when juxtaposed to the vast array of stars visible to the naked human eye. The stars observable on clear nights—the naked-eye-visible stars—constitute but a meager fraction of all the stars of our own "Milky Way" galaxy. Beyond this there is the 99.99 times greater array of the only-telescopically-visible 200 billion other already photographically identified galaxies, each consisting of an average of 100 billion stars, all coordinately operative within a 22-billion-light-year diameter sphere of Earth-planet-mounted instrumental observation and Earthians' photographic recording. This spherical-sweepout sphere of astronomical observation by minuscule humans on Earth describes a sphere of 123-plus-21-zeros-miles in diameter, all of which adds up at the present moment to each of the four billion humans on Earth having an exclusive personal quota of 25 billion stars or suns operatively available to their energetic regeneration, just awaiting his metaphysical know-how development, with the metaphysical assurance already established that the Universe is eternally regenerative. Energy crises are crises of ignorance induced by lethal self-interest.

326.07 This cosmic-accounting analysis discloses the omniuniversal orderliness that the scientist finds always to be underlying all of the only-ignorantly-apprehended illusions of randomness. This tells us that the seeming disorder of physical entropy is only superficial and explains why metaphysical thought

---

*Whether communication is by telephone hook-up or by wireless radio, what you and I transmit is only weightless metaphysical information. Metaphysical, information appreciative, you and I are not the telephones nor the wired or wireless means of the metaphysical information transmitting.

can always find the syntropic orderliness that cancels out all disorderliness. Disorderliness is nonthinking. Exclusively energetic brain, which stores the sensorial input data of all the special-case experiences, cannot find the synergetic interrelationships existing only *between* and never *in* any of the special-case systems considered only separately, any more than a library building in itself can find the unique interrelationships existing between the separate data that it houses. Only mind—the magnificent, weightless, metaphysical, pattern-seeking-and-apprehending function—has demonstrated the capability to intuitively apprehend and mathematically contrive the experimental means for identifying the significant, only-between-and-not-of, only mathematically expressible, eternal relationships, which altogether permit a seemingly inexhaustible variety of cosmic interrelationship patterning, and thereby human mind's capability to participate in the ceaseless complex of orderly intertransformings in such a manner as continually to abet the human mind's unique local-Universe functioning in maintenance of the cosmic integrity of eternal regeneration.

326.08 So long as humans progressively employ and develop their syntropic, energetic, metaphysical mind's capability to locally abet cosmic integrity, just so long will that metaphysical capability continue to operate in this particular local Universe's ecologically regenerative, planetary team aboard spaceship Earth. If, however, the entropic, energetically exploiting, antisynergetic, exclusively partisan profit motive—political or financial—continues to dominate and rule humans by force of arms, then the Earthian ecology team will become self-disqualifying for further continuance as a potentially effective local sustainer of cosmic-regeneration integrity. This is the net of what has now become metaphysically evidenced regarding the potential significance of humans aboard planet Earth. It is the nature of the contest between brain and mind. Brain is selfishly exclusive; mind is cosmically inclusive. Brain now commands the physical power to overwhelm humanity. But it is also the nature of mind's design science capability to render all humanity physically successful, thus eliminating human preoccupation with the struggle and thereby freeing all humanity to become metaphysically preoccupied with fulfilling its cosmic-regeneration functioning.

326.09 The physical Universe is an aggregate of frequencies. The human brain's senses are able to tune in directly only about one-millionth of the exclusively-mind-discovered and now experimentally proven and usable electromagnetic ranges of energetic reality. To this magnificent extent has the metaphysical brought humans into potential mastery of the physical. But thus far their physically most powerful political masters, misassuming ignorantly that there is fundamental inadequacy of life support, continue to exploit the fear that this induces in individuals, not for the individuals' sake but for concern and love for their dependents. Thus individuals concerned for the welfare of their dependents fearfully yield their economic-strategy mandates to the power-wielding brains of exclusively partisan ideologists.

326.10 Precession of Side Effects and Primary Effects

326.11 In recognizing the residual ignorance of variously dominant Earthian power-structure partisans of this particular cosmic-history moment, we do not impute malevolence or wrongdoing; we are only recognizing the checks-and-balances mechanisms of nonsimultaneously occurring and only partially overlapping Scenario Universe's vast variety of local conception-to-birth gestation rates. These gestations are the product of the associative energies' frequency synchronizations as well as of the concomitant rate of local disassociative energies, the dyings-off.

326.12 In the regenerative integrity of the cosmic design the locally supportive, human mind's understanding operative on planet Earth has to be midwife at the birth and experiential development of critical information inventorying; it must also sustain the total—Garden of Eden—ecological regenerativity as naturally accomplished by intense high-frequency information transmitted by the electromagnetics of chemical elements of the star Sun and receivingly translated by the Earth's land-borne vegetation and water-borne algae into photosynthetic sorting, reorganizing, and combining of the planet Earth's inventory of carbon, hydrogen, and other chemical elements. Some of these energies are redistributed to the biosphere, and some reenter directly into the integral metabolic multiplication and proliferation by all biological organisms of the hydrocarbon molecules and their concomitant environmentally supportive, chemical exchanging events. To function successfully in gales and storms while exposing adequate leafage to the Sun, the dry Earth-borne vegetation must send into the soil and rocks roots that draw water from the soil to cool themselves against dehydration by the heat of radiation exposure as well as to provide the vegetation with noncompressible, hydraulic structuring as well as with hydraulic distribution of the many eccentric, locally concentrated energy stressings of the vegetation's structure. Because the vegetation is rooted, it cannot reach the other vegetation to procreate. To solve this regenerative problem Universe inventively designed a vast variety of mobile creatures—such as birds, butterflies, worms and ants—to intertraffic and cross-pollinate the vast variety of vegetation involved in the biochemical refertilization complexities of ecology, as for instance does the honeybee buzz-enter the flowers to reach its honey while inadvertently cross-fertilizing the plants. Each biological specimen is structurally designed by the programmed codings of DNA-RNA and is chromosomically programmed to go directly to immediately rewarding targets while inadvertently, or unknowingly, producing (what are to it) "side effects" that inadvertently sustain the main objective of Universe, which is the sustenance of the synergetic circuitry of terrestrial ecology and thereby as well to sustain the cosmic regeneration.

326.13   Humans, like the honeybee, are born ignorant, preprogrammed with hunger, thirst, and respiratory drives to take in chemical elements in crystalline, liquid, and gaseous increments, as well as with procreativeness and parental-protectiveness drives. With their directly programmed drives humans inadvertently produce (what are to them) side effects, which results in their doing the right cosmic regenerative tasks for all the wrong reasons— or without any reason at all. This preliminary phase of preconditioned human reflexing, while lasting millions of years, is a gestative-phase behavior that becomes obsolete as humans' metaphysical mind discovers the principles of precession and discovers—only through vast, cumulative trial and error—the pattern experience of both terrestrial and cosmic ecology; whereafter humans will progressively recommit their endeavors in support of the recycling and orbitally regenerative effects, precessionally interproduced by all independently orbiting cosmic systems. This abrupt 90-degree reorientation constitutes the evolutionary stage through which humanity is now passing, wherein humanity will progressively exchange its exclusive preoccupation with self-preservation for that of supporting omniinclusive, cosmic integrity.

326.20   Pyramid of Generalizations

326.21   The physical Universe is characterized by local-system entropy, an ever-increasing, locally expansive randomness, and an ever-increasing diffusion, as all the different and nonsimultaneous transformations and reorientations occur. While the entrophy and disorderliness of physically exportive Universe increase and expand, we have the metaphysical Universe countering syntropically with energy importing, orderly sorting, and comprehensive contraction and storage of energy. In the metaphysically organized, syntropic-importing phase of cosmic-energy events we have the human mind digesting and sorting out all the special-case experiences and generalizing therefrom the persistent relationships existing *between* and not *in* the special characteristics of all the special cases. All the eternal, weightless principles apparently governing both the physical and metaphysical Universe are experimentally detected and digested into mathematical generalizations, of which only a very few are as yet known.

326.22   The whole process of generalizing generalizations forms a pyramid whose base consists of all the special cases of direct physical experiences. We can say, "We take a piece of rope and tense it," when we do not in fact have a rope in our hands. We have all had so many rope experiences that we can generalize our communication of the concept. This is a first-degree generalization. The discovery of always and only coexisting tension and compression is a second-degree generalization. Finding a whole family of always and only coexisting phenomena is a third-degree generalization, and conceiving therefrom "relativity" is a fourth-degree generalization.

326.23 In this pyramid of generalizations the human mind goes way beyond the biologicals in its development of the diminishing conceptual Universe—diminishing because, being progressively reduced and refined from a plurality of first-degree to a plurality of second-degree generalizations, and finally to as concise a form as $E=Mc^2$, which may some day be even more economically expressed as it is synergetically digested into ever more comprehensive and exquisite cosmic comprehension by metaphysically evolving human minds. So we find the metaphysical not only comprehending the physical—which should have been expected—but also encompassing and omni-accounting all the physical with the tetrahedron and thereafter reducing Universe's structural system myriadness to unity. The metaphysical, as with the circumferentially united, great-circle chord vectors of the vector equilibrium, masteringly coheres the physical by more effective use of the same quanta of energy. (See Sec. 440.08.)

326.24 Discovery of an acceptable hypothesis for explaining the role of humans as an essential metaphysical function of Universe, along with discovery of the mathematical proof of the tetrahedron as the true minimum limit structural system of Universe and its subdivision of Universe into macro- and microcosms as the sizeless conceptual system basis for generalizing general system theory, all develop as a consequence of our asking ourselves the question: How may we organize our self-disciplining on behalf of all humanity and in support of the integrity of eternal cosmic regeneration to deal comprehensively and capably with the maximum and minimum of limiting factors of the combined and complementary metaphysical and physical prime subdivisions of Universe?

326.25 The *theory of functions* holds for Universe itself. Universe consists at minimum of both the metaphysical and the physical. The inherent, uniquely differentiable, but constantly interproportional twoness of physical Universe was embraced in Einstein's one-word metaphysical concept, "relativity," and in a more specific and experimentally demonstrable way in the physicists' concept of complementarity.

326.30 Comprehensive Universe

326.31 Comprehensive Universe combines both the metaphysical Universe and the physical Universe. The local physical system is the one we experience sensorially: the conceptual metaphysical system is one we never experience physically but only consider in thought. As we discover in our grand synergetic strategy, we commence all problem-solving most advantageously at the supreme, terminally comprehensive level of Universe—that is, at the generalized-principle level. Thereafter we separate out from nonunitarily conceptual Scenario Universe one single, thinkable, experientially definable, holistic concept, which definition function inadvertently discloses the inherent polyhedral geometry of all conceptually sizeless thought referencing. We go on to find out through topological mathematics how to demonstrate that the difference between the cosmically total metaphysical and physical Universe and anyone experiential physical system, or any one conceptually thinkable system, is just one (positive and negative) tetrahedron, or one unity-of-twoness. This is to say that the difference between the finite (because an aggregate of finites) but nonunitarily conceptual total Scenario Universe (which we used to call infinity) and the physical Scenario Universe of energy with which physics deals, is just one finitely positive and one complementary, finitely negative tetrahedron. The exclusively conceptual metaphysical Universe is also proven to be a finite scenario because it too is an aggregate of locally terminal—ergo, finite—intellectual conceptioning experiences. The total Universe is just one (plus and minus) tetrahedron more than either the exclusively physical or exclusively metaphysical conceptualized Universe. (See Sec. 620.12.)

326.32   What man used to call infinite, I call finite, but nonunitarily conceptual—ergo, nonsimultaneously thinkable. What man used to call finite, I call definite, i.e., definable, conceptually—ergo, thinkably definable. The plus and minus, sizelessly conceivable, tetrahedral differences are all finitely and rationally calculable.

326.40   Metaphysical and Physical: Summary

Metaphysical cogitates reliably in respect to equilibrium.

Physical abhors equilibrium.

That which is communicated, i.e., understood, is *metaphysical.*

The means of communication is *physical.*

Metaphysical is unlimited and generalizable independent of time-space-sizing.

Physical is limited, experienceable, and is always special-case time-space-sized.

Metaphysical is unweighable, imponderable, and cannot move an electromagnetically or mass-attracted, levered needle.

Physical is always apprehensible by an instrument's needle leverage actuated by weight, pressure, heat, or electromagnetics.

Metaphysical discoveries clarify ever more comprehensively, inclusively, and are more economically communicable in their progressive description of the eternally changeless, and the rate of the more economic restatability continually accelerates.

Physical events are ever transforming and ever more acceleratingly entropic.

Humans are *metaphysical.* You and I are awareness, which is the identity of the weightless life.

The *physical,* automatedly rebuilding, information-gathering device we sensorially apprehend as our anatomy, employed by metaphysical mind, consists entirely of inanimate atoms and is therefore entirely physical.

Conceptioning is *metaphysical.*

Sensing is *physical,* and the sensed is physical and always inherently special-case.

Waves are *metaphysical* pattern integrities.

Metaphysical pattern integrity waves articulate as *physical* phenomena such as water, air, or electromagnetic fields, and thus communicate their nonsensorially apprehensible presence through the displacement of the sensibly detectable *physical* phenomena.

Symmetry is *metaphysical.*

Asymmetry is *physical.*

Equilibrium is metaphysical.

Disequilibrium is physical.

Conceptuality is *metaphysical* and weightless. Reality is metaphysically conceptualized information transmitted only through physical senses.

Realizations are special case *physical,* brain-sensed phenomena.

The *metaphysical* evolution of human awareness slows as it approaches the omniintegrated verity—physically unattainable by the only-physically-sensing, anatomical machine designed specifically only for limited operation with the highly specialized local Universe conditions of the Earthian biosphere.

Physical systems alone accelerate as they unravel entropically, ever approaching the speed of a noninterfered-with electromagentic, spherically expansive wave (186,000 mps.)

The generalized conceptions of *metaphysical* evolution tend ever to decelerate, simplify, consolidate, and ultimately unify.

The special case transformings of physical evolution tend ever to accelerate, differentiate, and multiply.

Generalization is *metaphysical:* mind function. Design is generalized metaphysical conceptioning.

Tools, artifacts, and all humanly contrived extracorporeal facilities designed by mind are always special-case, brain-processed, physically sensible phenomena.

Theoretical and applied science is conceptually *metaphysical.*

Applied science always results in special-case, brain-sensible, limited-longevity phenomena.

Metaphysically operative mind cannot design a generalized tool of unlimited and eternal capability, which unlimited and eternal capability is manifest only in purely metaphysical, weightless principles. Principles have no beginning-ending or other temporal limitations. Principles are truths.

We can invent only *physical*. The human mind, translating the generalized principles only through special-case, individual, brain-operated anatomical tools, can realize only phenomena of special-case, limited-capacity capability and durability.

326.50 Metaphysical and Physical: Cross-references

The design movement during the 1950s referred to as the Cranbrook experience often included both Cranbrook and the nearby University of Michigan College of Architecture and Design. Designers like the Eameses, the Saarinens, Harry Bertoia, and Carl Milles would interface with Fuller and form lifelong friendships. At the University of Michigan, young Saarinen, Yamasaki, Kahn, Birkerts, and occasionally Frank Lloyd Wright interacted with Bucky's ideas.

Bucky's assistant for the Dymaxion Deployment Unit, a converted grain bin called the Butler DDU was Walter Sanders, head of the Department of Architecture at the University of Michigan and later dean at the university. Here Bucky refined the 93-foot dome to enclose the Ford Motor Company's old Rotunda Building. Bucky's design was based on his octet truss, a 4-pound octahedral structure weighing only 2.5 pounds per square foot. It was an instant success. Recently, UM professor Joseph T. Lee donated a section of the octet truss to the Henry Ford Museum at Greenfield Village. Sanders, Oberdick, Eaton, Metcalf, Eero Saarinen's associate Glenn Paulsen (who would later become head of the Cranbrook Institute), and others would provide the basis of architectural studies at Michigan and stimulate Fuller's ideas. When Bucky was a visiting professor at Michigan, two of his students were Charles Correa and his wife, Monika, from Bombay, India. Other Michigan graduates—besides myself—included Charles Moore; Domino Pizza magnate Tom Monehan, a collector of Wright houses; and Raoul Wallenberg, later a Swedish diplomat and true humanitarian who would save the lives of thousands of Jews during World War II.

Charles Correa's essay on Bucky was originally written for a commemorative book that was not published. It is published here for the first time. Included also is Bucky's poem in his own hand, about his friend Edward Durell Stone and the architect's U.S. Embassy building in New Delhi, India.

Bucky

Charles Correa

In the early 1950s, when I was a student in the United States, Bucky was visiting professor at several architectural schools. I was lucky enough to be at two of them—first as an undergraduate at the University of Michigan and then a year later in the Master's class at MIT.

On both occasions, Bucky was a *tour de force*. He just knocked us out—with his ideas, his energy, and his superb and intuitive grasp of the forces that make the world around us. But perhaps the most extraordinary quality Bucky possessed—and one for which I think history will best remember him—was his ability to connect different things, different disciplines, different states of mind. "Only connect!" cried E. M. Forster. But it takes an intuitive and synergetic understanding of the forces that shape our environment, and our lives, to make this happen.

As visiting professor, Bucky would undertake a short project with the students, usually the design and construction of a specific structure. And as a background to this enterprise, Bucky gave wonderful lectures—at least one a day, each three to four hours long, traversing great stretches of time and space. These discourses provided the general theory from which Bucky could then launch the project he had in mind. At Michigan, I recall it was to design domes for the radar stations being built within the Arctic Circle. Bucky's idea was incredibly simple. As a young naval cadet, he had noticed that the air vents on ships were usually surmounted by small louvered globes, so that as the sea winds swirled the blades of the vents around, air was sucked up out of the innards of the ship: kitchens, boiler rooms, and so forth. What Bucky had conceived of for the Arctic Circle was not just his usual dome house, but a *revolving* dome house, with its surface made up of blades which could be angled to pick up the strong polar winds, and thus spin faster and faster until the whole dome itself became invisible. In short, one would be using the forces of Nature to make an invisible house.

What a brilliant idea! Can you imagine sitting in the viciously hostile environment of the Arctic Circle, warm as toast, with the comforts of life around, and with your domain stretching to the horizon in all directions, without any visible obstruction? Bucky went further. By adjusting the angle of the blades (as in an airplane propeller), the amount of air changes within the dome could be controlled—the ultimate Machine for Living.

Forthwith our class enthusiastically broke up into teams to make a full-scale working model—using an electric motor to propel aluminum blades attached around the perimeter of a bicycle wheel. Within a few days the working model was ready, and there was very great excitement indeed as the motor was started and the blades of the dome began to rotate. Since we couldn't test the contraption with real water, we threw uncooked rice toward the rotating blades, from the mezzanine level of our studio. Regardless of how fast we tried to get the blades to turn, just about every grain of rice went right through and hit the floor.

Was this a failure? No, strangely enough, Bucky had succeeded. Not in

The Mayflower

To: Tom Zung

Dymaxion Rating

Mepkin—
Conger Goodyear—
Museum of Modern Art
Edward ~~Darrell~~ Stone
Off to a flying start

Martinis,
Scotch and sodas,
Palava-- wartorn years
Edward Durrell Stone
Locked in his lower gears

Panama Canal Zone
Brussells
Beirut Hotel
Edward Durrell Stone
Doing very well

Rood-screened fenestration
New Delhi's
Sculptured lace
Edward Durrell Stone
Setting new era pace

Dignity
For Common Man
In Raleigh's white-green grace.
Edward Durrell Stone
In World Architects' First Place.

Buckminster Fuller

constructing a working model of his idea—obviously the crude technical means at our disposal ruled that out. Where he had won was in making you realize that such a concept was within the domain of the architect—and with that, he opened a door in your mind. It is an experience I have never forgotten.

Years later, after I had returned to Bombay, Bucky often visited India, sometimes alone, sometimes with family and friends. On each of these occasions it was a joy to meet him. For like a true student, Bucky had always found something new which he would share, and the deeply thoughtful and optimistic aspects of his nature continued to illuminate our lives. In this respect, I do not know why he bothered to call himself an architect—he encompassed an infinitely richer world than that meager label implies. But to the extent that Bucky was part of the architectural scene, I think he brought to our profession something invaluable: his work made one realize that the next important vector in the Modern Movement was not the silly conceits (strip windows and white walls) of the so-called International Style, but the *right to invent*. The right, as in Bucky's own ineffable phrase, to rearrange the scenery. The very same mandate by which the great nineteenth-century engineers, from George Stephenson or Isadore Brunel on, had started a process which produced some of the most profoundly moving images we know, far more potent than any Beaux Arts architect has ever accomplished: the suspension bridge high above the harbor, the railway rain streaking across the countryside, the ocean liner sailing out of port, the airplane coming in to land.

Like those great forerunners who preferred to create extraordinarily beautiful objects (rather than quibble over questions of style), Bucky could produce something as elegant as his Dymaxion cars—without bothering to mention their looks. Indeed, his life and work remind us all that the Modern Movement in architecture was much more than just a style—at its very best, it underscored perhaps the most fundamental human need of all: the need to rearrange the scenery. And in the process, invent a new future.

Part XXIX
Vital Statistics of Industrialization from *Untitled Epic Poem on the History of Industrialization*

Industrialization may be
delineated
in two important ways.
The first, most obvious
and most frequently employed,
is by **per capita annual** production
in **tons** or **dollars.**
The second is by dynamic measurement.

The first method shows the per capita resources,
rates of production,
and a representative sample
for instance
of manufactured articles of the U.S.A.
compared with the rest of the world.

But this first method constitutes
a **static** measurement of the economy.
Even when it shows growth
over a period of years,
it does not show the **reasons** for growth
or the **meaning** of growth.
For these we must turn to dynamic measurements.
In dynamic measurement
matter is combined with energy
by disciplined brainpower.

Dynamic measurement reveals
for instance
the entire trend of U.S.A. industry:—
which is

TO PRODUCE MORE WITH LESS—
for the SATISFACTION OF ALL MEN.

This trend appears first
in terms of what might be called
perishables.

Industrially speaking,
coal is a perishable,
inasmuch as it is **consumed** by industry
for the **release** of energy;
and the energy that can be released
from a ton of coal
has increased threefold
in the last twenty years alone.

Therefore, **statically** speaking,
a piece of coal is just a piece of coal;
but, dynamically speaking,
a piece of coal may become "three pieces"
in respect to performance
and in respect to man's derived satisfaction.

And if that piece
which has become three pieces
in potential power release,—
is not used until necessary,
it may through new science gains
become four pieces, dynamically.
in potential power release,—
Was this the meaning
in the parable of the
loaves and fishes?

This trend is even observable in food
which is also consumed
for the useful release of energy
to operate the "integral mechanisms" of man
which in turn **relay**
the initial brain operation
to the "external mechanism"

for the amplification of force
or increase in precision
of ultimate work to be done.

U.S.A. man now eats recordably less,
having less requirement of energy
as physical **work**
and less requirement of energy
as integral **heat**
to maintain the preferential
thermal conditions
to the best functioning
of the elaborate processes
of his integral mechanics,
in his now atmospherically better
mechanical extended and
controlled environment,
yet he jumps higher and runs faster
than his father
and circles the world in days
against his father's year.

The trend is also visible
for instance in the number of miles
yielded per pound of tire rubber,
increased fivefold in the last twenty years.
And one of the most fascinating manifestations
is to be found
within the oil industry
where the viscosity index of lubricating oil
remained at 100,
(that of the best natural Pennsylvania oils
misassumed to be the best
that would ever be known)
until 1930—
and has since shot nearly up to 200
by synthetic molecular design of researching man.
But the viscosity index
has also to do,
among other things,
with the life, pressure,

and temperature limitations
of the moving parts
that are being lubricated;
also of course with the life of the oil.
Thus it is seen that the dynamic gains
compound in their ramifications.

Another way to observe the dynamic trend
is in the constant reduction
of manufacturing **cost** per ton of bituminous coal;
cost per pound of electrolytic copper;
dollars per auto horsepower.
Or, to speak of cost in more human terms,
in constantly reducing
automobile, railroad, and airplane deaths
per passenger mile traveled,
despite both increased speed
and increasing population
of road and rail cars and planes
as well as of people;
by decreasing deaths per capita
caused by infectious diseases
or by industrial accidents;
and sum totally,
with **increasing** life expectancy
at all the various ages
of the human beings in the U.S.A.

The high point of the charts
illustrating this trend
shows graphically
what the American standard of living really is.

This is done by showing
how much more than a subsistence
is the advancing standard of living,
that is by charting
what the U.S.A. man can buy
over and above the absolute necessities
for a given amount of work hours
rendered to industry by him.

This will be seen to be
more than favorably comparable
with such excesses of other countries.

A second way in which industry
delivers more for less
is graphically illustrated in the economic behavior
of what might be called the "non-perishables,"
chiefly the industrial metals,
but also the organic materials,
such as fiber, rubber.

The central factor here
is **scrap** material
which is salvaged, separated,
and used over and over again.

The result of this recirculation is
that as time goes on
less **new** raw material is required
to make a given weight of product.

Scrap in this sense is not "shoddy"
for the pure chemical elements **iron** or **copper**
may be refined-out endlessly to ingot
from discarded use forms.
In fact steel scrap becomes
progressively enriched
to higher ultimate strength
with minor alloy constituents
not worth separating out.

Scrap has been a special tactical factor
in recent war economy
though of great economic importance
throughout all old industrial countries.

In the U.S.A.
for certain reasons other than war
scrap is highly important.

Here copper, for instance,
seems almost to have reached
what might be called a "scrap-point";
that is, the scrap supply
is almost sufficient to meet
all present domestic non-war needs,
and very little copper has, therefore,
to be mined for domestic use.
Also about 90 per cent
of U.S.A. domestic consumption of steel
is derived from scrap.
There is enough tin
in the U.S.A. recirculating cycle
to constitute a larger ore body than that in the
Malay Straits Settlement
whence it mostly originated.

Of course, practically speaking,
scrap both as "new"
and as combined old-and-new metal
is both domestically consumed and exported
so that these statements are only true
in the terms of a net fact accounting.

The most interesting scrap situations are
as already noted
those of copper and steel,
(though scrap is playing an important role
in all **non**-perishable categories),
for one reason among many,
that it renders basic monopoly impossible.

Scrap also places economic emphasis
upon business turnover
rather than upon **ownership.**

Thirdly,
industry delivers **more** for **less**
by causing a given amount
of non-perishable material
to do more work
or to be more effective.

Sometimes this is achieved
by combining it more efficiently with energy;
thus the average horsepower
and cruising speed of automobiles
have been constantly increased,
without adding weight.

In the U.S.A. 1920–1940
there has been a vast increase
in the proportion of energy consumption
to that of tonnage consumption
of industrial raw materials.

While one may find it difficult
to "see" energy as constituent in products—
in a pencil, a phonograph record—
the energy component is nevertheless there
and grows ever larger
by numerical proportion
to the tangible components.

Sometimes, if not always consciously
more for less is effected
by a better combination of materials
through exercise of brainpower.

Thus by **alloying** or synthesizing elements
under varying conditions,
new **stronger** alloys are evolved
whereby the same basic materials
acquire a strength
many times that indicated
by their isolated properties;
for instance, two per cent of beryllium
alloyed to copper
and thereafter jointly heat treated
jumps copper's tensile strength tenfold
and provides a non-fatiguing metal,
which, however, may happily
be worked before heat treatment
with all the facility
of pure annealed copper.

Thus also by **mental ingenuity**
a two-hundredfold increase
in messages per unit
of telephone wires
is effected.

Mavericks seem to be naturally attracted to each other. Bucky loved them, for they seemed to be more comfortable thinking outside the "rectangle." The "hip" Malcolm Forbes had four sons—Steve, CEO and publisher; Christopher (Kip), interested in the arts; Timothy, a producer of films; and Robert, all members of Forbes, Inc.—and a daughter, Moira. All are creative mavericks. Forbes demonstrates one piece of Fuller's advice to the young and others who would listen: "Do things that need doing." Regardless of one's personal political affiliation, one must respect Steve's tremendous efforts for what he believes. He has made some statements that will be long remembered in American political history. I commend him for his courage and tenacity in pursuing his goals as a presidential candidate.

Bucky Fuller, always apolitical, once remarked that if all politicians were jettisoned into space, there would be no appreciable effect on mankind, but if all scientists were lost, humanity would be in serious trouble. However, Fuller acknowledged that politics is necessary, akin to catching the common cold. In 1970 in an interview with *Playboy* magazine, Bucky is quoted as saying: "Politics exists and will always exist . . . but it must always be as accessory after the fact . . . an accessory after the fact of whatever the circumstances may be. . . . So I can see that we can get the political mood, the mood of man, to simply demand that their political parties on both sides merely yield in a direction that neither of them ever thought of before. The one will not yield to the other man's policy at all. He'll be yielding to the computer. And I find that anybody can yield to the computer. He can't lose face yielding to a machine." Bucky was always suspicious of "big government," so I believe he would have smiled at some of Steve's ideas. Here, Steve Forbes's contribution describes Bucky's vision of the future.

The Future According to Fuller

Steve Forbes

Where are America and the world headed in the new millennium? What are the obstacles and barriers that lie in our path, and how do we lead the world into a dazzling new era of economic prosperity, technological advancement, and individual opportunity?

These are the questions that captivated R. Buckminster "Bucky" Fuller for his eighty-seven years on Earth. The answers he discovered brought him twenty-eight patents, forty-seven honorary doctorates, the cover of *Time* magazine, and the presidential Medal of Freedom, as well as the interest of people all over the world, including me. His work is a treasure trove of original thinking, and any young person who discovers it is in for a treat.

Every young person wants to be inspired by someone who has a sense of the future. When I was growing up, my father seemed larger than life—a maverick businessman with a zest for life and an unquenchable sense of optimism about the future, his own and America's. I remember in the early 1960s, when Communist China's Chairman Mao began railing against the West as a bunch of "capitalist tools." Pop loved it. "That's exactly what we are—capitalist tools!" he would say. With a savvy sense of marketing, he seized Mao's line and made it his own. He made it the motto of *Forbes*. He began fashioning himself and the magazine as international ambassadors of free markets and free elections in a cosmic East-West battle for the future.

No wonder my father admired Fuller. Not your typical scholar or entrepreneur, Fuller was a brilliant thinker, a man constantly thinking about tomorrow—how to design, shape, and make the most of it.

America's first great futurist, Fuller coined the term "Spaceship Earth." He was convinced, even in the early part of the twentieth century, that life in the twenty-first century would revolve around a new global village in which breakthrough advances in transportation, communication, and business would make the world seem smaller and would dramatically expand individual opportunities for men and women everywhere.

In this sense, Fuller was light-years ahead of his time. He lived during the peak years of the Industrial Age, an age of bigness—big factories, big companies, big unions, big cities, and big government—an age in which most men and women were cogs in the giant corporate wheel. Yet Fuller envisioned the dawn of an Information Age, an age of individual freedom, choice, and opportunity.

An architect passionate about design issues, Fuller was arguably a rival of another great American architect and design futurist, Frank Lloyd Wright. Indeed, Wright once wrote to Fuller, "You are the most sensible man in New York, truly sensitive. [But] to say you are the most sensible man in New York isn't saying much for you, in that pack of caged fools."

Fuller's ambition was for buildings of the future to capture his sense of imagination and innovation. Why just pile brick upon brick or stone upon stone to construct a building? he wondered. Sure, people had done this for centuries. But was there not another way, perhaps a better way? Could not homes and office buildings be made in a way that would enclose far more space with far fewer materials at a far lower cost?

Such questions led Fuller in 1948 to invent the geodesic dome, accurately described by supporters as the lightest, strongest, most cost-effective structure ever devised. You have probably seen these spectacular creations without even realizing it. The U.S. Pavilion at the 1967 Montreal World's Fair showcased one of Fuller's geodesic domes. But perhaps the best-known is the dome that contains a marvelous ride into the future at Disney's Epcot Center in Orlando, Florida. Today there are more than 300,000 geodesic domes in use around the globe, and Fuller's design has even created the standard for American military radar domes, because their unique design allows them to withstand driving winds of up to 180 miles per hour.

A theorist, Fuller pioneered the concept of "synergetics" that has revolutionized the world of business. "Synergy" is a term that seems commonplace to us now. But it was Fuller who popularized it. He envisioned an interactive world where seemingly disparate disciplines would complement and reinforce one another, a world where emerging new technologies would allow people to do more with far less.

"For the first time in history, it is now possible to take care of everybody at a higher standard of living than any have ever known," Fuller said in 1980. "Only ten years ago the 'more with less' technology reached the point where this could be done. All humanity now has the option of becoming enduringly successful."

A writer and poet, Fuller penned twenty-three books before his death in 1983. In many ways, Fuller was the forerunner of more recent American futurists and trend-trackers such as Tom Peters *(In Search of Excellence),* John Naisbitt *(Megatrends),* Alvin Toffler *(Future Shock* and *The Third Wave),* and Arthur Clarke *(2001: A Space Odyssey).*

"He felt he was surfing history in a sense, trying to anticipate trends," his daughter, Allegra Fuller Snyder, told the *New York Times.* "Now some of those waves have come in. There's also a different kind of chemistry today, especially among young people, and that opens doors to his thinking once again."

Throughout his life, Fuller personified America's can-do philosophy. When his first geodesic dome failed, when he left Harvard (twice), when he experienced the devastating loss of a first daughter, Alexandra, he refused to give up on his dreams. He refused to give up on his passionate belief that tomorrow could be better than today if people applied themselves to creative, innovative, out-of-the-box-style thinking.

The measure of a man lies not in the degree of his perfection, but rather in his effort, in his vision, and in his work. This is key to understanding Fuller and his many contributions to human advancement. Fuller's ideas were alternately delightful, brilliant, implausible, misguided, and wacky. But they were always pushing the boundaries of conventional wisdom, and that in itself is immeasurably valuable.

What has happened to the lost art of dreaming? Where are the great thinkers of our time? Now that we have arrived in the twenty-first century— now that our 2001 space odyssey has begun—who is imagining life in 3001 and beyond? Are we raising young people to think and dream, to innovate and invent? In politics I fear we are becoming intimidated by positive change and innovation. Heaven forbid that such paralysis permeate the world of business, science, communications, architecture, art, literature, music, and beyond. And this is why we must be vigilant against the intrusion and depersonalizing effect of Big Government. For true innovation can only occur in an environment defined by personal freedom, and Big Government is the greatest threat to our personal freedom and the American spirit of enterprise.

One of the things Fuller missed—but would have loved, I think—was the birth and spectacular rise of the Internet here in America and around the globe. Think about it. The World Wide Web has become a revolutionary vehicle for the explosion of commerce and communication. It has created a world where individuals can prosper unimpeded (thus far) by the crushing taxes and regulation of government. A single mother can put her children to bed at night, and then using the Internet can pay her bills and balance her checkbook on-line, read up on the news she missed during the day, and send a note to her sister halfway around the country. On the Internet, a retired couple can now set up a used book business, take credit card orders on-line, use sophisticated databases to find out-of-print titles, talk to dealers around the world, and track their FedEx shipments—all from the comfort and security of their own home.

The Internet quite literally puts the world at students' fingertips. Knowledge is power. Wealth comes from within, from human capital like imagination and innovation. And new wealth is created all the time in a free market. At its best, learning is supposed to be about stirring the imagination and encouraging creativity. The new classroom is technology, and technology is an invaluable tool for gaining knowledge, not an excuse for wallowing in ignorance.

You and I are entering a new world—a world Buckminster Fuller would have loved—a world he actually foresaw long, long ago. Are we up to the challenge?

Buckminster Fuller

Can't Fool Cosmic Computer from *Grunch of Giants*

It must be remembered that, as clearly elucidated in *Critical Path* and *Operating Manual for Spaceship Earth*,* money is not wealth and that wealth is the organized technological capability to protect, nurture, educate, and accommodate the forward days of humans, whereas money is only a medium of exchange and a cash accounting system. Money has become completely monopolized by the supranational-corporation colossi, which inherently as legal abstractions ignore the problem of how to protect and nurture human lives.

In the very largest way of looking at planet Earth's socioeconomic-evolution events, we must observe that humans are designed with legs and not roots. Yesterday, humanity developed temporary roots as it cultivated its life-support food root-grown on the land. The metals made possible metal canning of food and mobilization of machinery. Today, all of human existence depends on the swift, world-around intercommunication system operating at 186,000 miles per second. We have transformed reality from Newton's "at rest" norm to an Einstein's 186,000-miles-per-second norm. Socioeconomically we have synchronized with the omni-intertransformative kinetics of the entire Universe.

Planetary economics has now shifted from a physical-land-and-metals capitalism to a strictly metaphysical, omniplanetary, omnicosmic-wealth know-how capitalism. The once noble and essential but now obsolete nations belonged to the rooted socioeconomic land-capitalism era of humanity. In reality, humanity is now

* The July 31, 1981, issue of *Publishers' Weekly* reported on the first U.S. publishers' trade fair in the People's Republic of China. Over six hundred publishers and eight thousand titles were represented, with several million English-reading Chinese attending the exhibit in six cities. Elaine Frumer, a member of the U.S. delegation reporting in an accompanying article titled "What the Children Looked At," listed *Operating Manual for Spaceship Earth* first on the list she compiled of the ten books the Chinese showed most interest in.

uprooted kinetically and occupying the whole planet. Capitalism is dumping its immobile real estate and depending on science to synchronize its affairs with the invisible realities, misassuming, however, that science knows what it is all about. To successfully dump all its real estate, capitalism has all but ceased "renting" and through enforced selling of "cooperatives" and "nothing else but condominiums" is forcing the citizenry into anchored exploitability, while it is always increasing the *corporate* deployability and mobile shift-about-ability around the world.

Since science and human inventiveness are continually learning how to produce ever more, and ever better performance, for each ounce of material, erg of energy, and second of labor and overhead time invested in any and all of industry's production functions, the real cosmic costs are always and only decreasing, and all price-increasing, as already noted, is corporate selfishness "gotten away with" by political-campaign obligations and by excruciatingly painful, behind-the-scenes corporate lobbyists' congressional bullying.

In 1940, two years before the U.S.A. entered World War II, the president of the Aluminum Company of America became interested in my use of corrugated aluminum in the Dymaxion Deployment Unit, which was the little, mass-producible, autonomous, one-room dwelling machine that I had developed for a group of Scottish leaders, intent on anticipating the wholesale bombing of England's industrial cities, who had proposed accommodation of the surely-to-be-displaced population in thousands of my Dymaxion Deployment Units to be installed on the Scottish moorlands. I was developing these deployment units at Butler Manufacturing Company in Kansas City. I was doing this by converting Butler's mass-production, twenty-foot-diameter, galvanized-steel grain bins into an autonomous, fifteen-hundred-dollar, well-insulated, fireproof, earthquake-proof, kerosene-ice-box-and-stove, Sears Roebuck–furnished, ready-to-move-into unit.

The Butler grain bins had been developed for the U.S. government "ever-normal-grainery" program. The Alcoa president hoped that I would switch from corrugated, galvanized iron grain bins to corrugated aluminum bins. He told me that the cost of aluminum would always decrease. He said that the main cost was that of the electrolytic refining of bauxite ore into aluminum. He said the cost of electricity for the process was always decreasing as we learned to produce ever more kilowatt-hours per each BTU of fossil fuel expended. Aluminum, he said, is one of nature's most abundant elements. The wholesale price of aluminum in 1940, when Alcoa's president made that statement to me, was twelve cents per pound. Today it is selling for over a dollar a pound, not because the Alcoa president was wrong in what he said but because massively organized selfishness has dishonestly changed the scoring system.

It is evident that the degree of technical "advantage" now attained by world-around industrial production capability, if realistically appraised and articulated, now shows that all humanity has just reached a state of comprehensive technical advantage adequate to provide a billionaire's level of living on an indefinitely sustainable base for all of the over four billion human passengers now aboard Spaceship Earth (see *Critical Path*). The world's economic accounting system, if properly entered into the world's computers, will quickly indicate that comprehensive economic success for all humanity is now realizable within a Design Science Decade. All it takes is shifting from weaponry to livingry production.

History's unprecedentedly large and invisible supranational Grunch of Giants being too supra- and infra-visibly large to be sensitively comprehended, it is difficult to surmise and accredit that the almost omni-computerized giant may be evolution's agent of most effective establishment of a world-embracing socioeconomic system most logically suited for the mass-production and distribution of its products and services to economically successful humanity. It could well be that the total-world-involved, supranational giant corporations' computer operations might, to their corporate directors' astonishment and to popular surprise, lead the Grunch into profitable discard of all that is not true, as for instance that anybody owns anything. Commonly acknowledged operational custodianship and popular reaccreditation of the integrated world-around technology management may supplant "ownership" with Hertz and telephone-renting.

The way that the giant can be successfully led into doing so is for a substantial majority of humanity, and eventually all humanity, realistically to comprehend the falsity of the greater part of the inventory of academic premises and axioms upon which the thus misconditioned reflexing of "educated" society is based. For instance, there is no God-validated deed to property of any kind whatsoever. There are no solids. There are no things—only systemic complexes of events interacting in pure principle. There is no up or down in Universe. There is no cubic structure. There are no straight lines. There is no one-, two-, or three-dimensionality. There is only four- or six-dimensionality, etc. As we eliminate that which isn't true, we inadvertently admit into reality that which is true. As world society divests itself of that which experimental evidence demonstrates to be untrue and embracingly enters into its computer the mathematical formulae of all that can be experimentally proven to be true, all the socially, selfishly malignant characteristics of the giant may vanish and the omni-pro-social-advantage-producing capabilities may prevail and flourish.

FIFTEENTH-TO-NINETEENTH-CENTURY
GIANTS' OATH

Fee-fie-fo-fum
I smell the blood of a Britishman

Be he alive or be he dead
I'll grind his bones
To make my bread.

TWENTIETH-TO-TWENTY-FIRST-CENTURY
GIANTS' OATH

We-Fort-five-hun
Steal kudos and credit American
Be it live checking
Or savings "dead"
We knead their dough
For dividend bread.

Each of the giants of today's great Grunch is a quadrillionfold more formidable than was Goliath. Each is entirely invisible, abstract, and completely ruthless—not because those who run the show are malevolent but because the giant is a non-human corporation, a many-centuries-old, royal-legal-advisor-invented institution. The giant is a so one-sidedly biased abstract legal invention that its exploitation by the power structures of thirty generations has made the XYZ corporations and companies seemingly as much a part of nature as the phases of the Moon and clouds of the sky. Corporations operate on an unnatural economic basis that makes a successful Las Vegas roulette bet a trifling success. If you bet your money on the fortunate corporation, your bet is paid and repaid to you quarterly, continually, ad infinitum, often more copiously each year. Assuming an investment of 100 shares at a cost of $2,750 in I.B.M. in 1914 when the company was first formed, it would have grown by June 1959 to 59,320 shares valued at $19,308,900, plus $1,089,000 in cash and stock dividends and an additional $101,906 from rights and privileges.

As I have frequently recalled, the grossly mis- or under-informed, 95-percent-illiterate world society of 1900 misassumed the existence of a fundamental and dire inadequacy of life-support to be operative on our planet, wherefore it concluded, and the political-economic system as yet maintains, that it has to be only one or the other of the planet Earth's two great political ideologies which can survive, and that all the people governed by the loser must perish—"there is not enough for both."

The now-predominantly-literate world population of 1981 has developed an intuitive awareness of the illogicality and even madness of all political systems.

All the foregoing inadequate life-support misassuming by both political parties and all the major religious organizations, as earlier noted, has resulted in both sides having jointly spent six and a half trillion dollars in developing the present capability to destroy all humanity within one hour. Humanity at large is logically intuiting that the same sum spent in the direction of improving the lives of the presently deprived many might readily have brought about better results than race suicide.

The awareness of the emergence of a new world society has been only intuitive, because it is actualized only by a superficial knowledge of the overall integrated effects of an almost entirely nonsensorially contacted invisible reality of electronics, chemistry, metallurgy, atomics, and astrophysics. The epochal events of humans landing on the Moon, satellite-relayed "instant" around-the-world information, and an exclusively direct Sun-powered, Paris-to-London flight are altogether reorienting world-around humanity's intuitive thinking into the realization that we can now do so much with so little that we can indeed take care of everybody.

There is a deep urge on the part of vast number of world "youth"—irrespective of their years—to do something right now about their intuition, which develops an impatience and ever more volatile group psychology. There is therefore an urge toward open physical revolt even amongst some of those who do know there is a bloodless, design-science, revolutionary option to attain socioeconomic success for all. The hotheads want to yield to their impatience. To those who urge us to join forces in bloody revolution, I reply as follows.

Before humans could be designed to occupy it, planet Earth had to be designed. Before planet Earth could be designed, the solar system had to be designed. Before the solar system could be designed, galaxies had to be designed. Before special-case galaxies could be designed, special-case macro-micro Universe, all its atoms and molecules, gravity, and radiation had to be designed. Before *any* realizable designing was possible, it was cosmically necessary to discover and employ the full family of eternally coexistent and synergetically inter-augmentative, only-by-mathematical-equations expressible, intercovarying, generalized principles governing the generalized design of eternally regenerative scenario Universe. And before all recognition of the eternal generalized principles and their inherent design-science functions, it was further necessary to have:

1. The design of an eternally regenerative, radiationally expansive and gravitationally contractive, everywhere and everywhen complexedly intertransforming, non-simultaneously episoded, scenario Universe.
2. The generalized design of galaxies of entropic matter-into-energy as radiation-exporting stars and generalized star systems of planets serving syntropically, as radiation-into-matter importing planets.
3. The design of planet Earth as the Sun-orbiting, biosphere-protected, and oxygen-atmosphere-equipped incubator of DNA-RNA design-controlled biological life and of that life's photosynthetic conversion of entropic radiation into syntropic, orderly hydrocarbon molecules and a vast variety of

hydraulically compressioned, crystallinely tensioned, exquisitely structured biological species omni-inter-regenerating as an ecological omni-life and human-thinking support-system.

4. The eternal mathematics—numbering and structuring. The eternally extensive mathematical spectrums of frequencies, wave lengths, and harmonic intervals.

Only thereafter could those human beings progressively re-evolutionize exclusively by trial-and-error enlightenment, from their born-naked state of absolute ignorance, to discover their scientific-principle-apprehending minds and thereby (now for the first possible moment in history) to glimpse humanity's semi-divine functioning-potential as local-Universe critical-information-gatherers and local-Universe problem-solvers in support of the integrity of eternally regenerative Universe.

Only now for the first time can we human beings effectively revolutionize society in an adequate degree to fulfill what I have identified in *Critical Path* as being the number one objective of humanity's inclusion in comprehensive evolution.

The generalized principles have always—eternally—existed and have always been available for each special-case, revolutionary design-realization. The special-case design revolution has always had to precede the extra-special-case local social revolution, whether it be by inventing guns which overwhelm archery or inventing the wireless telegraph which transmits messages halfway around the world at six million miles per hour, making utterly obsolete the Pony Express and its concomitant "Wild West" socioeconomic behavior-patterning. The greatest evolution-producing revolutions are complex and take the longest to be realized

What I hoped I had made clear in *Critical Path* is that the inherently half-century-long design-science-revolution phase of attainment of universal economic success has been successfully completed and now needs only the *bloodless* socioeconomic reorientation instead of the *political revolution* to exercise humanity's option to "make it" for all.

I hoped that *Critical Path* made clear that fifty-three years ago I anticipated today's transition of humanity from 150 nations operating independently and remotely from one another to an omni-integrated world family, with all that socioeconomic transition's conditioned-reflex stressings and shockings.

I hoped that *Critical Path* made it clear that the world-data integration I initiated more than a half-century ago has kept growing into a comprehensive record of the invisible design-science revolution being achieved only by ever more performance, realized for ever less energy, weight, and time units invested per each increment of accomplished livingry functioning; and that more than a half-century ago I reduced my design-science, human-environment-augmenting structures and technologies to full-scale, physically working demonstrations of their advancement of technological advantage to economically accommodate an around-the-world pulsatively deploying-and-converging, kinetic society.

I hoped that *Critical Path* made it clear that if I or some other individual had not taken these anticipatory initiatives more than a half-century ago, the comprehensively integrated physical-gaining-for-all now to be immediately attained simply through inaugurating the mass-production phase of its already developed prototypings, then the design-science revolution could not now be realizable—it would take another fifty years to do the critical path work, and we now have only fifty months within which to exercise our option to convert all Earthian industrial productivity from killingry to livingry products and service systems.

I hoped that *Critical Path* made it clear that lacking the accomplishment of the design-science revolution, while also undergoing the transition into a one-world amalgamation of humanity which we are now experiencing, humans would have been catalyzed only into a world-around social revolution of the same bloody historic pattern of revengeful pulling down of the advantaged few by the disadvantaged many.

I hoped that *Critical Path* made it clear that the accomplished design-revolution's prototypes and developmental concepts now make possible for the first time in history a *bloodless* social revolution successfully elevating all humanity to a sustainable higher living-standard than ever heretofore enjoyed by anyone.

I hope that *Critical Path* made it clear that despite the reality of humanity's option to make it for all humanity, my own conclusion as to whether humanity will do so within the critical time and environmental development limits is—that it will remain cosmically undecided up to the last second of the option's effective actuation, knowing that beyond that imminent moment lies only the swift extinction of humans on planet Earth.

The critical path I committed myself to in 1927 was and as yet is that of applying all technology and science directly to accomplishing the mass-produced components for advanced livingry for all humanity, instead of continuing to invest the advanced science and technology inadvertently falling out of the weaponry industry into the livingry tools industries. This critical path was inherently a fifty-year path.

I will now summarize the last few pages quickly.

The social revolution potential now can for the first time in history realize economic success for all and a comprehensive world enjoyment that involves not revengefully toppling the economically successful minority but elevating all humanity to a sustainable higher level of existing and interacting than any humans have heretofore either experienced or dreamed of.

The now-potentially-to-be-omni-successful social revolution could never before in history have been realized. Until 1970 there had always been enough physical resources but not enough metaphysical resources (of experience-won know-how) on our planet to render the physical technology capable of taking care of everyone at a sustainable, eminently successful level of physical well-being—bloodlessly accomplished and sustainable without the co-existence of either a human slave or working class. Until 1970 it had realistically to be either you *or* me, not enough for both. Since 1970 it has become realistically you *and* me—all else is automated acceleration to human-race extinction on planet Earth.

As we know, when the on-foot soldiers at Crécy stuck their pikes into the ground, points slantingly upward and forward, to impale the bellies of the advancing charges of cavalry, it was social revolution, brought about by design-science revolution. Thus armed with their newly-science-designed pikes, the long-overwhelmed many on foot began to gain emancipation from being overwhelmed by the horse-mounted few. Design did it.

Both the word *revolution* and the words *social revolution* have many meanings, from that referring to the mechanical revolutions of a wheel to the social changing of life-styles which occurred when the horse-mounted and -carriaged few, with their many on-foot servants, stable boys, grooms, coachmen, and the vast slums they drove through were almost entirely design-science-obsoleted by the automobile—mounted many covering vastly more miles and having only a diminishing self-servant-class functioning as some of the riders became the auto-production workers, gas-station attendants, etc.

Greater justice and economic improvement for the many is not always the result of social revolution. The Europeans' guns' overwhelming of the American Indian bow-and-arrow weapons was in most ways a retrogressive social revolution implemented by design-science revolution. It is always the design revolution that tips the social scales one way or the other. However, sum totally the combined design and social revolutions ultimately favor the many. Between 1900 and today, 60 percent of humans in the U.S.A. have attained a standard of living far in advance of those of the greatest potentates of 1900 while concurrently doubling the life-span of that fortunate 60 percent.

Never before in all history have the inequities and the momentums of unthinking money-power been more glaringly evident to so vastly large a number of now literate, competent, and constructively thinking all-around-the-world humans. There's a soon-to-occur critical-mass moment when the intuition of the responsibly inspired majority of humanity, in contradistinction to the angered Luddites and avenging Robin Hoods, faced with comprehensive functional discontinuity of nationally contained techno-economic system, will call for and accomplish a world-around reorientation of our planetary affairs. At this critical moment will occur a realization by the responsibly inspired majority that the adequate capacity of the invisible technology to sustainingly support all humanity depends on all the resources, physical and metaphysical, being always and only employed for all of world-around humanity as a completely integrated techno-economic system operating entirely on its daily income principally of Sun-emanating energy. The integrated world-techno-economic system purpose is in contradistinction to a union of 150 autonomously operating nation-states, as with the United Nations. All this can now be comprehensively commonwealth-accounted in time-energy work units. All this can provide regenerative-initiative accommodating access of human individuals to the ever multiplying commonwealth-techno-economic facilities. The degree of individual-initiative computerized access to the commonwealth facilities will be predicated on the demonstrated performance and sustained integrity of the individual's ever-forwardly-anticipatory designing competence.

I have been a deliberate half-century-fused inciter of a cool-headed, natural, gestation-rate-paced revolution, armed with physically demonstrable livingry levers with which altogether to elevate all humanity to realization of an inherently sustainable, satisfactory-to-all, ever higher standard of living.

Critical threshold-crossing of the inevitable revolution is already underway. The question is: Can it be successfully accomplished before the only-instinctively-operating fear and ignorance preclude success, by one individual, authorized or unauthorized, pushing the first button of chain-reacting all-buttons-pushing, atomic, race-irradiated suicide?

The only happily promising recourse of each human individual is to our highest intellectual facilities and their mutual, ego-deflated, unselfishly loving preoccupation with comprehensivity and our employment of the most powerful tools of all:

(A) the family of generalized scientific principles governing the operational design of eternally regenerative Universe itself;

(B) comprehending and effectively employing synergetics, with the books *Synergetics* and *Synergetics 2* presenting the comprehensive omni-image-able mathematical coordinate system employed by nature, thus avoiding the mentally debilitating, vast-majority-of-humanity-excluding quasimathematical coordinate system employed by present-day science;

(C) comprehending the major objectives and operating strategies of the major opposing power structures of world politics, their present status quo and probable future trending;

(D) comprehending the fundamentals of economics, of wealth vs. money, of the principal features and functioning of industry, banking, and securities;

(E) comprehending the educational system in general as well as the discovery of the shortcomings of science, engineering, and education in general;

(F) synergetically comprehending "what it is all about," as propounded in *Critical Path* and this book, *Grunch of Giants,* and discovering what our options are to confront imminent race disaster; and

(G) the individual discovery of God by a vast majority of human individuals—not the discovery of religions, but the discovery that each and every individual has an always-instantly-open, no-intermediary-switchboard-authority-to-contend-with, no-interference-of-any-kind, direct "hot-line to God": i.e., the weightless, nonphysical communication occurring teleologically between the differentially limited, weightless, nonphysical, temporal, special-case mind of the individual human and the comprehensively integrated, macro-micro unlimited, weightless, eternal, generalized mind of God.

In the course of his life, R. Buckminster Fuller published over twenty-five books and literally thousands of articles. With his last half-dozen books, Bucky came to know and work with Michael Denneny, an editor with Macmillan Publishing who later moved to St. Martin's Press, where he continues to work.

Denneny was to become an editor "familia" with Bucky, working on Fuller's magnum opus, *Synergetics* and *Synergetics 2,* as well as on *And It Came to Pass—Not to Stay.* After Denneny moved to St. Martin's Press, together they published *Critical Path* and *Grunch of Giants* and reprinted *Tetrascroll: Goldilocks and the Three Bears,* a story Fuller used to tell his daughter, Allegra. Denneny's last editorial association with Bucky was with *Inventions: The Patented Works of R. Buckminster Fuller.*

Bucky's Apologia

Michael Denneny

My task was not to preach about God
but to serve God in silence about God.

Inventions, published in 1983, is the last book R. Buckminster Fuller prepared before his death at the age of eighty-eight. Although he died a few months before it was published, Bucky oversaw all stages of the book except for the final reading of the galleys. Appropriately enough, it is a classic apologia in the tradition of Plato's *Apology,* an explanation to his fellow citizens of the course of his life, a rendering of accounts, and a justification for what might seem—in both cases—a strange and easily misunderstood story. As such, it seems to me the most original and profound philosophical report we have seen in the twentieth century, a work that invites both intellectual and moral reflection.

But the book as it exists is not a pure rendering of Bucky's original vision; it is rather a compromise between that vision and the exigencies of the practical world, which, as his editor, I had to take into account. Bucky originally saw the book as composed of four parts: the long autobiographical essay he called "Guinea Pig B," supported, as by a tripod, by the catalog of his twenty-eight patented works, the texts of his forty-seven honorary doctorate citations, and the Basic Biography prepared by his office. We agreed that the many drawings in the patent applications should be included, since the efficacy of any invention depends on the correct concrete details, as well as at least one photograph of each finished object. But this meant the book as he originally envisioned it would have been horrendously expensive, putting it out of reach of the people he most wanted to reach and defeating his purpose.

Bucky was totally reasonable about deleting the "claims" section from each patent. Required by the Patent Office for purely legal reasons, these sections merely reiterate information in the patent itself, often in several different ways, and including them would have led to excessive repetition to no point. Much more difficult was persuading him to delete the Basic Biography, a dense work of some seventy-two double-column printed pages, which was sort of like a curriculum vitae gone mad. Only when I suggested adding at the very beginning of the text a footnote stating that this material was to be read in conjunction with the Basic Biography and informing the reader that it could be ordered from the Fuller Institute for five dollars did he—reluctantly and only after a number of discussions—acquiesce. But when it came to deleting the honorary doctorate citations, or cutting them down to just the names of institutions that had so honored him, he was adamant. I can still remember vividly the lunch discussion for which I had prepared so carefully. After warning him that I thought the reader would not get the point of their inclusion and would skip them, and feared reviewers would see including them as a monumental ego trip, I hastened to add that I had finally understood why he wanted them in and how they functioned, in conjunction with the other two legs of the tripod, to support "Guinea Pig B." At which point he looked at me with a radiant smile and said, "Well, dear boy, then you write an introduction to that section, explaining it to them!" I figured winning two out of three points with Bucky was not bad.

"Guinea Pig B" (B stands for Bucky) started as a speech Bucky gave to the Senate of the United States, an appropriate forum for this most American of thinkers. It is his final, most sustained, and most moving attempt to explain to his fellow citizens what his life had been about, and it circles around the central, pivotal event in that life: his decision at the age of thirty-two not to commit suicide. In 1927, Buckminster Fuller was a total "failure," a college dropout, penniless, unable to earn a living and support his wife and newborn daughter. A man on the verge of suicide, he had walked out on one of the long rocky breakwaters that extend from the beaches of Chicago into Lake Michigan, where he underwent a profound religious experience that was to shape the course of his activity for the next sixty years. Committing "egocide," he turned away from any effort aimed only at his own personal advantage and decided to commit his life to doing what needed to be done for the advantage of all humanity. But was there anything that could be done for humanity by a single individual, without the aid of political movements, religious organizations, or the state? Bucky decided to find out. He decided to make of his attempt an experiment, Guinea Pig B, and to thoroughly document this "search and research project designed to discover what, if anything, an unknown, moneyless individual, with a dependent wife and newborn child, might be able to do effectively on behalf of all humanity that could not be accomplished by great nations, great religious or private enterprise, no matter how rich or powerfully armed."

This event in 1927 is the kernel of Bucky's lifelong activity, the central story of the man's life, the moment he returned to again and again, interpreting the event, amplifying it, searching for its meaning. Indeed, "Guinea Pig B" almost reads like a dialogue between the eighty-eight-year-old Bucky and the thirty-two-year-old man standing on the shores of Lake Michigan. Over a gap of sixty years the older Bucky speaks to (and for) the younger man as he recounts the philosophical/religious drama taking place on that breakwater, step by philosophical step ("I said," "I then said to myself," "As a consequence of the foregoing, I then said"). Whether the young Bucky thought these precise thoughts is not the point. Sixty years of reflection and experiment allowed the elder Bucky to give his younger self the amplified philosophical meaning that may have been present in the original event only implicitly. Like Socrates at his trial explaining how he had spent a lifetime trying to understand the Delphic oracle's odd and whimsical declaration that he was the wisest of men, the elder Bucky can now give a fuller account of what the younger Bucky did that night on the shores of Lake Michigan. In his decision to commit egocide rather than suicide, to choose life over death, to commit his efforts to the betterment of all humanity rather than attempting to "earn a living" for himself and his family, Bucky was in effect throwing in his lot with God. Like Pascal, he was making his own wager. "If I am doing what God wants done . . . then I do not have to worry about not being commissioned to do so by any Earthians and I don't have to worry how we are going to acquire the money, tools and services necessary." What happened on the breakwater in 1927 was that Bucky made a classic leap of faith, plunging into his new commitment to use his energy and intellect only for the betterment of all humanity as confident as any swimmer that he would be supported and buoyed up by the waters of life. And as he felt his life drawing to a close, he prepared this report on the result of his half-century experiment, "Guinea Pig B."

And here is where the tripod comes in, the three columns which support his essay, each of which is fundamental to understanding Bucky's life and thought.

First, the record of his patents. "Ideas," Bucky once said to me, "are a dime a dozen. The trick is to render them into concrete artifacts that will actually be of some use to humanity." Not preaching to mankind but inventing solutions to mankind's problems—solutions that could be implemented in concrete artifacts that would change the way people lived rather than their opinions—was his mission. "It was to be with these artifacts alone that I was committing myself to comprehensively solve all humanity's physio-economic problems." "The prime public record of my more-than-half-century's fulfillment of my commitment has been realized in the working artifacts themselves"—beginning with the 300,000 geodesic domes built around the world. Of what other thinker can it be said that 300,000 structures built on the basis of his philosophy are still standing? However one may consider this man and his philosophy, a close perusal of his inventions should give anyone pause. Thinkers like Michel Foucault might suggest that the philosopher's mission is not to articulate theory but to invent intellectual tool kits people could use in the struggle for their own liberation, but no one has taken this line of thought quite so far as Bucky.

Second, the citations for the forty-seven honorary doctorates he was awarded. From the very beginning, Bucky realized that for his experiment to work, "you are going to have to do all your own thinking." Many have used those words, but few have meant them as radically as Bucky. And in an era of increasing specialization, when the validity of intellectual achievement is judged in the first instance by an ever narrowing group of one's scientific or scholarly peers, it is particularly interesting to consider the academic recognition granted to a man who was a relentless and thoroughgoing comprehensivist. Bucky's utter determination to think for himself, to disregard the conditioned thinking of our inherited scientific disciplines in favor of direct contact with an intuitioned reality, unavoidably exposed him to the danger of being labeled a crackpot by the scientific and educational establishment. Bucky felt such social pressure in favor of intellectual conformism caused many young people to falter in their determination to think for themselves, and he hoped that his example might strengthen their commitment to absolute intellectual integrity. After all, these citations were written by members of various university and college faculties, precisely the knowledge establishment which dismisses totally independent thinkers as crackpots. Thus these informed assessments of his work can serve as critical appraisals of the historical relevance, practicality, and relative effectiveness of his only apparently foolish decision, to disregard conventional thought. Like the record established by his patents, these public recognitions are an integral part—akin to experimental data—of his half-century commitment to see what a single individual can accomplish without organizational backing, including the organizational backing of our knowledge establishment.

And finally the Basic Biography prepared by his office, which is not in the final book. This is an extraordinary document which lists all of Bucky's "primary and secondary social functions," academic appointments, awards, publications, exhibitions, museum shows, major domes, and keynote addresses— i.e., as much of his life as could be objectively listed, quantified, and tallied. Bucky's fondness for this document has mystified many people, but in a way it follows logically enough from his discovery that "I seem to be a verb" (the title of one of his books). To Bucky, "R. Buckminster Fuller" was not a noun but a verb, not a thing like other things but a series of activities. After all, this is a man who had committed "egocide"—the ego, the I, is a name, a noun (okay, a pronoun)—to devote himself to solving the basic problems of humanity, and that activity of problem solving can indeed be described as a series of "primary and secondary social functions."

These three legs of the tripod work in a way akin to experimental data to support the conclusions in "Guinea Pig B." If you want to see what the single individual, thinking totally for himself, can do on behalf of all mankind, here is a test case. Guinea Pig B is a fully documented case study of what one individual was able to achieve in the twentieth century, "an encouraging example," Bucky hoped, "of what the little, average human being can do if you have absolute faith in the eternal cosmic intelligence we call God."

Buckminster Fuller

Introduction: Guinea Pig B from *Inventions*

I am now close to 88 and I am confident that the only thing important about me is that I am an average healthy human. I am also a living case history of a thoroughly documented, half-century, search-and-research project designed to discover what, if anything, an unknown, moneyless individual, with a dependent wife and newborn child, might be able to do effectively on behalf of all humanity that could not be accomplished by great nations, great religious or private enterprise, no matter how rich or powerfully armed.

I started out fifty-six years ago, at the age of 32, to make that experiment. By good fortune I had acquired a comprehensive experience in commanding and handling ships, first as a sailor in Penobscot Bay, Maine, and later as a regular U.S. naval officer. The navy is inherently concerned with not only all the world's oceans, but also the world's dry land emanating exportable resources and import necessities and the resulting high seas commerce. The navy is concerned with all vital statistics. I saw that there was nothing to stop me from thinking about our total planet Earth and thinking realistically about how to operate it on an enduringly sustainable basis as the magnificent human-passengered spaceship that it is.

Planet Earth is a superbly conceived and realized 6,586,242,500,000,000,000,000-ton (over 6.5 sextillion tons) spaceship, cruise-speeding frictionlessly and soundlessly on an incredibly accurate celestial course. Spaceship Earth's spherical passenger deck is largely occupied by a 140-million-square mile "swimming pool," whose three principal widenings are called oceans.

Upon the surface of the "swimming pool," humanity is playing high-profit gambling games with oil-loaded ships. The largest of all such ships in all history is a quarter-mile-long tanker of 580,000 tons. At top speed it can cross the 3,000-mile-wide Atlantic Ocean in six days. That 3,000-mile, six-day tanker distance is traveled every two and one-half minutes by the eleven-quadrillion-times-heavier

Spaceship Earth, which has been moving at this fast rate for at least seven billion years with no signs of slowing or "running out of gas." As it travels around the Sun at 66,000 m.p.h., it also rotates at an equatorial velocity of 1,000 m.p.h.

The units of time and energy expenditures as "matter" or "work" necessary to structure, equip, and operate all the transcendent-to-human contrivings, biological and chemical organisms and equipment, and their "natural" operational events, including the time-energy units invested in creating and operating volcanoes, earthquakes, seaquakes, and tornadoes, as well as to accomplish this fully equipped and complexedly passengered planet Earth's 66,000-m.p.h. cosmic-highway-traveling speed, stated in the terms of time and energy expended per each ton-mile accomplished at that speed, produce a numerical figure of a staggering magnitude of energy expending.

This staggering energy-expenditure figure for operating planet Earth is in turn utterly belittled when compared to the sum of the same units of time-energy expenditures for structuring, equipping, integrally operating, and moving all of the asteroids, moons, and planets as well as the stars themselves of each of all the known approximately 100 billion other star systems of our Milky Way galaxy, as well as of all the asteroids, moon, and planets, and stars of all the approximately 100 billion star systems of each of all the other two billion galaxies thus far discovered by Earthians to be present and complexedly interacting and co-intershunting with Spaceship Earth in our astro-episode neighborhood of eternally regenerative scenario Universe.

All of these "really real" cosmic energy expenditures may be dramatically compared with, and their significance considered in respect to the fact that, the total of all energy used daily—95 percent wastefully—by all humans for all purposes aboard Spaceship Earth amounts to less than one-millionth of 1 percent of Spaceship Earth's daily income of expendable energy imported from the Universe around and within us.

A vast overabundance of this Earthian cosmic energy income is now technically impoundable and distributable to humanity by presently proven technology. We are not allowed to enjoy this primarily because tax-hungry government bureaucracies and money-drunk big business can't figure a way of putting meters between these cosmic energy sources and the Earthian passengers, so nothing is done about it.

The technical equipment—steel plows, shovels, wheelbarrows, boilers, copper tubing, etc.—essential to individuals' successful harvesting of their own cosmic energy income cannot be economically produced in the backyard, kitchen, garage, or studio without the large-scale industrial tools' production elsewhere of industrial materials and tools-that-make-tools, involving vast initial capital investments. If big business and big government don't want to amass and make available adequate capital for up-to-date technological tooling, people will rarely be able to tap the cosmic energy income, except by berry-, nut-, mushroom-, or apple-picking and by fishing.

Volumetrically, S.S. Earth is a 256-billion-cubic-mile spherical vessel of 8,000-miles beam (diameter), having at all times 100 million of its approximately 200 million square miles of spherical surface always exposed to the Sun—with the other hemisphere always in nighttime shadow—bringing about enormous atmospheric, temperature, and pressure differentials and all their resultant high-low weather-produced winds that create all the waves thunderously crashing on all the trillions of miles of our around-the-world shorelines. The atmospheric temperature differentials in turn induce the electromagnetic potential differentials that transform the atmosphere into rain- and lightning-charged clouds from Sun-evaporated waters of the three-fourths-ocean-and-sea-covered Earth.

Gravity's ability to hold together the planet itself, its waters, its biosphere, and other protective mantles, as well as to pull the rain to Earth, thus combines with the photosynthetic capability of the Earth's vast vegetation to harvest solar radiation and store its energies in a manner readily and efficiently convertible into alcohols. The alcohols (four types) constitute the "Grand Central Reservoir" of cosmic-radiation- and gravity-generated energy in its most immediately-convertible-into-human-use state—for instance, as high-octane motor fuels, synthetic rubber, and all other products misnamed "petrochemical products" by their exploiters, whose petroleum is in reality a time-, pressure-, and heat-produced by-product of alcohol.

It was reliably reported in February 1981 that many thousands of individu-

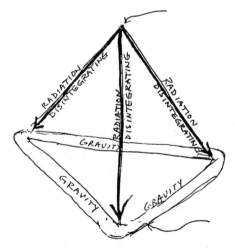

Radiation is disintegrating because it is held together only at this end of each of its energy-manifesting vectors. Disintegrating, the vectors can be angularly aimed, ergo focused. Gravity is inherently integrated as a closed system with no ends, ergo is an inherently closed system having twice the coherence integrity of equally energy-vectored radiation.

als in the U.S. have developed their private distilling equipment and are producing their own alcohol directly from the vegetation's photosynthesizing of the Sun's radiation into hydrocarbon molecules, which may be converted into the alcohol with which these moonshiners are now successfully operating their automobiles.

For exploratory purposes we will now tentatively adopt my "working assumption," later described in detail, that humans are present on Spaceship Earth only because they have an ultimately-to-become-operative, critical function to perform in Universe—a function of which humanity, in general, is as yet almost totally unaware. Assuming for exploratory purposes that this cosmic function of humanity is in due course demonstrably proved to be valid leads to the discovery that this ultimate functioning has involved the investment of important magnitudes of cosmic energy over a period of billions of years, to develop a celestial "incubator planet" having a regeneratively sustained, exactly controlled environment, successfully accommodating the complex development and maintenance of the hydraulically structured (60 percent water) and 98.6-degrees-Fahrenheit-tuned human organisms. We will now also assume that this biosphere's omni-controlled environment can successfully nurture, grow, and develop such an abundance of ecological life support as to be able to accommodate the only-by-shockingly-wasteful-trial-and-error-education of humans in preparation for their ultimately-to-be-employed-and-maintained semi-divine functioning in Universe, as local Universe information harvesters and local Universe problem solvers in support of the integrity of eternally regenerative scenario Universe, provided, however, that, if after the billions of years of their development, those humans living right now can and do pass their final exam for graduation into this semi-divine functioning.

We have next to assume that this crucial test may well lead to humans erroneously pressing the buttons that can now release so much destruction as—within only a few days after tomorrow—to terminate further human life aboard this planet.

Since destruction of humans in Universe would by our working assumption seem to be cosmically undesirable and to be accomplishable only by the anti-intelligent use of the mega-mega-concentrates of energy that humans have learned only recently how to produce and explosively detonate, it is vitally worth our while to stretch our conceptual faculties to understand the physical potentials and possible mystical significance of our most comprehensive inventory of cosmological and cosmogonical information.

The omni-conserved, nonsimultaneously intertransforming energies of eternally regenerative Universe consist most simply of a plurality of omnimagnitude syntropic convergences here and entropic divergences there. The syntropic convergences integrate as matter. The entropic divergences disintegrate as radiation. The convergently associative function is gravitationally integrative and inherently nondivisible and nondifferentiable until convergently realized as matter.

The radiated disintegration of matter, on the other hand, is inherently differentiable and subdivisible, ergo assignable to a plurality of distinctly separated, vastly remote interminglings with illions of other systems' entropically separated-out atomic constituents, thereafter individually to intermingle tentatively and in progressive syntropy as one of myriads of entirely new star systems.

Radiation is inherently focusable and shadowable; gravity is inherently nonfocusable and shadowless.

All the stars are atomic-energy-generating "plants." The star Sun is a hydrogen-into-helium intertransformative regenerator of radiation. The Sun operates internally at a heat of 26,000,000° Centigrade—a scale on which 0° represents the freezing point and 100° the boiling point of water.

The cosmic design problem—with the solution of which we are now concerned—was that of employing a safe way of providing the right amount of exclusively star-emanating energy for developing and maintaining the critical-function-in-Universe-serving humans on the only planet having the right environment for both protecting and ecologically supporting those delicately, intricately designed humans, together with their physically and metaphysically critical functionings.

The next star nearest to planet Earth—beyond the Sun—is 300,000 times farther away from the Earth than is the Sun. In solving the "humans on Earth" problem, it was therefore necessary to employ the cosmic facilities in such a manner as to transmit the appropriate amount of the specific radiation constituents in nonlethal concentration from the Sun, the star nearest to the human-incubating planet Earth—the planet having not only the most propitious environmental condition for humanity incubating, but also maintaining the *exact* vast complex of close-tolerance-of-physical-error limits within which humans could survive.

Humans cannot impound enough energy by sunbathing to keep them alive and operative. Planet Earth's safe importing of Sun radiation must first be impoundingly accomplished by vegetation in energy quantities adequate to supporting not only the vegetation and the humans but all the other myriads of species of life altogether constituting the regenerative ecologic system. To do this, the angular fan-out concentration of energy intensity must be accomplished by attaining adequate distance from the Sun, and thereby to arrive at protoplasmically tolerable increments of energy exactly sustained for conversion by botanical photosynthesis into hydrocarbon molecules, which thereafter can be assimilated metabolically by all other living biological organisms.

COURTESY OF THE BUCKMINSTER FULLER INSTITUTE

This means that the Sun's surface energy radiation must be transformingly programmed to angularly deconcentrate during its eight-minute, 92-million-mile passage from Sun to Earth to arrive in nonlethal increments and at non-lethal temperatures.

The farther away a cosmic radiation source is, the less concentrated the radiation. Only about one-billionth of the radiation given off omnidirectionally by the Sun is so aimed as to impinge directly upon the surface of planet Earth.

With all the space of Universe to work with, the great designing wisdom of

Universe seems to have found it to be essential that the energy organized as the predominantly water-structured substance that is all biological life be maintained at a distance apparently never greater than 95 million miles and never less than 91 million miles away from the atomic-energy-generating plant of the Sun's initial magnitude of radiation concentration.

Sad to say, those of the present Earthian power-structure's scientists who assert that they can safely bury atomic radiation wastes within the ever-transforming structure of planet Earth are rationalizing information critical to human continuance aboard planet Earth. They and their only-selfishly-motivated masters are gambling the future of all humanity to win only the continuing increase of their personal economic power control for the few remaining years of only their own lives.

While the totality of atomic wastes now Earth-buried or ocean-sunk is as yet of a feasibly rocketable total bulk and weight for a plurality of blastings-off, we may still send such atomic wastes back into the Sun where they can once more become safely exportable.

Humans who think they are better designers than God are the most to be pitied of liars—the liars who are believingly convinced by their own only-self-conveniencing fabrications.

On rereading what I have just said about humans falsifying critical information, I am retrospectively shocked at my making those two negative citations.

Firstly, my positive information, which if comprehended can help toward realization of physical success for humanity, when followed by my negativism, reduces the credibility of my wisdom and thereby reduces the value of my positive information.

Secondly, we observe from experience that nature has its own checks and balances, accelerators and brakes, temporary side-tracking, overload circuit breakers, self-starters, transformers, birth and death rates, and complex over-all evolutionary gestation rates.

Nature's biggest, most important problems take the longest to solve satisfactorily because they can be solved only by lessons humanly learned through trial-and-error mistake-making; and, most important, by those who make the mistakes and their self-recognition and public acknowledgment of their errors and their only-thereby-learned-from positive clues to effective solutions of evolutionary problems—in the present instance, the problem of how best to abruptly terminate further atomic energy development for human use as fissionally or fusionally generated aboard planet Earth. The problem is one of immediate and direct concern to each individual of our four billion humans, as well as to all the potential many yet to be born and to God. Its satisfactory solution can be arrived at only through major design-science initiative-takings that produce far superior technologies to render spontaneously obsolete the previous undesirable technology.

Vividly, I recall an occasion when I was about ten and my father heard me call my own brother a fool. My father said, "Bucky, I do not take everything I read as being reliable, not even when I read it in the Bible. But I find by experience that statements in the Bible are far more often reliable than are declarations printed elsewhere. There is a statement in the Bible that you should remember: he who calls his brother a fool shall be in danger of hell's fire."

In these critical times let us no longer make the mistake of identifying as fools those with whom we disagree.

The late twentieth century's confused, fearful human chitchat about an Earthian energy crisis discloses the abysmal state of ignorance within which we Earthians now struggle. All this results in realization of the almost absolute futility of the disintegrated ways in which humanity's present leaders cope with this one-and-only-available Spaceship Earth and its one-and-only-available Universe and with the problems of human survival, let alone attainment of physical success for all humanity within the critical time limit.

For all those who wished to observe them—I being one such person—all the foregoing concepts were already apparent in 1927, though in far less obvious degree. The sum of all these facts made it clear to me in 1927 that no matter how much I could, did, or might as yet accomplish, as a human problem-solver employing only an artifact-designed industrial-production revolution on Spaceship Earth—seeking by techno-economic obsolescence, rather than by political reform, to make physically obsolete all ignorantly incapacitating reflexing of humanity—I could not possibly make more of a mess-of-it-all than that being made by the behind-the-scenes absolutely selfish world power-structures' puppeting of the 150 "sovereign" prime ministers, their national legislatures, and their in-turn-puppeted generals and admirals, and the latter's omni-intercompetitive commanding of our one-and-only spaceship, with each five-star "admiral general" looking out only for his own sovereignly-escutcheoned stateroom and all the starboard-side admirals trying to find a way of sinking all the port-side admirals without the winning admirals getting their own feet wet, let alone being drowned.

I saw that there was nothing to stop me from studying—hopefully to discover, comprehend, and eventually employ design-wise—the integrated total family of generalized principles by which nature operates this magnificent, human-passengered, spherical spaceship as entirely enclosed within an external set of physically unique, spherically concentric environmental zones altogether producing the critically complex balance of intertransformative energy conditions essential to maintaining an omniregenerative planetary ecology—all accomplished in local Universe support of eternally omniinterregenerative Universe itself—by means of planet Earth's syntropic, bio-chemical capability to photosynthetically convert stellar radiation (primarily that of the Sun) into hydrocarbon-structured vegetation that in turn is con-verted as "food" into all manner of biological proliferatings, ultimately—after aeons of enormous heat and pressure treatment produced by deep-Earth burial—to be converted into fossil fuels.

This Earthian energy impounding and conserving altogether constitutes a cosmic accumulation of energy ultimately adequate—billions of years hence— to produce "critical mass" for self-starting its own "all-out" atomic energy generators and thus itself becoming a radiation-exporting star.

I saw that this planet Earth's organic-biochemical interstructuring is (1) tensionally produced only by triple-bonded, no-degrees-of-freedom, crys-talline interarrayings of atomic events, (2) compressionally structured only by double-bonded, flexibly jointed, pressure-distributing and omni-stress-equalizing, hydraulic interarrangements of atomic events, and (3) shock-absorbingly structured, single-bondedly and pneumatically, by gaseous interarrangement of atomic events.

I saw that approximately one-half of all the mobile biological structuring con-sists of water, which freezes and boils within very close thermal-environment limits, the physical accommodation of which limiting requirements is uniquely maintained in Universe only within the biosphere of Spaceship Earth—that is, so far as human information goes.

And I repeat for emphasis that I saw in 1927 that there was nothing to stop me from trying to think about how and why humans are here as passengers aboard this spherical spaceship we call Earth. Return to this initial question has always produced for me the most relevant and incisive of insights. There-fore I hope all humanity will begin to ask itself this question in increasingly at-tentive earnestness.

I also saw that there was nothing to stop me from thinking about the total physical resources we have now discovered aboard our ship and about how to use the total cumulative know-how to make this ship work for everybody— paying absolutely no attention to the survival problems of any separate na-tions or any other individual groupings of humans, and assuming only one goal: the omni-physically successful, spontaneous self-integration of all hu-manity into what I called in 1927 "a one-town world."

I knew at the 1927 outset that this was to be a very long-distance kind of search, research, and development experiment, probably to take at least one-half a century to bring to fruition, with no capital backing. At any rate, I want you to understand now why I had no competition undertaking to solve all human physio-economic problems only by an environment-improving, artifact-inventing-and-developing revolution, which inadvertently produced its recognition by the media, which incidental news publishing is the only reason you know about me—especially since I have been only inadvertently producing news-provoking artifacts for fifty-six years. All that news has failed to induce any sincerely sustained realistic competition with my efforts and on my economic premise of non-money-making—but hopefully sense-making—and only-by-faith-in-God-sustained objectives.

Because it was a very large undertaking, I didn't know that I would be here to see it through for all those 56 years. I was born in 1895. The life insurance companies' actuarial life-expectancy for me was 42 years. I was already 32 when I started the project. I have been amazed that things have worked out to the extent that they have, that I am as yet vigorously active, and that I have been able to find so many relevant things that a little individual can do that a great nation and capital enterprise can't do.

Yet I am quite confident there is nothing that I have undertaken to do that others couldn't do equally well or better under the same economic circumstances. I was supported only by my faith in God and my vigorously pursued working assumption that it is God's intent to make humans an economic success so that they can and may in due course fulfill an essential—and only mind-renderable—functioning in Universe.

Assuming this to be God's intent, I saw that if I committed myself only to initiating, inventing, and full-scale prototyping of life-protecting and -supporting artifacts that afforded ever more inclusive, efficient, and in every way more humanly pleasing performances while employing ever less pounds of materials, ergs of energy, and seconds of time per each accomplished function, a young public's enthusiasm for acquisition of those artifacts and youth's increasing satisfaction with the services thereby produced might induce their further development and multiplication by other significance-comprehending and initiative-taking young humans.

I saw that this ever-multiplying activity could lead ultimately to full-scale, world-around, only-for-industrial-mass-production-prototyped artifacts. I saw that this mass-production-and-distribution of livingry service could provide an adequate inventory of public-attention-winning-and-supporting artifacts, efficient and comprehensive enough to swiftly provide the physical success for all humanity.

This would terminate humanity's need to "earn a living," i.e., doing what others wanted done only for others' ultimately selfish reasons. This attending only to what needs to be done for all humanity in turn would allow humanity the time to effectively attend to the Universe-functioning task for the spontaneous performance of which God—the eternal, comprehensive, intellectual integrity usually referred to as "nature" or as "evolution"—had included humans in the grand design of eternally regenerative Universe.

It was clear to me that if my scientifically reasoned working assumptions were correct and if I did my part in successfully initiating, and following-through on realizing, the previously recited potential chain-reactive events, I would be supported by God in realistic, natural, but almost always utterly surprising-to-me ways. I therefore committed myself to such initiations, realizations, and followings-through.

Because I knew at the 1927 outset of the commitment that no one else thought my commitment to be practical or profitable, I also knew that no one would keep any record of its evolvement—should it be so fortunate as to evolve. Since I intended to do everything in a comprehensively scientific man-

ner in committing myself to this very large-scale experiment (which, as already stated, sought to discover what a little, unknown, moneyless, creditless individual with dependents could do effectively on behalf of all humanity that—inherently—could never be done by any nations or capital enterprise), I saw clearly that I must keep my own comprehensive records—records being a prime requisite of scientific exploration. This I have done. It has been expensive and difficult both to accomplish and to maintain. It is comprehensive and detailed. I speak of the record as the "archives." They consist of:

A. The "Chronofile," which in 1981 consisted of 750 12" x 10" x 5" volumes. These volumes contain all my correspondence, as well as sketches and doodles made during meetings with others, and also back-of-envelope and newspaper-edged notes, all maintained chronologically—in exact order of inbound and outbound happenings—all the way from my earliest childhood to the present keeping of such records as induce discovery of what to avoid in future initiatives

B. All the drawings and blueprints I have been able to save of all the design and full-scale artifact-inventing, -developing, and -testing realizations

C. All the economically retainable models

D. All the moving picture and television footage covering my work

E. All the wire and tape recorded records of my public addresses

F. All the affordable news-bureau and clipping-service records of articles or books written by others about me or my work

G. All the posters announcing my lecturing appearances as designed and produced by others

H. A large conglomeration of items (for instance, over 100 T-shirts with pictures of my work or quotations of my public utterances) produced and distributed by students at many of the 550 universities and colleges that have invited me to speak; collection of awards, mementos, etc.

I. All the multi-stage copies of the manuscript and typescript versions of my twenty-three formally published books and many published magazine articles

J. Over 10,000 4" x 5" photo negatives and over 30,000 photographs, all code-listed, covering my life and work; also 20,000 35-mm projection slides

K. My own extensive library of relevant books and published articles

L. All my financial records, including annual income tax returns

M. All the indexes to the archival material

N. All the drafting tools, typewriters, computers, furniture, and file cabinets for an office staff of seven

O. A large collection of framed photos, paintings, diplomas, cartoons, etc.

P. Biographical data, published periodically (approximately every three years), summarizing all developments of my original commitments

Q. The "Inventory of World Resources, Human Trends and Needs"

R. The World Game records

The archives' collected public record now consists of over 100,000 newspaper and magazine articles, books, and radio and television broadcasts about me or my work, unsolicitedly conceived and produced by other human beings all around the world since 1917.

The prime public record of my more-than-half-century's fulfillment of my commitment has been realized in the working artifacts themselves—the 300,000 world-around geodesic domes, the five million Dymaxion World Maps, the many thousands of copies of each of my twenty-three published books—and, most important of all, within the minds and memories of the 30,000 students I have taught how to think about how to design socially needed, more efficiently produced artifacts.

I do not now employ, and never have employed, any professional public relations agents or agencies, lecture or publishing bureaus, salespeople, sales agencies, or promotional workers. As indicated earlier, I am convinced that nature has her own conceptioning, gestation, birth, development, maturization, and death rates, the magnitudes of which vary greatly in respect to the biochemistry and technological arts involved. The most important evolutionary events take the longest.

Since maintenance of the updating and safety of the archives is as yet my responsibility, they are not open to the public, though scholars from time to time are allowed to view them and be shown items of special interest to them. Because I avoid employing any professional agencies, the magnitude of my development is not kept track of and publicly reported by any of the professional agency associations. For instance, I am not included in the annual statements appearing in the news regarding the public speakers most in demand. Therefore every three years or so my office updates my "basic biography," as it is called, to be distributed to those who ask for information.

Because we are now entering upon the 1927-initiated half-century period of realization, it is now appropriate to make public exposure of the record in order to encourage youth to undertake its own mind-evolved initiatives.

I am therefore publishing herewith:

1. A compilation of my U.S. patents, with photographs of each of the realized artifacts covered by each patent.

2. The compendium of honorary doctorate citations. Since I did not graduate from any college or university and since I have not amassed riches and made generous financial gifts to the schools, I am confident that none of the doctorates were conferred upon me as a financial benefactor but only in recognition of my on-campus academic activities, as a visiting lecturer, a research project initiator and director, or as an appointed professor. Here my work could be intimately judged for its educational value, wherefore the awarding of honorary doctorate degrees to me constitutes an objective assessment of

the magnitude and validity of my working knowledge and of its usefulness to the educational system and society. Like the public record established by my patents, these doctorates can serve as critical appraisal of the historical relevance, practicality, and relative effectiveness of my half-century's experimental commitment to discover what, if anything, an individual human eschewing politics and money-making can do effectively on behalf of all humanity.

I hope that the record so documented (and your hoped-for-by-me close examination of it) will serve as an encouragement to you as individuals to undertake tasks that you can see need to be attended to, which are not to the best of your knowledge being attended to by others, and for which there are no capital backers. You are going to have to test a cosmic intellectual integrity as being inherently manifest in the eternally generalized scientific and only mathematically expressible laws governing the complex design of Universe and of all the myriads of objective special-case realizations.

I am presenting all these thoughts and records because I think we are coming to a very extraordinary moment in the history of humanity, when only such a spontaneous, competent, and ultimately cooperative design-science initiative-taking, on the part of a large majority of human individuals, can ensure humanity's safe crossing of the cosmically critical threshold into its prime and possibly eternal functioning in the macro-micro cosmic scheme.

All humans have always been born naked, completely helpless for months, beautifully equipped but with no experience, therefore absolutely ignorant. This is a very important design fact. I do not look at such a human start as constituting a careless or chance oversight in the cosmic conceptioning of the intellectual integrity governing the Universe. The initial ignorance of humans was by deliberate cosmic (divine) design.

We know that before humans are born forth, naked from their mothers' wombs, their protoplasmic cells are chromosomally (DNA-RNA) programmed to produce each human in incredibly successful regenerative detail.

If we study both the overall integrated system and the detail-design features of our own physical organism—for instance, our optical system and its intimacy with our brain's nervous system—we realize what miraculously complex yet eminently successful anticipatory-design-science phenomena are the humans and the cosmic totality of our complex supportive environment.

If I accidentally scratch-cut myself when I am 3, nature goes instantly to work and repairs the cut. I don't know, even now at 88, how to repair my own cell-structured tissue, and I certainly didn't know how at 3 years of age. We understand very little. Obviously, however, we are magnificently successful products of design in a Universe the complexity and intricacy of whose design integrity utterly transcends human comprehension, let alone popularly acceptable descriptions of "divine design."

Therefore the fact that we are designed to be born naked, helpless, and ignorant is, I feel, a very important matter. We must pay attention to that. We are also designed to be very hungry and to be continually rehungered and rethirsted and multiplyingly curious. Therefore, we are quite clearly designed to be inexorably driven to learn only by trial and error how to get on in life. As a consequence of the design, we have had to make an incredible number of mistakes, that being the only way we can find out "what's what" and a little bit about "why," and an even more meager bit regarding "how" we can take advantage of what we have learned from our mistakes.

Suppose a hypothetical 3-billion-B.C. "you," being enormously hungry, ate some invitingly succulent red berries and was poisoned. The tribe concluded, "Berries are poisonous. People can't eat berries." And for the next thousand years, that tribe did not do so. Then along came somebody who showed them they could safely eat blueberries.

At any rate it took a long, long time for humanity to get to the point where it was inventing words with which to integrate and thereby share the lessons individually learned from error-inducing experiences. Humans could help one another only by confessing what each had found out only by trial and error. Most often "know-it-all" ego blocked the process.

After inventing spoken words, humanity took a long time more to invent writing, with which both the remote-from-local-community and the dead gave the remote-from-one-another beings information about their remote-in-time-and-space error-discovering and truth-uncovering experience. Thus does evolution continually compound the wisdom accruing to error-making and -discovering experiences, the self-admission of which alone can uncover that which is true.

Informed by the senses, the brains of all creatures unthinkingly reflex in pursuing only the sensorially obvious and attractive wilderness trails, waterways, and mountain passes along which may exist fallen-leaves-hidden pitfalls.

This fact made possible the effectiveness of the pitfall trap—falling into the pit at 90 degrees to the 180 degrees line of sight spontaneously followed in pursuit of the obvious. This straightaway reflexing is frequently employed by nature to give humans an opportunity to learn that which humanity needs to learn if humanity is ever to attain the capability to perform its cosmically assigned, spontaneous, intellectually responsible functioning. If ego is surmounted, mind may discover and comprehend the significance of the negative event and may thereby discover the principles leading not only to escape from the entrapment but discovery of what is truly worthwhile pursuing, but only as a consequence of the mind's exclusive capability to discover and employ not just one principle but the synergetically interoperative significance of all the human mind's thus-far-discovered principles.

Up to a very short time ago—that is to say, up to the twentieth century—humans were only innately wise but comprehensively information-ignorant. We have had to discover many errors to become reasonably intelligent.

For instance, when I was young those humans who were most remote from others had to travel a minimum of six months to reach one another; approximately none of them ever did so. That is why Kipling's "East is East, and West is West, and never the twain shall meet" seemed so obviously true to the vast millions who read or heard his words. That has all changed. Today, any one of us having enough credit to acquire a jet air-transport ticket can—using only scheduled flights—physically reach anybody else around the world within twenty-four hours, or can reach each other by telephone within minutes.

The furthestmost point away from any place on our approximately 25-thousand-mile-circumferenced Earth sphere is always halfway around the world, which is 12,500 great-circle miles away in any direction from the point at which we start. Flying the shortest distance from the exact North Pole to the South Pole, any direction you first head in will be "due south." If you keep on heading exactly south you will find yourself following a one-great-circle meridian of longitude until you get to the South Pole. Furthermore, to reach your halfway-around-the-Earth, furthest-away-from-you point—which is always 12,500 miles away—flying in a Concorde supersonic transport, cruising efficiently at Mach 2 (approximately 1,400 m.p.h.), and including stopover refueling times, you will reach that furthestmost halfway-around-the-world-from-where-you-started point in half a day.

Humanity, which yesterday was remotely deployed by evolution, is now being deliberately integrated to make us all very intimate with one another and probably ultimately to cross-breed us back into one physically similar human family.

When I was young we were extremely ignorant about other people. I was told that people even in the next town were very dangerous: they "drink whiskey and have knives. . . . you had better not go over there." I was 7 years old when the first automobile came into Boston. When the owner of one of those excitingly new 1902 automobiles drove me over into one of those "dangerous" next towns, I could see no one who appeared to be more "dangerously threatening" than any of my very nonthreatening hometown neighbors.

I was 8 when the Wright brothers first flew. I was 12 when we had the first public wireless telegraphy. I was 19 when we had our first "world" war.

I was born in an almost exclusively walking-from-here-to-there world, a Victorian world in which we knew nothing about strangers. The assumption was that all strangers were inherently very unreliable people. When I was young, 95 percent of all humans were illiterate. Today over 60 percent of the three times as many humans present on Earth are literate. All this has happened unpredictedly in only one lifetime. The majority of the older humans of today are as yet apprehensive of strangers and pretend nonrecognition while the majority of those 30 years and younger tend to welcome strangers, often with open arms.

Something very big has been going on in my particular generation's lifetime where all the fundamental conditions of humanity are changing at an ever-accelerating acceleration rate.

When I was young, not only were 95 percent of human beings illiterate but their speech patterns were also atrociously difficult to understand. I had two jobs before World War I. The men I worked with were very skilled but their awkwardly articulated, ill-furnished vocabularies were limited to about 100 words—50 percent of them blasphemous or obscene. Primarily they let you know how they felt about matters by the way in which they spit—delightedly, amusedly, approvingly, or disgustedly. They were wonderfully lovable and brave human beings but that swearing and spitting was the most articulate and effective expressive language they had. Their pronunciations varied not only from town to town and from one part of a town to another but also from house to house and from individual to individual.

Something extraordinary has happened. Only within our last-of-the-twentieth-century time, approximately everybody has acquired a beautiful vocabulary. This did not come from the schooling system but from the radio and TV, where the people who secured the performing jobs did so by virtue of their common pronunciation, the clarity of their speech, and the magnitude of their vocabularies. They no longer spoke with the myriad of esoteric pronunciations of yesterday. People were introduced to a single kind of language. This brought about primary common-speech patterns. The necessity for pilots of airplanes operating daily around the world from countries all around the world to have a common language has swiftly evolved into a common language. Olympic Games, athletics in general, and frequently televised world affairs have all been accelerating the coming of a to-be-evolved world language. That the language most commonly used in 1983 is English is unfortunate and untrue. What we call English was not the language of long-ago-vanished Angles and Jutes. It is the most crossbred of all the world-around languages of all the world-around people who on their ever westwardly and mildly northwestwardly colonizing way have historically invaded or populated England. "English" now includes words from all the world's languages and represents an agglomeration of the most frequently used and most easily pronounceable words.

Since throughout at least three million years people did not understand languages well, they were relatively ignorant. Their group survival required leaders. Thus through the comprehensively illiterate ages we came historically to require powerful leaders—sometimes as physical warriors, sometimes as

COURTESY OF THE BUCKMINSTER FULLER INSTITUTE

spiritual warriors, sometimes as unique individual personalities. Statesmen steered great religious organizations or great government organizations that led the affairs of illiterate, uninformed humanity.

We have come now to a completely new moment in the history of humans at which approximately everybody is "in" on both speech and information.

Humanity has been coming out of a group-womb of permitted ignorance made possible only by the existence of an enormous cushion of natural physical resources with which to learn only by trial and error, thus to become somewhat educated about how and why humans happen to be here on this planet. Despite a doubling of world population during my lifetime, humanity has at the same time gone from 95 percent illiterate to 60 percent illiterate.

Until I was 28 we Earthians knew only about our own Milky Way galaxy. In 1923 Hubble discovered another galaxy. Since that time we have discovered two billion more of them. That explosively accelerated rate of expansion of our astronomical-information-acquisitioning typifies the rate of popular increase of both scientifically general and technically special information during my lifetime.

We are living in a new evolutionary moment in which the human is being individually educated to do the free-standing human's own thinking, and each is thereby separately becoming extraordinarily well informed. We have reached a threshold moment where the individual human beings are in what I consider to be a "final examination" as to whether they, individually, as a cosmic invention, are to graduate successfully into their mature cosmic functioning or, failing, are to be classified as "imperfects" and "discontinued items" on this planet and anywhere else in Universe.

We are at a human examination point at which it is critically necessary for each of us individually to have some self-discovered, logically reasonable, experience-engendered idea of how and why we are here on this little planet in this star system and galaxy, amongst the billions of approximately equally star-populated galaxies of Universe. I assume it is because human minds were designed with the capability to discover from time to time the only-mathematically-stateable principles governing the eternal interrelationships existing between various extraordinary phenomena—a capability possessed by no known phenomenon other than humans.

How and why were we given our beautiful minds with their exclusive access to the scientific principles governing the operational design of eternally regenerative Universe?

As an instance of the human mind's capability, we have the integration of the unplanned, only evolutionarily combined works of Copernicus, Kepler, and Galileo informedly inspiring Isaac Newton into assuming hypothetically that the interattractiveness of any two celestial bodies varies inversely as of the second power of the arithmetical distances intervening—if you double the arithmetical distance intervening, you will reduce the interattractiveness to one-fourth what it had been. Astronomers applied his hypothesis to their celestial observations and discovered that this mathematical formulation explained the ever-mobile interpositioning behaviors of all known and measuringly-observed celestial bodies. Thus, Newton's hypothesis became adopted as a "scientifically generalized law."

In dramatic contradistinction to the brain's functioning within only directly nerve-sensing limits, the human mind has the capability, once in a great while, to discover invisible, soundless, unsmellable, untasteable, untouchable inter-relationships eternally existing between separate, special-case cosmic entities, which eternal interrelationships are not manifest by any of the interested entities when considered only separately and which interrelationships can be expressed only mathematically and constitute eternal cosmic laws—such as Newton's discovery that the interattractiveness existing invisibly between any two celestial bodies always varies inversely as the second power of the arithmetically expressed distances intervenes those bodies. Such Universal laws can be expressed only mathematically. Mathematics, we note, is *purely* intellectual. Altogether these laws manifest the eternal intellectual integrity of Universe that I speak of as "God."

Human intellect (mind) has gradually discovered a number of these extraordinary, generalized, non-sensorially-apprensible eternal principles. We human beings have been given access to at least some of the design laws of the Universe.

We don't know of any other phenomenon that has such a faculty and such an access permit, wherefore we may assume that we humans must be cosmically present for some very, very important reason.

As far as I know, we humans haven't thought or talked very much about how and why we are here as either a desirable or a necessary function of Universe. We have talked a great deal about the great mystery of being here. But the majority of our public talking centers on the egotistical assumption that human politics and the wealthy are running the Universe, that the macrocosmic spectaculars are an accessory amusement of our all-important selfish preoccupations, and that Universe's microcosmic invisibles are exclusively for corporate stockholders' money-making exploitation—though always individually discovered or invented but only industrially developed, funded by humanity's taxes-paid military defense expenditures after production rights are transferred to prime corporations.

We humans are overwhelmed because we are so tiny and the Earth is so big and the celestial systems so vast. It is very hard for us to think effectively and realistically about what we feel about the significance of all we have learned about Universe.

At any rate, we are now at a point where we have to begin to think realistically about how and why we are here with this extraordinary capability of the mind. Our remaining here on Earth isn't a matter of the cosmic validity of any Earthian economic systems, political systems, religious systems, or other mystical-organization systems. Our "final examination" is entirely a matter of each of all the individual human beings, all of whom have been given this extraordinary, truly divine capability of the mind, individually qualifying in their own right to continue in Universe as an extraordinary thinking faculty. If all of humanity as a cosmic invention is to successfully pass this final examination, it would seem to be logically probable that a large majority of all Earthian individuals make and do the passing for themselves and for the remaining numbers.

Are we individually going to be able to break out from our institutionally misconditioned reflexing? Are we going to be able to question intelligently all the things that we have been taught only to believe and not to expose to the experiential-evidence tests? Are we going to really dare to make our own behavioral strategy decisions as informed only by our own separate experimentally or experientially derived evidence? The integrity of individually thinking human beings—as mind—vs. brain-reflexing automatons is being tested.

My own working assumption of why we are here is that we are here as local-Universe information-gatherers and that we are given access to the divine design principles so that we can therefrom objectively invent instruments and tools—e.g., the microscope and the telescope—with which to extend all sen-

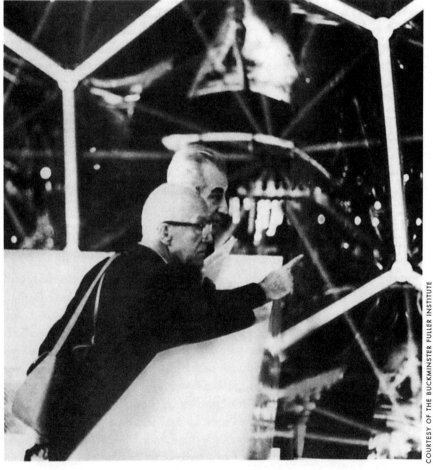

sorial inquiring regarding the rest of the to-the-naked-eye-invisible, micro-macro Universe, because human beings, tiny though we are, are here for all the local-Universe information-harvesting and cosmic-principle-discovering, objective tool-inventing, and local-environmental-controlling as local Universe problem-solvers in support of the integrity of eternally regenerative Universe.

To fulfill our ultimate cosmic functioning we needed the telescope and microscope, having only within the present century discovered that almost all the Universe is invisible. I was two years old when the electron was discovered. That began a new era. When I entered Harvard my physics text had a yellow-paper appendix that had just been glued into the back of the white-leaved physics book. The added section was called "Electricity." The vast ranges of the invisible reality of Universe constituted a very new world—for humanity, a very different kind of world.

When I was born, "reality" was everything you could see, smell, touch, and hear. That's all there was to it. But suddenly we extended our everyday doings and thinkings, not linearly but omnidirectionally into the vast outward, macro-ranges and inwardly penetrating to discover the infra-micro-tune-in-able ranges of the invisible within-ness world.

We began to discover all kinds of new chemical, biological, and electromagnetic behaviors of the invisible realm so that today 99.999 percent of the search and research for everything that is going to affect all our lives tomorrow is being conducted in the realm of reality nondirectly contractable by the human senses. It takes a really educated human to be able to cope with the vast and exquisite ranges of reality.

So here we are—as human beings—majorly educated and individually endowed with developable capabilities of getting on successfully within the invisible ultra-macro and infra-micro world.

Here we have found that each chemical element has its own electromagnetic spectrum wavelengths and frequencies that are absolutely unique to that particular element-isotope. These wavelengths and frequencies are nondirectly tune-in-able by the human senses but are all spectroscopically differentiable and photographically convertible into human readability.

Repeating to emphasize its significance, we note that exploring with the spectroscope, photo-telescope, and radio telescope, humans only recently have traveled informationally to discover about two billion galaxies with an average of 100 billion stars each, all existing within astronomy's present 11.5 billion light-years' radial reach in all directions of the vastness around us.

Taking all the light from all those galaxies we have been able to discover spectroscopically—and have inventoried here on board planet Earth—the relative interabundance of each of the unique categories of all the chemical elements present within that 11.5 billion-light-years realm around us. That little Earthian humans can accomplish that scale of scientific inventorying makes it possible to realize what the human mind can really do. One can begin to comprehend what God is planning to do with the humans' cosmic functioning.

In our immediate need to discover more about ourselves we also note that what is common to all human beings in all history is their ceaseless confrontation by problems, problems, problems. We humans are manifestly here for problem-solving and, if we are any good at problem-solving, we don't come to utopia, we come to more difficult problems to solve. That apparently is what we're here for, so I therefore conclude that we humans are here for local information-gathering and local problem-solving with our minds having access to the design principles of the Universe and—I repeat—thereby finally discover that we are most probably here for local information-gathering and local-Universe problem-solving in support of the integrity of eternally regenerative Universe.

If our very logically and experientially supported working assumption is right, that is a very extraordinarily important kind of function for which we humans were designed ultimately to fulfill. It is clearly within the premises of the divine.

We note also that when nature has a very important function to perform, such as regenerating Earthian birds to play their part in the overall ecological regenerating, nature doesn't put all her birds' eggs in one basket. Instead she provides myriads of fail-safe alternate means of satisfying each function.

Nature must have illions of alternate solutions in this cosmic locality to serve effectively the local information-gathering and local problem-solving in local support of the integrity of an eternally regenerative Universe—should we humans fail to graduate from our potential lessons-learning games of ever more exquisite micro-invisible discrimination and ever greater political, economic, educational, and religious mistake-making.

When in 1927 I started the experiment to discover what a little individual might be able to do effectively on behalf of all humanity, I said to myself, "You are going to have to do all your own thinking." I had been brought up in an era in which all the older people said to all the young people, "Never mind what you think. Listen to what we are trying to teach you to think!" However, as experience multiplied, I learned time and again that the way things often turned out evidenced that the way I had been thinking was often a more accurate, informative, and significant way of comprehending the significance of events than was the academic and conventional way I was being taught, so I said, "If I am going to discard all my taught-to-believe reflexings, I must do all my own thinking. I must go entirely on my own direct experiential evidence."

I said, "It has been an impressive part of my experience that most human beings have a powerful feeling that there is some greater and more exaltedly benign authority operating in the Universe than that of human beings—a phenomenon they call 'God.' There are many ways of thinking about 'God' and I saw that most children are brought up to 'believe' this and that concerning the subject. By the word *believe* I mean 'accepting explanations of physical and metaphysical phenomena without any supportive physical evidence, i.e., without reference to any inadvertently experienced information-harvesting or deliberate-experimental-evidence expansion of our knowledge.'"

I then said to myself, "If you are going to give up all the beliefs and are going to have to go only on your own experiential evidence, you are going to have to ask yourself if you have any experience that would cause you to have to say to yourself there is quite clearly a greater intellect operating in the Universe than that of human beings."

Luckily I had had a good scientific training, and the discovery of those great, eternal, scientific principles that could only be expressed mathematically—mathematics is purely intellectual—made me conclude that I was overwhelmed with the manifests of a greater intellect operating in Universe than that of humans, which acknowledgment became greatly fortified by the following experiential observations.

Scientific perusal of the personal diaries, notes, and letters written by three of history's most significant scientific discoverers—just before, at the time of, and immediately after the moment of their great discoveries—finds that the written records of all of them, unaware of the others' experiences, make it eminently clear that two successive intuitions always played the principal role in their successful accomplishing of their discoveries: the first intuition telling them that a "fish was nibbling at their baited hook" and the second intuition telling them how to "jerk" their fishing line to "hook" and successfully "land" that fish.

The records of these same three scientists also make clear that to start with there must be human curiosity-arousal followed by an intuition-excitation of the human that causes the individual to say, "There is something very significant going on here regarding which I have yet no specific clues, let alone comprehension, because special-case evidences of generalized principles are always myriadly and only complexedly manifest in nature." To qualify as a scientifically generalized principle the scientifically observed and measured special-case behavioral manifests of the generalized interrelationship principles must always be experimentally redemonstrable under a given set of explicitly and implicitly controlled conditions.

Scientifically generalized principles must be repeatedly demonstrable as being both exceptionless and only elegantly expressible as simple, three-term mathematical equations, such as $E = mc^2$.

Constantly intercovarying, ever experimentally redemonstrable interrelations, which have no exceptions, are inherently "eternal."

As a consequence of the foregoing, I then said, "Brain is always and only coordinating the information reported to us by our senses regarding both the macro outside world around us and the micro world within us, and recording, recalling, and only reflexively behaving in response to previous similar experiencing—if any—or if 'none' to newly imagined safe-way logic. No one has ever seen or in any way directly sensed anything outside the brain. The brain is our smell-, touch-, sound-, and image-ination tele-set whose reliability of objective image formulation has been for all childhood so faithful that we humans soon become convinced that we are sensing directly outside ourselves, whereas the fact is that no one has ever seen or heard or felt or smelt outside themselves. All sensing occurs inside the brain's 'television control zone.' The brain always and only deals with temporal, special-case, human-senses-reported experiences. Mind is always concerned only with multi-reconsidering a host of special-case experiences that have intuitively tantalizing implications of ultimately manifesting the operative presence of an as-yet-to-be-discovered eternal principle governing an invisible, unsmellable, soundless, untouchable eternal interrelationship and that complete interrelationships' possibly-to-be-discovered, constantly covarying interrelationships' rate of change. Such principles, whether discovered or not, are intuitively held to always both embrace and permeate all special-case experiences."

I then said to myself, "I am scientifically convinced that the thus-far-discovered and proven inventory of unfailingly redemonstrable generalized principles are a convincing manifest of an eternal intellect governing the myriad of nonsimultaneously and only overlappingly occurring episodes of finite but non-unitarily conceptual, multi-where and multi-when eternally regenerative scenario Universe.

"In so governing the great Universe's integrity, cosmic intellect always and only designs with the generalized principles that are inherently eternal.

"That all the eternal principles always and only appear to be comprehensively and concurrently operative may prove to be an eternal interrelationship condition, which latter hypothesis is fortified by the fact that none of the eternal principles has ever been found to contradict another. All of the eternal principles appear to be both constantly interaccommodative-intersupportive and multiplyingly interaugmentative.

"When you and I speak of *design,* we spontaneously think of it as an intellectual conceptualizing event in which intellect first sorts out a plurality of elements and then interarranges them in a preferred way.

"Ergo, the 'eternal intellect'—the eternal intellectual integrity—apparently governs the integrity of the great design of the Universe and all of its special-case, temporal realizations of the complex interemployments of all the eternal principles."

So I said, "I am overwhelmed by the evidence of an eternally existent and operative, omnicompetent, greater intellect than that of human beings."

Consequently, I also said in 1927, "Here I am launching a half-century-magnitude program with nobody telling me to do so, or suggesting how to do it." I had absolutely no money and my darling wife (who has now been married to me for 66 years) was willing to go along with my thinking and commitments. I said to myself, "If I, in confining my activity to inventing, proving, and improving, and physically producing artifacts suggested to me by physical challenges of the a priori environment, which inventions alter the environment consistently with evolution's trending, whereby I am doing that which is compatible with what universal evolution seems intent upon doing—which is to say, if I am doing what God wants done, i.e., employing my mind to help other humans' minds to render all humanity a physically self-regenerative and comprehensively intellectual integrity success so that humans can effectively give their priority of attention to the ongoing local Universe information-gathering and local problem-solving, primarily with design-science artifact solutions, which will altogether result in comprehensive environmental transformation leading to conditions so favorable to humans' physical well-being and metaphysical equanimity as to permit humans to become permanently engaged with only the by-mind-conceived challenges of local Universe—then I do not have to worry about not being commissioned to do so by any Earthians and I don't have to worry how we are going to acquire the money, tools, and services necessary to produce the successively evolving special-case physical artifacts that will most effectively increase humanity's technological functioning advantage to an omnisuccess-producing degree."

This became an overwhelming realization, for it was to be with these artifacts alone that I was committing myself to comprehensively solve all humanity's physio-economic problems.

I then said, "I see the hydrogen atom doesn't have to earn a living before behaving like a hydrogen atom. In fact, as best I can see, only human beings operate on the basis of 'having to earn a living.' The concept is one introduced into social conventions only by the temporal power structure's dictums of the ages. If I am doing what God's evolutionary strategy needs to have accomplished, I need spend no further time worrying about such matters.

"I happen to have been born at the special moment in history in which for the first time there exists enough experience-won and experiment-verified information current in humanity's spontaneous conceptioning and reasoning for all humanity to carry on in a far more intelligent way than ever before.

"I am not being messianically motivated in undertaking this experiment, nor do I think I am someone very special and different from other humans. The design of all humans, like all else in Universe, transcends human comprehension of 'how come' their mysterious, a priori, complexedly designed existence.

"I am doing what I am doing only because at this critical moment I happen to be a human being who, by virtue of a vast number of errors and recognitions of such, has discovered that he would always be a failure as judged by society's ages-long conditioned reflexings—and therefore a 'disgrace' to those related to him (me)—in the misassuredly, eternally-to-exist 'not-enough-for-all,' comprehensive, economic struggle of humanity to attain only special, selfish, personal, family, corporate, or national advantage-gaining, wherefore I had decided to commit suicide. I also thereby qualified as a 'throwaway' individual who had acquired enough knowledge regarding relevantly essential human evolution matters to be able to think in this particular kind of way. In committing suicide I seemingly would never again have to feel the pain and mortification of my failures and errors, but the only-by-experience-winnable inventory of knowledge that I had accrued would also be forever lost—an inventory of information that, if I did not commit suicide, might prove to be of critical advantage to others, possibly to all others, possibly to Universe." The realization that such a concept could have even a one-in-an-illion chance of being true was a powerful reconsideration provoker and ultimate grand-strategy reorienter.

The thought then came that my impulse to commit suicide was a consequence of my being expressly overconcerned with "me" and "my pains," and that doing so would mean that I would be making the supremely selfish mistake of possibly losing forever some evolutionary information link essential to the ultimate realization of the as-yet-to-be-known human function in Universe. I then realized that I could commit an exclusively "ego" suicide—a personal-ego "throwaway"—if I swore, to the best of my ability, never again to recognize and yield to the voice of wants only of "me" but instead commit my physical organism and nervous system to enduring whatever pain might lie ahead while possibly thereby coming to mentally comprehend how a "me"-less individual might redress the humiliations, expenses, and financial losses I had selfishly and carelessly imposed on all the in-any-way-involved others, while keeping actively alive in toto only the possibly-of-essential-use-for-others inventory of my experience. I saw that there was a true possibility that I could do just that if I remained alive and committed myself to a never-again-for-self-use employment of my omni-experience-gained inventory of knowledge. My thinking began to clear.

I repeated enlargingly to myself, "If I go ahead with my physical suicide, I will selfishly escape from my personal pain but will probably cause great pain

to others. I will thereby also throw away the inventory of experience—which does not belong to me—that may be of critical evolutionary value to others and even may be said to belong not to me but only to others.

"If I take oath never again to work for my own advantaging and to work only for all others for whom my experience-gained knowledge may be of benefit, I may be justified in not throwing myself away. This will, of course, mean that I will not be able to escape the pain and mortification of being an absolute failure in playing the game of life as it has been taught to me."

I then found myself saying, "I *am* going to commit myself completely to the wisdom of God and to realization only of the advantages for all humanity potentially existent in what life has already taught and may as yet teach me." I found myself saying, "I *am* going to commit myself completely to God and to realization of God's apparent intent to assign semi-divine functioning to an as-yet-to-qualify-for-such-functioning humanity. To qualify for such local-Universe's evolutionary adjustings, humans themselves must intelligently discover and spontaneously employ their designed-in potentials and themselves realize the sustainable success of their evolutionary scheduled physio-economic potential."

From that time, 56 years ago, I have had absolute faith in God. My task was not to preach about God, but to serve God in silence about God. Because such commitment to faith is inherently a "flying blind" commitment, I have often weakened in my confidence in myself to comprehend what it might be that I was being taught or told to do. Because I am a human and designed like all humans to learn only by trial and error, I have had many times to do the wrong things in order thereby to learn what next needed to be done. Making mistakes can be and usually is a very dismaying experience—so dismaying as to make it seemingly easier to "go along with unthinking custom."

If I had not in 1927 committed "egocide," I would probably have yielded long ago to convention and therewith suicide of my "only-for-all-others" initiative.

Friends would say, "You are being treacherous to your wife and child, not going out to earn a living for them. Come over here and we will give you a very good job." When, persuaded by their obvious generosity and concern, I did yield, everything went wrong; and every time I went "off the deep end" again, working only for everybody without salary, everything went right again.

I was convinced from the 1927 outset of this new life that I would be of no benefit whatsoever to the more than two billion humans alive in 1927 if I set about asking people to listen to my ideas and endeavoring to persuade them to reform their thinking and ways of behaving. People listen to you only when in a dilemma they recognize that they don't know what to do and, thinking that you might know, ask you to advise them what to do. When they do ask you and you have only a seemingly "good idea" of what they might do, you are far less effective than when you can say, "Jump aboard and I'll take you where you want to go"—or "Jump aboard that vehicle and it will take you to where you want to go." This involves an inanimate artifact to "jump aboard."

Quite clearly I had to address not only the specific but also the comprehensive problem of how to find ways of giving human beings more energy-effective environment-controlling artifacts that did ever more environment-controlling with ever less pounds of materials, ergs or energy, and minutes of time per each realized functioning, until we attained the physically realized techno-energetic ability to do so much with so little that we could realize ample good-life support for everyone, hoping that under those more favorable physical circumstances humans would dare to be less selfish and more genuinely thoughtful toward one another, instead of being lethally and subversively competitive for a share in the existing misassumed-to-be-fundamental inadequacy of life support, perpetually to be extant, on our planet. Every time I recommitted myself so to do, everything went well again.

So we—my wife and family—have for 56 years realized a series of miracles that occur just when I need something, but not until the absolutely last second. If what I think I need does not become available I realize that my objective may be invalid or that I am steering a wrong course. It is only through such nonhappenings that I seem to be informed of how to correct both my grand strategy and its constituent initiations.

I can't make plans of how to invest that which I don't have or don't know that I am going to have. I cannot count on anything. During all these last 56 years I have been unable to budget. I simply have to have faith and just when I need the right-something for the right-reasoning, there it is—or there they are—the workshops, helping hands, materials, ideas, money, tools.

Throughout the last 56 years, I have been able to initiate and manage a great many physical-artifact developments, well over $20 million direct-expenditures' worth of artifact-prototype-producing-and-testing physical research and development. For the last twenty years my income has averaged a quarter of a million dollars a year and my office overhead, travel, invention research and development, and taxes have also averaged a quarter of a million dollars. I am always operating in proximity to bankruptcy but never going bankrupt. While I owe all the humans of all history an unrepayable debt of only-by-experience-winnable knowledge, I don't owe anybody any money and have never consciously and deliberately gained at the expense of others.

I have no accrued savings of earnings. Income taxes take away even the most meager cushioning of funds. I have no retirement fund. I am on nobody's retirement roll. My wife's and my own social security combine to $9,000 per year. The tax experts tell me that the base of the U.S. government taxing theory relates to capitalism and its initially-amassed-dollars-investments in physical production facilities and the progressive depreciation of those physical properties. I operate in a differently accounted world in which there is nothing that I can depreciate wherewith to accrue further-initiative-funding capability. The "know-how" capital capability with which I operate is always appreciating at an exponentially multiplying rate.

There is always an inventory of important follow-through tasks to be accomplished and a number of new, highly relevant critical initiatives to be taken.

Over the years there have also had to be a number of important errors to be made and important lessons to be learned. From the outset of my 1927 commitment and the first twenty subsequent years, there were individuals who were altruistically inspired to support my commitment. For instance, in 1928 a lawyer friend of mine gave me his services for nothing. Seeing the advantage for me of incorporating my activity, he took a few shares in my corporation to pay for his very prodigious services with a hope of incidental personal profit (see my book 4D *Timelock*). Fortunately, the individual investments of those who sought to help me were as monetarily small as they were altruistically large, for their direct profits were never realized and their investments seemingly lost, unless they felt greatly rewarded, as many did, to discover that they had helped to launch an enterprise that as years went on seemed to be ever more promising for all humanity.

The last such "ultimately lost," friendly financial backing occurred during my design and production of the Dymaxion Dwelling Machine's mass-production prototype for Beech Aircraft in Wichita, Kansas. It was there, in

1944, 1945, and 1946, that I produced in full working scale the Dymaxion House, which weighed the three tons that I had estimated and published it would weigh when I designed it in 1927–29. This Beech-Dymaxion realized weight of three tons proved the validity of both my structural and economic efficacious theory and its important technological advance over its conventional, equi-volumed and -equipped counterpart residence of 1927, which weighed 150 tons. It was the 1927 design-initiative discovery that I could apparently by physical design reduce fifty-fold the weight of materials necessary to produce a home, given certain operating standards, that gave me a "rocket blast-off" as an increase of confidence in my theory of solving humans' economic problems by producing ever more performance with ever less energy investments. The confirmation realized at Beech Aircraft seventeen years later was a second-stage rocket acceleration of my only-by-artifact problem-solving initiative.

Produced on the premises of Beech Aircraft, this 1944 mass-production dwelling machine prototype, along with its 100 percent spare parts, was ordered and paid for by the U.S. Air Force. The prototype, along with its 100 percent spare parts, cost $54,000 and after World War II was turned over by the government to a privately organized two-hundred-thousand-dollar corporation formed by about 300 subscribers—averaging a $666 gamble. This corporation hoped to organize the mass production of the physically realized and three-times-refined government-paid-for prototype. Beech Aircraft had its production engineering department plan the tooling and complete an estimate of producing the Dymaxion Dwelling Machine at a rate of 20,000 per year. Beech then made a firm offer to produce them at that rate for $1,800 each, delivered in Wichita minus the kitchen equipment and other electrical appliances to be provided by General Electric on a rental basis of $200 a year. Beech required, however, that a ten-million-dollar tooling cost be provided by outside finance. This was not raised because there existed no high-speed, one-day, "turnkey"—no marketing, distributing, and installing service industry. Prepayment checks for 37,000 unsolicited orders had to be returned. The hopefully-into-mass-production gamble of the private corporation occurred despite my two-fold warning that (1) experience had by then taught me that the gains accruing to my work apparently were distributable only to everybody and only as techno-economic advantage profits for all humanity and (2) that while I was producing an important prototype dwelling machine suitable for mass production at low cost, there existed as yet no distribution and maintenance service industry (and that the latter would require a development period taking another third of a century).

No one has ever won their direct monetary gains by their investment-for-profit bets on my artifact-inventing and -developing concepts. I have always

been saddened by this because the backers' motives were most often greatly affected by a personal affection for me. Fortunately, this gambling on the financial success of my work despite my warnings seems altogether to have ceased a third of a century ago.

With several hundred thousand geodesic domes, millions of my world maps and books, and structural and mechanical artifacts now distributed and installed around the world, I have realized new and ever greater degrees of technical advantaging of society and its individuals. I have continually increased the knowledge of the means of accomplishing across-the-board and world-around advances in technological abilities to produce more economically effective structures and machinery that do ever more work with ever less pounds of materials, ergs of energy, and seconds of time per function—all accomplished entirely within the visible structural inventions and invisible (alloying, electronic, etc.) realms of technology. Twelve years ago I had learned enough to be able to state publicly that humanity had now passed through an evolutionary inflection point whereafter, for the first time in history, it is irrefutably demonstrable that a ten-year design revolution that employs all the physical resources of humanity (now majorly invested in killingry technology) and transforms weaponry scrap into livingry technology, can, within ten design-science-revolution years, have all humanity enjoying a higher standard of living—interminably sustainable—than any humans have ever experienced, while concurrently phasing out all further everyday uses of fossil fuels or atomic energy. We can live handsomely using only our daily income of the Sun's and gravity's multi-way-intergenerated energies.

This omni-humanity eco-technical breakthrough opportunity involves the successful inauguration of the industrial mass production of the dwelling machines and all other geodesic and tensegrity, air-deliverable, move-in-today, environmental controls.

The technology for producing the dwelling machines, their air deliverability, their energy harvesting and conserving, and their prolonged autonomy of operation, has now reached the service industry launching stage. The dwelling machine service industry will not sell houses but will only rent them (as with rental cars or telephones). Much of the dead U.S. automobile-manufacturing industry can and probably will be retooled to produce the dwelling machines that will be needed to upgrade the deployed phases of living of four billion humans. The environment-controlling service industry will provide city-size domes for protecting and housing humanity's convergent activities and the dwelling machine will accommodate all humanity's divergent activities.

My part in this half-century-long technological development has all been fulfilled or occurred as an absolute miracle. From what I have learned I can say—at this most critical moment in all the history of humanity—that whether we are going to stay here or not depends entirely on each and all of our individually attained and maintained integrities of reasoning and acting and not on any politically or religiously accepted or power-imposed socioeconomic-credo-system or financial reforms.

At the outset of my 1927 commitment to exploring for that which only the individual could do effectively for all humanity, while depending entirely on the unpromised-to-him backing of his enterprise only by God, it became immediately evident that if indeed the undertaking became affirmatively supported by God, it would entail many extraordinary physical and metaphysical insights regarding both human and cosmic affairs. I asked myself, "(1) Can you trust yourself never to turn to your own exclusive advantage the insights entrusted to you only for the realization of benefits for all humanity and Universe itself? (2) Can you also be sure that you will never exploit your insight by publicly declaring yourself to be a special 'son of God' or a divinely ordained mystic leader? (3) Can you trust yourself to remember that you qualified for this functioning only because you were an out-and-out throwaway?

(4) Can you trust yourself to reliably report these facts to others when they applaud you for the success of the experiment with which you were entrusted?"

Fortunately, I can, may, and do report to you that I have never broken that trust nor have I ever been tempted to do so.

I do not deem it to be a breaking of that trust when—entirely unsolicited by me or by my family or by any of those working with me on my staff—individuals or organizations outside my domain come spontaneously to me or my staff to employ me as a speaker, author, architect, or consultant. It is these unsolicited, uncontrived, spontaneous short-engagement employments of me in one role or another, plus—on rare occasions—an unsolicited outright gift to me of money, materials, tools, working space, commissions for designs, orders for specific products, etc., which altogether uncontrived employments and unsought gifts I have classified for you as the "miracles," always unforeseeable-in-advance, which have financed or implemented my technical initiatives.

I hope this book will prove to be an encouraging example of what the little, average human being can do if you have absolute faith in the eternal cosmic intelligence we call God.

The archives of R. Buckminster Fuller have been hailed by archive authority Nicholas Olsberg as one of the most important twentieth-century collections of one man's ideas, subjects, and thinking. For fifteen years the Buckminster Fuller Institute served as custodian for this important resource.

Recently, in 1999, Allegra Fuller Snyder, chairwoman of BFI, commented, "I am so gratified that the Fuller Archive will make its permanent home at Stanford University, whose vision and goals are so closely aligned with my father's thinking. He always felt California represented the cresting wave of human evolutionary patterns, and Stanford is right at that interface of cutting-edge thinking and technology. I am particularly excited that with the move to Stanford we will be broadening accessibility to the scholarly community, and especially to the students and new thinkers who I feel are so significantly gathered at Stanford."

Michael Keller, University Librarian and Director of Academic Information Resources at Stanford, has stated that he regards the Fuller Archive as "one of the most important acquisitions" during his thirty-year career—important because of its usefulness for research and teaching across multiple disciplines.

The R. Buckminster Fuller Archives

Allegra Fuller Snyder, Professor Emerita

What are archives? I'm sure as we all look around our desks at the undistinguished piles of paper, some of which eventually find their way into more orderly files, we have no thought of considering this material as an archive. Yet most archives are born from these chaotic beginnings. Most archives come about through hindsight. We come to realize, or others do, that certain parts of those endless papers may be unique and may document what others will find of use and interest.

What is fascinating about my father and his archives is that it didn't happen like this at all. I think the seeds were sown through an extraordinary event that occurred when he was four years old, an age when we are already very much aware of the world around us. At four he received his first pair of glasses. Until that moment his eyes had been so farsighted that all about him was but a blur. With this event he realized that the world he had existed in for four years was a very different one from the world that he now encountered when he put his glasses on. It made a fundamental change in his thinking as well as his seeing.

He became aware that change is critical to our understanding and experience. And that change, and the understanding of change, has something to do with the lens we are wearing. It was many years later that he came to recognize that "99.9 per cent of all that is new, transpiring in human activity and interaction with nature, is taking place within the realms of reality which are utterly invisible, inaudible, unsmellable, untouchable by human senses." That "Man is only impressed by the things he can see. We can't see the hands of the clock move. We can't see the tree grow. We can't see the stars move. We can't see the atoms move. We find man really not accrediting all the continual evolutionary change." It is only with new glasses, with lenses provided by electromagnetic sensors, micro- and macrophotography, telescopes and microscopes, that these new realities begin to exist for us.

From that moment, he was trying, in his youthful ways, to understand and document change. He seemed to feel his life process would hold a demonstration of that change. He was born in 1895, "a suburban New Englander, born in the Gay Nineties," and stepped into a new century almost immediately. He was aware that life was fundamentally changing as he was living it. 1895 was the year automobiles were introduced, and wireless telegraph and the automatic screw machine invented. It was the year X-rays were discovered, and shortly thereafter both the North and South Poles were discovered. He knew somehow that it was important to document those changes. "Most children like to collect things. At four I started to collect documents of my own development." But his own development was counterbalanced with the changes he was continuing to note occurring around him. As a freshman at Harvard in 1913 he felt that "the subway, which then opened to connect Cambridge and Boston by a seven-minute ride, was a harbinger of an entirely new distance-time relationship of humanity and its transforming environment."

He recalled, "By 1917 I was convinced that a much greater environmental transformation was beginning to take place in our generation's unfolding experience." And he determined to employ his "already rich case history" as a method for understanding that change. A year later he determined how; in 1918–1919 he had the task of organizing official records for the U.S. Navy, and he was asked to organize these records chronologically. "My experience before the Navy was that people kept static kinds of files in terms of names and topics, but in the Navy important records were kept chronologically. I thought it might be interesting if I took my own private papers and put them into chronological order. I did so; and I asked my mother for any papers she had regarding me and I put them all into order too."

He called the results of his effort the *Chronofile*.

The Chronofile did chronicle the early events of his life, but in 1927, this chronicle was to become what he regarded as his critical "tool." The concept of tools or artifacts was central to my father's emerging thinking. He understood tools as "externalizations of originally integral functions, aiding the evolutionary process and the regeneration of the species." He saw this tool, the Chronofile, as externalizing his thinking.

He determined to create this tool "by methodical and chronological inventorying of all the human communications in which I was personally involved." This was to be a very rigorous record, a scientific experiment, and he saw himself as "Guinea Pig B" as he committed himself to this very large-scale experiment. "I saw clearly that I must keep my own comprehensive records—records being a prime requisite of scientific exploration." This was a documentation of process. He was not concerned with an end product. He was interested in not looking back but in seeing forward, which prompted him later to realize, "I seem to be a verb, an evolutionary process."

He learned through the Chronofile "to see myself very objectively . . . to see myself as others see me." The Chronofile "persuaded me, ten years after its inception, to start my life as nearly new as is humanly possible to do." "In 1927 I gave up entirely the idea of trying to use my capabilities to develop special economic and physical advantage for [my family] and instead committed myself to the proposition that if those whom I love were indeed the kind of human beings I thought them, their true happiness could only develop through an awareness that our efforts were always in the direction of progressively increasing advantage for all humans without any biases whatsoever."

The details of the Chronofile also charted for him some of the major changes he was seeking to understand, such as "humanity's epochal graduation from the inert, materialistic 19th into the dynamic 20th century"—a change which, he said, "terminated Sir Isaac Newton's normally 'at rest' world of myriadly and remotely isolated hybrid cultures, to which change was anathema; and opened Einstein's normally 'dynamic' omni-integrating world culture to which change has come to seem evolutionarily inevitable." The Chronofile became "the scientific documentation of the emergent realization of the era of accelerating acceleration of progressive ephemeralization."

I remember well when the first bound volumes of the Chronofile took their central place on our bookshelves, in our small apartment on 88th Street in New York. They almost overwhelmed us, but they were very impressive, with their brown leather bindings and soft green covers. The bindings displaying the words "DYMAXION CHRONOFILE, R. Buckminster Fuller," stamped in gold. They arrived just after he had published his first book, *Nine Chains to the Moon,* in 1938, and totaled sixty-seven volumes. To an eleven-year-old they were an important presence.

These turned out to be the only bound volumes, for the incoming data began to be almost as uncontainable as was water for the Sorcerer's Apprentice. The Chronofile alone could not incorporate all the relevant work, so interrelated collections began to emerge, including drawings and blueprints; models; moving picture and television footage; wire- and tape-recorded records; photo negatives, photographs, and 35mm projection slides; news clippings; and posters. Of key importance were the multistage copies of his manuscripts and typescript versions of published books and many published magazine articles. There were also such special and specific projects as the "Inventory of World Resources, Human Trends and Needs" and the World Game records. And not to be overlooked were all his financial records. Finally, in the 1950s, indexes of the archival material were added, which were carefully maintained until his death in 1983. All of the above materials are now known as the R. Buckminster Fuller Archive.

Maintenance of the archives—their safekeeping and updating—was solely the responsibility of my father and, later, his small staff. He found he could not maintain a sufficient staff to open them to the public. Their full significance could not, therefore, be realized at the time. In 1976 they were formally given to me, but my commitment was to seeing that they remained at my father's office, where he and his staff could have easy access to them at all times. This was the situation at the time of his death.

After his death I was confronted with a quandary and a challenge. I wanted the archives to remain the "tool" that my father had always understood them to be: not static, but dynamic, and something that inspired a sense of process in others. But who could make the best use of these tools, and how? My son and I developed the Buckminster Fuller Institute, which I felt initially might make use of this tool; but the reality of a very limited staff, who were overwhelmed by the tasks of maintenance and utilization, soon made it clear that this was not a reasonable answer. Over the years I had the pleasure of visiting such people and organizations as Jonas Salk, with his wonderful vision (never realized) of a Salk Institute for the Humanities; the Canadian Centre for Architecture; the Ford Museum in Dearborn; and Princeton University. Could any of them make full use of the archives? Would any of them be interested in them and willing to understand them as tools for action rather than static documents and memorabilia?

Some of the things I was concerned about:

—That there be an acknowledgment of my father's unique thinking/learning process. I was looking for a place that would respect a person like Bucky

Fuller, who was self-taught and had no formal credentials in any area, even though later he was to hold forty-seven honorary doctorates.

—One of the most important criteria was that the archives always be kept as an integrated whole—given the importance of "comprehensivity" in my father's work and thinking. And this meant finding the archives a final home that was itself comprehensively oriented, not limited by specialization. My father was constantly looking for connections, breaking boundaries between named areas of specialization, and urging consideration of cross-connections between fields of interest.

—That the place itself have a sense of history and continuity, which would be a guarantee for the future as well as for respect of the past. My father's sense of change always reflected the dynamic between past (a sense of history) and future. He saw them as equally important.

—That there be a sense of compatibility with my father's strong vision for the future. I wanted the repository for the archives to be a place of creative imagination, with the capabilities to guide and steer the utilization of the archives in a way that would maximize their impact as a resource for education and inspiration. I hoped this would be a place that would assist all of us in "being optimally effective for all of humanity" as the Chronofile had initially inspired my father to be.

—I was looking for a place that supported both research and education comfortably. Most institutions of higher learning declare they are research oriented because their concept of education is limited. I find that research is education, a fundamental aspect thereof, and its outcome enlightens the educational process. In the longest run I wanted to feel that the institution holding the Archives would respect the idea that education is as fundamental to the human process as breathing, and as continuous.

In talks with Sir Harry Kroto (Nobel Prize co-laureate for the discovery of the buckyball) about this subject, he shared with me his feelings that research agendas in physics and chemistry are driven by an applied focus. He has been critical of this approach, feeling strongly that pure research should be encouraged, the results being totally unpredictable, perhaps opening the doors to something completely new and very important. There is no question in my mind that thinking should be unencumbered by expectation and outcome, but I find sometimes research (as articulated by higher education) is without passion. To engage in an exquisite study that ignores any relation to a larger whole, "the big picture," is a concern to me. But this brings us full circle, to the recognition that application and utilization are less effective when education and research do not inspire them. It was my conclusion that the Archives should be available to researchers, scholars, and students at all levels, and certainly to the public at large.

In August 1999, the Archives were placed in the concerned and loving care of Stanford University and its Special Collections. With this action all my hopes and visions for the Archive are in process of becoming a reality.

Stanford Press Release

The following are excerpts from the Stanford press release issued on July 20, 1999:

The Stanford University Libraries announced the acquisition of the R. Buckminster Fuller Archive. The collection comprises the personal papers and working records of Buckminster Fuller, the architect, engineer, inventor, philosopher, author, cartographer, geometrician, futurist, teacher and poet, as assembled during his lifetime and maintained since his death in July 1983.

The Fuller Archive is known as a collection of incomparable comprehensiveness—truly an archive of 20th-century man—and may well be the most extensive known personal archive in existence. "I regard this archive as one of the most important acquisitions during my career," said Michael Keller, university librarian and director of Academic Information Resources at Stanford, who evaluated the collection while the university considered its acquisition. He said the Archive is particularly important because it is useful for research and teaching across multiple disciplines. These kinds of primary resources, "the personal papers, the early manuscripts and notes, the correspondence and the record of the critical development and reception of ideas, are invaluable documents for research of our programs in the history of science, technology and ideas," Keller said.

Fuller received numerous honorary degrees and awards, including the Queen Elizabeth Royal Gold Medal for Architecture, the Gold Medal Award of the National Institute of Arts and Letters, and the Presidential Medal of Freedom, the highest civilian award given by the U.S. government. The Buckminster Fuller Archive occupies approximately 2,000 linear feet of archival shelf space, spans almost eight decades, and thoroughly records almost every aspect of Fuller's life and works. Included in the Archive are 900 published and unpublished manuscripts; a "Chronofile" of more than 200,000 letters; approximately 4,000 hours of film, video, and audio tapes; 15,000 photographs documenting projects; 500 boxes of project reports and research; more than 150,000 research news clippings; the Inventory of World Resources, Human Trends and Needs; models; and memorabilia.

Fuller meticulously and self-consciously constituted the Archive and, just as meticulously, organized it, making obvious efforts to gather documentation that related to his life and career to supplement the papers that were generated by his own activities. The result is as complete a record of one private individual's life and work as could ever reasonably be found, Keller said.

The value of the Fuller Archive must be seen in relation to Fuller's achievements and to his association with other creative figures of importance to modern cultural life and thinking, ranging from Gropius to Einstein to Gandhi, Keller said. The collection is said to include every substantive piece of paper, film, photography, or tape that passed across Fuller's desk in his lifetime, resulting in the most complete possible documentation of his activities and associations, his thoughts and his projects.

"The collection is a significant resource for the vast range of ideas, subjects, and people with whom Fuller was concerned, and the Stanford University Libraries are its ideal new home for further study and exploration," said Assunta Pisani, associate university librarian for collections and services at Stanford. She noted that the collection will be of tremendous research value to faculty and students in a large number of disciplines, ranging across the humanities, the social sciences and the sciences.

Work to organize and fully describe the collection will begin as soon as possible, said Robert Trujillo, head of the Department of Special Collections at Stanford, where the Fuller Archive will be located. Trujillo expects that the collection will become available to scholars within a relatively brief period of time.

Working with Buckminster Fuller from *The Artifacts of R. Buckminster Fuller*

Don L. Richter

It was my privilege to study with R. Buckminster Fuller in 1949 at the Institute of Design in Chicago. It was "Bucky" who first stimulated my interest in the development of new structural systems. He also made me aware that to continue to build structures with slow, inefficient technology with a huge waste of precious raw materials was a luxury society could no longer afford: the impact of man on his environment is becoming more important than the impact of the environment on man. Fuller faced such issues during his early studies of great circle and geodesic dome geometry. These developments were concerned primarily with dome framing techniques, and, more particularly, with the geometry of such frames.

The Early Work of D. L. Richter

It was my conviction that while a geodesic framing system had many advantages from the standpoints of structure and production, it represented only part of the enclosure solution. Given a good structural dome frame, some form of cover is still required to make it a shelter enclosure. Furthermore, if this cover, or skin, is made tough enough to act as a secure barrier between interior contents and exterior environment, it might also be used to carry a share of the structural burden. I reasoned that if it were shaped properly, the skin might replace the frame totally and become the complete structure.

This concept led to a series of independent studies on the effects and factors involved in compound curved shell shapes. The research studies expanded to include many different forms of shaped surfaces, such as nonspherical domes, involute and hyperbolic paraboloid shells, as well as many other shapes without names. These dome models are indicative of the early research into compound-curved shapes developed in 1949.

It was recognized that while such research discoveries are vital, it is equally important to develop them into real products for a real world. This pioneering design work has resulted, over the past twenty-five years, in many dome structures that are being used throughout the world, from Alaska to the Antarctic. The first opportunity to employ these new stressed-skin principles combined with a space truss system and geodesic dome geometry came in 1956, when I was with the research and development division of Kaiser Aluminum. The problem posed was to design and fabricate a 150-foot dome of all aluminum construction for erection in Hawaii.

The structural system of space truss and stress-skin development for the Hawaii Dome Project was patented and erected by Kaiser Aluminum. The first of many such domes was designed for the Hawaiian Village Convention Center built in 1957. Several years were required to develop the Hawaiian geodesic dome structures for use in more severe climates. To facilitate their realization, we designed permanent tooling and trained personnel in the new construction techniques.

Temcor Domes

Following the success of the Hawaiian projects, I formed Temcor with two others in Torrance, California. Temcor was fortunate to have R. Buckminster Fuller as a member of our board of directors until his death in 1983. In Torrence, we designed a full line of pentagonal and hexagonal geodesic domes in sizes up to 232 feet in diameter. Our success in the myriad problems related to design fabrication and erection is perhaps best exemplified by the construction of a triple geodesic dome sports complex in 1973 in upstate New York. It comprises an interconnected cluster of three hexagonal-shaped domes, each having a clear span of 232 feet and a height of 62 feet at the apex.

The more than 33,000 gold-anodized aluminum panel and strut components for the three domes were engineered and fabricated to precise tolerances at Temcor's manufacturing plant in Torrance and shipped to Elmira, New York, for assembly. They were erected at ground level around the base of a lifting tower and raised into place. After a dome was totally assembled around the tower, it was hoisted and secured to previously prepared supports, and the lifting tower was then removed.

It is rather significant that aluminum geodesic domes designed and manufactured by Temcor have been selected in competition with nongeodesic domes of steel, concrete, wood, and fiberglass plastic—an excellent proof of the coordinated design, fabrication, and erection indicated by Bucky Fuller many years earlier.

Temcor also manufactures two other types of patented dome structure that employ geodesic great circle geometry: the Crystogon structures and Poly-Frame domes, each of which has its own unique applications.

Temcor Crystogon

Crystogon domes employ triangular panels of acrylic secured to a framework of extruded aluminum. The transparent or translucent plastics are nonstructural: they do not carry loads from the aluminum dome frame. The inherent stiffness of the geodesic frame has been proven to be particularly important for this product. Unlike the Expo '67 dome in Montreal, Temcor Crystogon domes have a single layer of geodesic framing plan, yet they can be built in spans up to 250 feet.

Temcor Crystogon domes are now employed in a variety of applications including arboretums and aviaries, skylights in restaurants, all-weather swimming pool covers, and other public buildings. Perhaps the most dramatic Crystogon installation to date is the 150-foot diameter Tropical Botanical Center Exhibition Building in Des Moines, Iowa.

Temcor PolyFrame

In contrast to the Crystogon, the Temcor PolyFrame dome employs wide-range beams of extruded aluminum that are covered with flat, triangular aluminum panels. As with the Crystogon, the nonstructural panels are designed primarily to provide protective covering and contribute only secondarily to the strength of the dome frame.

The PolyFrame domes are widely used for bulk storage covers, wastewater treatment plants, sports facilities, and special structures such as the famous South Pole PolyFrame and Long Beach *Spruce Goose* domes. Its advantage as a tank cover is obvious when one considers that a concrete dome for a typical 150-foot diameter tank weighs about 360 tons. A steel cover for the same tank weights about 175 tons. But an aluminum PolyFrame dome weighs only 18 tons—just five percent of the concrete and ten percent of the steel ones. Part of this efficiency is the direct result of using geodesic geometry.

Dome Geometry Comparison

To illustrate the merits of geodesic structures, one could compare them with the computer analyses of three other dome framing systems. In this assessment, all four domes were made as nearly equal as their individual geometries would permit. All four domes analyzed had the same base diameter, the same spherical radius, and the same number of gusset, or nodal, points interconnected by struts following their four different geometries: geodesic, lattice, Lamella, and Schwedler.

A true geodesic geometry does not employ concentric lesser circles. This is the key feature to look for when determining whether the dome geometry is fully geodesic. The great circle arcs in the geodesic dome example of the accompanying figure extend from the base ring on one side of the dome to a corresponding point on the other side. The space truss formed by the three sets of intersecting great circles yields surprisingly uniform, almost equilateral triangles.

The plan view of the lattice geometry dome has framing struts that follow intersecting, spiral-like patterns that connect the base ring to the concentric inner rings and the apex of the dome. None of the framing lines in the lattice dome follows a great circle arc.

A typical Lamella geometry also has the horizontal lesser circles concentric with the base ring. The Lamella dome has an advantage over the previous two systems, because the clutter of members intersecting at the apex has been reduced. The triangles formed between rings and struts are also more nearly equal in size.

In the plan view geometry of the Schwedler, or radial rib, dome, you will notice that a group of frame members extend from the base tension ring to the apex in great circle arcs. The base ring and the three concentric inner rings are lesser circles. Virtually all the commonly used nongeodesic dome frame systems have a number of such lesser circles in concentric rings. In the Schwedler dome, the diagonal framing members required to stabilize the dome have been shown as dashed lines. These diagonals are required to make the dome a completely triangulated three-dimensional space truss.

Although other geometric configurations and variations have been studied, these four basic types are the most representative. For our computer analysis of the four types, the following equalizing criteria were used:

1. All domes are the same overall size—that is, 100 feet in diameter and 14 feet high.
2. All domes have sixty-one nodes (strut connections).
3. All domes used tubular struts of the same diameter to make up the frame.

The results of our computer analysis are shown in Table 1. Although the Schwedler dome without diagonals looks rather good under uniform symmetrical load, the unbalanced loading caused by simulated snow and wind is far more important. Under such loading, the Schwedler dome without diagonals would only support 12 p.s.f. on one-half the dome, with six pounds distributed over the other half. By adding the diagonals to the Schwedler dome, its strength was more than doubled. This increase in strength was accomplished by reducing the tubular frame wall thickness and without increasing the total weight of the dome.

TABLE 1

	Schwedler Radial Rib				
	Without Diagonals	With Diagonals	Lattice	Lamella	Geodesic
Weight in pounds	10,000	10,000	10,000	10,000	10,000
Number of nodal points	61	61	61	61	61
Max. uniform load p.s.f. evenly distributed	45	29	25	32	45
Max. unbalanced load p.s.f. right side p.s.f. left side	12 6	28 14	20 10	28 14	40 20
Relative strength with unbalanced load	30%	70%	50%	70%	100%
Inches deflection with unbalanced load of 12 p.s.f. and 6 p.s.f.	$17^1/_2$	$1^1/_2$	$1^5/_8$	$^7/_8$	$^7/_8$

The lattice dome will support 20 p.s.f. under the unbalanced condition, the Lamella, 28 p.s.f. By contrast, the geodesic configuration will support a full 45 p.s.f. as a symmetrical load over the total surface, or a nonsymmetrical loading of 40 p.s.f. on one side and 20 p.s.f. on the other.

South Pole PolyFrame

A special application of the PolyFrame dome, and one that has received worldwide attention, is the U.S. Navy's South Pole Station. This huge, all-aluminum dome is 50 feet high and 164 feet in diameter, and serves as a giant weather break to protect the Navy's science headquarters, communications center, and crew's quarters.

The lightweight Temcor dome was selected for this difficult task because it could be broken down readily into coded components for air shipment and assembly under frigid working conditions. Moreover, the PolyFrame dome is inherently capable of withstanding heavy winds and drifting snow. The South Pole Dome was designed to withstand winds of 125 miles per hour, uniform loads from ten feet of snow on top, and, even more importantly, unbalanced loadings of thirty feet of snow on one side.

PolyFrame Covers the *Goose*

Probably the most significant verification of geodesic efficiency in dome structures is the Temcor PolyFrame dome erected to house and exhibit Howard Hughes' *Spruce Goose* aircraft in Long Beach, California. Housing without damaging the world's largest wooden airplane of World War II vintage presented some very difficult structural and logistics problems. The *Spruce Goose* has a wing span of 320 feet and a tail section that stands 100 feet above the ground. The Temcor dome was selected to be the permanent home of the flying boat instead of steel, wood, concrete, or fabric structures. The PolyFrame dome, including its foundations, saved the builder well over a million dollars—a great verification of the fact that quality need not be sacrificed to reduce costs when good design and geodesic principles are employed.

The 415-foot-diameter aluminum dome required to house the flying boat rests directly on the foundation at ground level and rises 130 feet at the center to clear the aircraft. Unique erection procedures were required to realize this dome. Notice the large temporary opening left in the side of the dome to allow passage of the flying boat. One view clearly shows the special geodesic geometry as modified for the construction requirements.

Bucky Fuller expressed to me complete satisfaction with the design during a visit to the site of the Temcor *Spruce Goose* dome. The dome represents the state of the art in efficient structures, as it requires only 0.75 ounce of material per cubic foot of enclosed volume.

These and other construction concepts brought forth by Bucky Fuller will become more feasible with each passing day. My work and success with Temcor is only a beginning.

Describing himself as a layman, E. J. Applewhite can also be described as Bucky's first teenage admirer. He has known Fuller and the family for more than sixty years and was involved with the Wichita Dymaxion house. He worked with the CIA as Deputy Inspector General and Chief of the Inspection Staff. Applewhite spent five years with Fuller, working on two of his major publications, *Synergetics: Explorations in the Geometry of Thinking* and *Synergetics 2: Further Explorations in the Geometry of Thinking*. Applewhite has also been a member of the board of the Buckminster Fuller Institute.

Applewhite's contribution was originally published in *The Chemical Intelligencer* (volume 1, 3, pp. 52–54) in 1995 and is reprinted by permission of the author. It is such an interesting article that I asked to reprint it in its entirety. It tells the story of the discovery of C_{60} by the 1996 Nobel Prize laureates Harold W. Kroto, Robert F. Curl, and Richard E. Smalley and their naming the molecule after Bucky.

The Naming of Buckminsterfullerene

E. J. Applewhite

Systematic chemical nomenclature has always been corrupted—or enhanced, depending on your point of view—by the prevalence of eponyms. The fact that C_{60} was named buckminsterfullerene could be construed as (a) an erratic departure from the etiquette of attributing discoveries to individuals, (b) trivial, or (c) the validation of an intuitive vision of a designer of geodesic domes. H. W. Kroto said that the newly discovered carbon cage molecule was named buckminsterfullerene "because the geodesic ideas associated with the constructs of Buckminster Fuller had been instrumental in arriving at a plausible structure."[1] It is becoming, in Fuller's case, that he made no claim; the honor was bestowed by others. The Israeli poet Yehuda Amichai once described naming as "the primary cultural activity," the crucial first step anyone must take before embarking on thought. John Stuart Mill declared, "The tendency

[1]H. W. Kroto, *Nature* 1987, 329, 529.

has always been strong to believe that whatever received a name must be an entity or being, having an independent existence of its own."

When Harry Kroto, Robert Curl, and Richard Smalley, the experimental chemists who discovered C_{60}, named it buckminsterfullerene, they accorded to Richard Buckminster Fuller (1895–1983), the maverick American engineering and architectural genius, a kind of immortality that only a name can confer—particularly when it links a single historical person to a hitherto unrecognized universal design in the material world of nature: the symmetrical molecule C_{60}. Smalley's laboratory equipment could only tell them how many atoms there were in the molecule, not how they were arranged or bonded together. From Fuller's model they intuited that the atoms were arrayed in the shape of a truncated icosahedron—a geodesic dome. Only after a novel phenomenon or concept is named can it be translated into the common currency of thought and speech.

This newly discovered molecule, a third allotrope of carbon—ancient and ubiquitous—transcends the historical or geographical significance of most named phenomena such as mountains of the moon or Antarctic peaks and ridges. Cartographers named two continents for Amerigo Vespucci, because he asserted (as Columbus did not) that the coasts of Brazil and the islands of the Caribbean were a landmass of their own and not just obstacles on the route to Asia. C_{60} is a far more elemental discovery, it is more ancient, and it pervades interstellar space. Fuller has no reason to envy Vespucci.

Buckminsterfullerene was discovered by chemists who were not looking for what they found. Kroto was looking for an interstellar molecule. Smalley said he hadn't been very interested in soot, but they agreed to collaborate. Smalley's laboratory at Rice University had the exquisite laser-vaporization and mass-spectrometry equipment to describe the atoms of newly created molecules. Scientific experimenters investigate nature at a level where revelation is often unpredictable and sometimes capricious. This is a phenomenon that Fuller (who was not a scientist, but a staunch defender of the scientific method) generalized into the dogmatic statement that all true discovery is precessional. For Fuller, the escape from accepted paradigms is precessional. (Vespucci precessed; Columbus did not.) Fuller had a lifelong preoccupation with the counterintuitive, gyroscopic phenomenon of precession. He defined precession, quite broadly, as the effect of bodies in motion on other bodies in motion. Every time you take a step, he said to me many times, you precess the universe.

For that matter, one may say that Kroto and Smalley in recognizing the shape of the C_{60} molecule made a precessional discovery. Earlier, Osawa, in a paper published in Japanese in 1970, had described the C_{60} molecule with the truncated icosahedral shape; so had Bochvar and Gal'pern in 1973 when they published a paper in Russian on the basis of their calculations. They all recognized the novelty of the molecule and conjectured that its structure should afford great stability and strength. However, neither Osawa nor Bochvar and Gal'pern had experimental evidence, nor did they consider their result important enough to follow up their finding with further work or to convince others to do so. Curiously, in 1984 a group of Exxon researchers made an experimental observation of C_{60} along with many other species. They failed, however, to discern the shape of this species and did not recognize its special importance. These precursors to Kroto and Smalley apparently lacked the requisite—precessional—insight to appreciate the significance of what they had found. Kroto and Smalley's precessional insight was best manifested by their decision to give a name to the C_{60} molecule of the truncated icosahedral shape.

As a longtime close friend of the Fuller family, as his collaborator on his *Synergetics* (1975) and *Synergetics 2* (1979), and as a trustee of the Buckminster Fuller Institute (BFI), I rejoiced vicariously in the molecular celebration of his name. I preserved the copy of its first publication in *Nature* (November 1985), with the C_{60} molecule on its cover, and, with the compulsion of an archivist, I documented the proliferation of reports on this molecule in the professional literature for some while thereafter. While I sensed that Professors Kroto and Smalley had granted the name for perhaps trivial reasons, I felt that there was a greater resonance between C_{60} and Fuller's writings and design philosophy than the mere congruence of the topology of that molecule and Fuller's geodesic domes. Fuller did not develop his peculiar geometry in order to build a dome. Of course, he delighted in building domes and built a great many of them (though all were replicable, no two of his prototypes were the same), and he succeeded admirably in containing a greater volume of space in an enclosed stable structure than any architect or engineer before him had ever done. (He had a dozen or so patents relating to his domes.) But I knew that Fuller was one of the most celebrated but least understood original thinkers of his day. Fuller did not develop his original great-circle coordinate geometry in order to build domes; he built domes because otherwise people would not understand the geometry—which rejected the *XYZ* coordinate system of standard mensuration. He advanced synergetics as nothing less than a new way of measuring experience and as a new strategy of design science which started with wholes rather than parts.

Although I felt that it was presumptuous for me, as a nonscientist, to address Kroto and Smalley on Fuller's behalf, I nevertheless offered them copies of Fuller's *Synergetics* books and drew their attention to collateral aspects of Fuller's work that might be relevant to their major discovery. I was careful to disavow any claim for priority of discovery on Fuller's behalf. He did not anticipate C_{60}, but its discovery did validate his intuitions that geodesic design plays a more significant role in nature's arrangements than had hitherto been recognized. Fuller would have been less surprised than any of us to learn that the sixty-atom array possessed an extraordinary property of stability. Although he regarded the hydrogen atom as the simplest—and hence the most beautiful—design in nature, Fuller had a lifelong interest in the carbon atom, and, in many of his writings and lectures, he celebrated J. H. van't Hoff's 1874 concept of the tetrahedral configuration of carbon bonds.

Some years later, on March 21, 1991, on a visit to Houston, I had the opportunity to call on Professor Smalley in his laboratory at Rice University and pay him homage, specifically on behalf of the Fuller family and the BFI—expressing our gratification in the luster that he and Professor Kroto had added to Fuller's name. He greeted me with a hospitality, a sympathy, and an enthusiasm matching the cordiality of the correspondence I had initiated with Professor Kroto at the University of Sussex in Brighton. A sense of destiny permeates his large, comfortable office; he told me I was sitting on the very couch where he and Kroto had christened the new molecule on September 9, 1985. He told me about how he and his colleagues had sat up all night making models out of Gummi Bear jelly beans and paper cutouts of pentagons and hexagons. I recalled that Fuller as a child had made models out of toothpicks and dried peas, and he had always felt that geometry should be taught as a hands-on laboratory discipline. Smalley said that he had overcome any initial reservations he might have had to Kroto's proposal to name C_{60} buckminsterfullerene. For one thing, the standard IUPAC name for the molecule was impossibly awkward and difficult to read, much less speak. When I asked him why he found the name so appropriate, he said that it was because it conveys in a single word so much information about the shape of the molecule, and he found a happy congruence in the fact that its twenty letters match the twenty faces of the icosahedron—a letter for each facet. All even-number carbon cluster-cage molecules are now termed fullerenes. The root name Fuller lent itself to generic applications with the various other conventional suffixes, producing not just fullerenes, but fulleranes, fullerenium, fullerides, fullerites, fulleroids, fulleronium, metallofullerenes, and so forth. Colloquially—even affectionately—they are subsumed as buckyballs.

As Smalley escorted me out of the laboratory complex on that steaming hot March afternoon (Houston is like that), I was exhilarated by his convicition that C_{60} is one of the most stable and photoresistant molecules known to chemistry, and also probably the most proliferating, and possibly the oldest. A new branch of organic chemistry indeed—and countless textbooks had instantly been rendered out of date.

After a few letters objecting to the name "buckminsterfullerene" had appeared in the columns of *Nature,* Harry Kroto gallantly defended its choice on the grounds that no other name—none of the forms of the classic Greek geometers—described the essential three properties of lightness, strength, and the internal cavity that the geodesic dome affords. To the protest that nobody had ever heard of Fuller, he submitted that the name would have educational value. A fine exercise of onomastic prerogative.

Fuller was not a chemist. He was not even a scientist, and made no pretension of adhering rigidly to an experimental and deductive methodology, and he did not follow the rules of submitting published papers to peer review. But he had an extraordinary facility for intuitive conceptioning. Jim Baggott, in his superb account *Perfect Symmetry: The Accidental Discovery of Buckminsterfullerene,*[2] quotes Fuller in an epigraph: "Are there in nature behaviors of whole systems unpredicted by the parts? This is exactly what the chemist has discovered to be true." Baggott goes on to describe how Fuller had derived his vector equilibrium (cuboctahedron, in conventional geometry) from the closest packing of spheres of energy. What he had was a principle that led to the design of geodesic structures capable of a strength-to-weight ratio impossible in more conventional structures. Fuller had a highly generalized definition of the function of architecture that put him outside the scope of the academicians' view of their discipline. Bucky said that "architecture is the making of macrostructures out of microstructures."

Baggott concludes: "Fuller's thoughts about the patterns of forces in structures formed from energy spheres had led him to the geodesic domes.... That his geodesic domes should serve as a basis for rediscovering these principles in the context of a new form of carbon microstructure has a certain symmetry that Fuller would have found pleasing, if not very surprising."

[2]J. Baggott, *Perfect Symmetry: The Accidental Discovery of Buckminsterfullerene* (Oxford: Oxford University Press, 1994).

The Mind of Buckminster Fuller from *Synergetics Dictionary*

Sample Entries Edited by Fuller

RBF DEFINITIONS

Acceleration:

Physics Recognizes two and only two uniquely differentiable motion patterns.

"The angular acceleration is ~~rotary~~ circumferential.

~~The~~ Linear acceleration is radial *with respect*

to the ~~circular~~ acceleration generating operator observing. Angular Acceleration. As when a weight on the end of a string is swung in a circular pattern around and above one's head, and

2. Linear acceleration

As when the string holding the weight in circular orbit is released and the weight flys off on a radial course ~~away~~

– Cite RBF to EJA + BOLB, 3200 Idaho, ?, 17 Feb '72

directly away from the weight and string operator.

RBF DEFINITIONS

Cycle:

"Convergence to frequency magnitude is tunability.

As with all wave phenomena, tunability is in terms of

whole cycles ~~to (cycled with ? ?)~~ CONVERGING TO a vertex.

Three intervals plus three events = tetra.

Four intervals plus four events = octa.

Five intervals plus five events = icosa.

There are no other fundamental cycles.

- Cite RBF holograph, Synergetics Notes, 1955
- Sketch by RBF, Santa Barbara, 10 Feb'73

RBF DEFINITIONS

Dymaxion Airocean World Map: (a)

"I am attaching a copy of the world map published by LIFE on
my new universal-hinging projection. I have taken off the
global map onto this new projection in several other ways, for
instance, with the North Pole and again the magnetic North
Pole, and the pole of the ecliptic as centers of triangles
instead of squares. And another takeoff, particularly useful
for navigational purposes, is that in which the vertexes of
squares and triangles coincide at the poles.

"The new projection method is also extremely useful in relating
the astronomical map to the land map of the world. This is
because the spherical angles are all proportionately or
symmetrically reduced when translated to plane geometry and
vice versa; furthermore, every point on my plane geometry
projection is vertically above the universally deployed center
of the earth. All interior points retain their symmetrical
positioning whether graphed in spherical or plane geometry.
Therefore points in the astronomical projections may be made
to occur vertically above points on the earth when they are
actually in zenith, with the triangulation of astronomical
positions usefully related by direct graphical method to the
terrestrial map."

- Cite RBF Ltr. to Gilbert Grosvenor, Wash., DC; 29 Apr'43

RbF DEFINITIONS

Dymaxion Airocean World Map: (b)

"The article in LIFE did not describe any of the mathematical
properties of my projection method. I am sure that you would
be interested to have it pointed out that the triangular sections of my projection method represent those unique spherical
triangles whose several vertexes are each coincident with a
vertex of another identical triangle of a system of eight
triangles, altogether forming a spherical triangular lattice
of great circle arcs of 60° completely enclosing the sphere.
This spherical triangular lattice (with equilateral spherical
quadrangle interstices) represents the surface coincidence with
a sphere of a unique system of tetrahedral segments of a sphere,
all of whose apexes coincide at the center of the sphere. It
happens that these particular equiangular spherical triangles
of the infinite number between 180° and 60° are the only
spherical triangles whose chords together with their interior
vertexial radii form a united system of lines describing
uniform, unit size, equilateral 60° triangles whose interior
apexes coincide with the center of the sphere.

"There is no set of spherical triangles which uniformly subdivides all the surface of a sphere (as with the eight 90° equiangular triangles or the faces of an icosahedron) whose central"

- Cite RBF Ltr. to Gilbert Grosvener, Wash., DC; 29 Apr'43

RBF DEFINITIONS

Dymaxion Airocean World Map: (c)

"60° apexes also coincide at the center of the sphere. The
apexes of all other spherical segment tetrahedra either fall
beyond the center or fall short of the spherical center.
This particular spherical triangle and tetrahedral unit which
I have used is the only exception.

"It happens, however, that this symmetrical subdivision of
the surface of the sphere by my eight spherical tetrahedra
leaves a void of six spherical squares whose chords and radii
form spherical pyramids whose apexes also coincide with the
center of the sphere. Thus this system provides uniform and
symmetrical chords and radii, any right angle or diagonal
subdivision of which on the spherical surface must be the
intersection of a plane passing through the center of the
sphere and is therefore a great circle. Thus it is possible
by employing these unique spherical equiangular triangles
and 'squares' (quadrangles) to provide a quadrangular grid of
great circles in the square and unique symmetrical triangular
grid of great circles in the triangle (great circle phenomena
not found in any other symmetrical spherical triangle) both
symmetrically and uniformly subdividing the enclosing
boundaries that allows of universal plane geometry projection"

- Cite RBF Ltr. to Gilbert Grosvener, Wash. DC; 29 Apr'43

RBF DEFINITIONS

Dymaxion Airocean World Map: (d)

"in the terms of the same uniform and symmetrical subdivisions
without defractions of angles of transferred data along the
hinges of the necessarily sectional projections, required for
universal direction of unwrapping of the spherical map.

"All the interior structural geometry of the model thus devised
consists of universally symmetrical equilateral and equiangular
inside truss structure, united individually at their external
vertexes and all joined internally at a universal vertex
center, represents the unique stabilized, nonredundant
four-dimensional force diagram of any dynamically radiant or
convergent spherical organization. It provides a mathematical
module system 'tri-' and 'bi-'secting central angular unity
and graphic model of the decimal twelve, or duodecimal
system, essential to mathematical facility in radionics. It
relates simple geometry to dynamic graphical requirements of
electronics.

"The respective interior triangular and quadrangular great
circle grids which terminally intercept the enclosing sides
of the eight spherical triangles and six spherical squares in
mutually uniform linear intervals may be collapsed to plane"

- Cite RBF ltr. to Gilbert Grosvener, Wash. DC; 29 Apr'43

RBF DEFINITIONS

Dymaxion Airocean World Map: (e)

"surface grids uniformly subdivided by interior triangles and
squares. This collapsing may be accomplished by 'loosing' the
unit apex centers of the tetrahedrons and quadrahedrons while
holding the vertex positions of the squares or triangles and
allowing the radii to 'dangle' parallel to one another with
their loosed terminals in one place.

"Uniform subsidence of the spherical arc segments of the major
spherical triangles and squares of the spherical projection
lattice into plane geometry sections of squares and triangles
is accomplished by concentric shrinking to the chordal plane
in such a manner that the right-angle relationship of all
interior points in respect to the enclosing sides remains
intact. It is the retention of the interior perpendicularity
of points to enclosing sides that makes the hinging of the
triangles and squares possible in a manner that, at the same
time, does not disproportionate or refract the contours of
areas partially occurring on adjacent triangles or squares.

"It is also this method of uniformly progressive concentric
correction by subsidence from spherical segment to plane
geometry which provides the unique characteristic of this"

- Cite RBF Ltr. to Gilbert Grosvenor, Wash. DC: 29 Apr'43

RBF DEFINITIONS

Dymaxion Airocean World Map: (f)

"method of projection which distinguishes it from all other
methods. The unique characteristic referred to is that the
projected diagram retains true measurement, shape, direction,
and distance throughout all of the enclosing boundaries of
the segments with mathematically controlled distortion
'massaged' to the center of the projection areas. All other
projections are true in measure, shaping, and direction only
at an interior point or along one side or along one or several
separated lines or arcs crossing the projection with progressive distortion articulated outwards towards one or more of
the enclosing edges of the projected diagram. In other words,
my new projection is uniformly corrected by internalization
while all other projections are corrected by some systematic
externalization of error. This allows of true external association of my projection units, which is impossible in all
other methods demonstrated to date.

"Only in the case of the azimuthal or gnomonic projections
where correction is radiantly distributed does this exterliza-
tion of correction allow of uniform relationship of one
portion of the spherical projection to another; but in the
cases of the azimuthal or gnomonic hemispheres, there is only"

- Cite RBF Ltr. to Gilbert Grosvenor, Wash., DC; 29 Apr'43

Often referred to as Mr. Cleveland, Herbert E. Strawbridge first met Bucky in 1967 when he and his wife, Marie, along with their daughter, Holly, attended an Energetics Society symposium sponsored by Constantinos A. Doxiadis. Other attendees included Margaret Mead, Marshall McLuhan, Lawrence Halprin, Arnold Toynbee, Edmund Bacon, and Jonas Salk.

In Cleveland, where Fuller had an architectural office, Strawbridge implemented some of Doxiadis's ideas with his formation of the Northern Ohio Urban Systems Research Corporation (NOUS), a regional planning study with global impact. He is now a retired CEO of the Higbee's Department Stores. Strawbridge initiated the revitalization of Cleveland's Flats area, which has become the city's entertainment district. Partially thanks to his effort, Cleveland is now known as a "comeback city" that has recovered after having fallen on hard times.

Panayis Psomopoulos, the secretary of the World Society for Ekistics; Dr. Wesley W. Posvar, President Emeritus of the University of Pittsburgh; and Strawbridge and others help keep alive the Ekistics Society to provide a forum for discussing world problems at conferences. Currently Strawbridge is president of the educational John P. Murphy Foundation and the Kulas Foundation.

Ekistics and R. Buckminster Fuller

Herbert E. Strawbridge

One of Bucky Fuller's many great contributions to mankind was his helping to found and thereafter being an active participant in the World Society for Ekistics. In fact, he acted as president of the society from 1975 to 1977, thus defying his own personal rule never to serve as an officer of any of the many international organizations in which he was involved. His involvement with the World Society for Ekistics began in 1962 and lasted until his death in 1983. His lovely wife, Anne, was his constant companion in the many meetings of the society and thus earned the status of sweetheart of the society.

The reader might wonder what the World Society for Ekistics is, and why

Bucky Fuller was so interested. First, a definition of the word "ekistics" is in order. It is a coined word with Greek roots and means "the science of human settlements." The idea that human settlements fitted into a science came through the thoughts of a Greek genius by the name of Constantinos A. Doxiadis. In 1963, "Dinos" Doxiadis convened thirty-four of the world's great thinkers in Athens to discuss, in symposium format, the problems facing human settlements around the globe. Bucky Fuller was among that first group of twenty, and with the exception of only one other person and Doxiadis himself he was the only person to attend all twelve of the annual symposia that were held.

Bucky was a natural for this exercise in creative thought that was meant to improve the habitat of mankind, as his very nature was "ekistical." He lived, breathed, and thought ekistically. And during the various symposia, he could stimulate and be stimulated by the other participants. They were his peers. Persons such as Margaret Mead, Marshall McLuhan, Arnold Toynbee, R. Llewelyn Davies, and Barbara Ward Jackson attended one or more of the symposia, along with about three hundred others. The majority who attended more than one symposium became strong advocates, just like Bucky, of ekistics. The number of participants in each of the symposia ranged from the original twenty to almost a hundred, many of whom were accompanied by their spouses or other traveling companions.

In preparing this paper, I asked the secretary of the World Society for Ekistics, Panayis Psomopoulos, for any material he had in the files relative to Bucky and his personal participation. I am now the proud possessor of numerous papers and photographs which clearly show Bucky's enormous contribution and enthusiastic participation. As all that knew Bucky can attest, he was a man never without creative thoughts or inspiring words, and he could and did express those thoughts most readily. Because of his tendency to talk for several hours when he warmed up to a subject, and because each session of the symposia was confined to only a few hours in which numerous individuals wished to submit comment, most of Bucky's contributions came in the form of submitted papers. The majority of those submitted papers are several thousand words long and crammed with the great thoughts that occupied his mind. Those papers are preserved as he presented them, which is most fortunate, since if he had only made verbal comments they would have been summarized or omitted entirely.

My personal exposure to Bucky came during the ninth (1967) and tenth (1972) symposia, to which I had been invited as the president of the Northern Ohio Urban Systems Research Corporation (NOUS). In 1975, NOUS engaged Dinos Doxiadis and his team of consultants to devise a twenty-year master plan for development of the northeastern section of Ohio. Doxiadis had developed an ekistical model for how urban communities should develop to provide their citizens with excellent living conditions as well as protect their environment from overambitious extension of suburban sprawl. The advanced thinking of Doxiadis's planning efforts included extensive computerization modeling, which at the time was almost unique. The NOUS experience was meaningful but, unfortunately, was cut off too soon because of the inability to convince a sufficient number of industrial, commercial, and political leaders that urban sprawl was a costly process that didn't have to be endured. Today, urban sprawl is a hot subject for all Americans and even some from other nations, and in the northern Ohio region the NOUS studies are being dusted off and updated.

At my very first symposium, Delos Nine (the symposia were named after the famous Greek island of Delos because in ancient Greece, the island was considered the seat of all wisdom), it was obvious that Bucky Fuller was the pet of all of the prior attendees. Their greetings to him were of the warmth reserved for those special friends that are always welcomed and loved. His demeanor was that of one who loved being where he was and of one who loved his fellow participants.

I was a newcomer and was surely out of place in the company of so many grand world figures, but Bucky soon turned his charm my way and enfolded me into his circle of friends. A situation was to develop during one of the excursions that were part of the charm of the symposia that highlighted Bucky's marvelous sense of humor and lack of self-importance. All several hundred participants, traveling companions, student observers, and staff were bused to a small village in the center of the island of Rhodes for a dinner and a performance of native dance. The entire village's central square had been taken over, with multiple picnic tables set up and a large bandstand for the orchestra. On the streets surrounding the square, the village's citizens had set up their chairs to observe the festivities. By good fortune, my wife and I were at the same table as Bucky and Anne. Also at that table were Stanley Winkleman and his wife, Peggy, plus several other couples. Stanley Winkleman was from Detroit and was the head of a chain of women's apparel stores; he had been very active in interracial relations, and that was why he was part of the symposium. Needless to say, Bucky was by far the most renowned person at the table.

As the dinner progressed, I saw one of the village's citizens get up from his chair and make his way through the square and toward our table. I commented to Bucky that one of his great admirers was approaching. In the few seconds before the gentleman arrived at the table, there was some joking about Bucky's being known everywhere, even in a remote village on Rhodes. When the Greek gentleman did get to the table, he said, "Are you Stanley Winkleman?" The switch was perfect. The table exploded with laughter, and Bucky said, "See, Stanley is more famous than I!" The dumbfounded Stanley

Winkleman finally said, "How do you know me?" The answer was that the Greek had been a towel boy at Stanley's club in Detroit for years and had retired to his little hometown in Greece. The essence of that story was the joy that Bucky got out of the episode. It really showed the very human side of one of the truly world-famous people, R. Buckminster Fuller.

The World Center of Ekistics publishes a scholarly paper called *Ekistics*, and in one of the issues (Vol. 50, No. 302, September/October 1983) the editor, Panayis Psomopoulos, paid a most fitting tribute to the memory of Bucky and Anne Fuller. I include the last paragraph of that memorial statement because it summarizes the great life of a truly amazing man:

> This unusual addition to the "Anthropocosmos" page is, we believe, most appropriate. In Greek, *anthropos* and *cosmos* make up the concept of "Anthropocosmos." *Anthropos* is the human being and *cosmos* the world, but also the ornament and the pride and the beauty of the universe. With their lives, Bucky and Anne embodied the ideal of Anthropocosmos, humanity at its most sacred, here to bring us light and beauty and thought.

It was a joy to be in the presence of Bucky Fuller, and it was a great mental exercise to be present and to hear, and later to read, his uniquely great thoughts. The world and mankind were and will continue to be exceptionally well served by this extraordinary individual.

The World Society for Ekistics is still in existence and continues to serve as one of the great contributors to discussions of world problems at almost annual conferences held all over the globe. Bucky Fuller has been followed as president by some of the greatest thinkers from around the world; the present holder of the chair is Dr. Wesley W. Posvar, President Emeritus of the University of Pittsburgh. These individuals have all had one common characteristic—an abiding faith that the world can be made a better place in which all humans may have a better living standard.

Integrity from *Cosmography*

The many and seemingly unrelated topics we have reviewed are intimately interrelevant and seem to be operating synergetically. Einstein's greatness evolved from his synergetic concern for all experimentally verified data regarding the Universe and its progressively beknownst to us, ever more complexedly and nonsimultaneous physical intertransformings—associatively as matter and disassociatively as radiation.

At the very moment humanity has arrived at that evolutionary point where we do have the option for everyone to "make it," I find it startling to discover that all the great governments, the five great religions, and most of big business would find it absolutely devastating to their continuance to have humanity become a physical, metabolic, economic success. All the political, religious, and money-making institutions' power is built upon those institutions' expertise in ministering to, and ameliorating, the suffering, want, pain, and fears resultant upon the misassumption of a fundamental inadequacy of life support on our planet and the consequent misfortune of the majority of humans.

Some religious bodies battle politically and morally against abortion, which inherently eliminates their most lucrative raison d'être—humans and more humans and then concomitant adversity and suffering, and their need of ministry.

The institutionalized catering to want and suffering gives us a sense of the almost certainly fatal dilemma we are in. Another relevant threat to human continuance in Universe is our world education systems' deliberate cultivation of specialization, despite the fact that each individual human is born physically and metaphysically equipped to function as a natural comprehensivist, with a unique mind designed to ascertain and comprehend the generalized design principles governing interrelationships. Surely if nature had wished humans to be specialists, she would have given them the special integral equipment for so doing, as she has given to all other creatures.

How did it come that the educational system was organized to counter this innate proclivity, environmental versatility, and multifacted capability?

We have observed for aeons herds of wild horses led by a king stallion. Every once in a while an unusually big and powerful young stallion is born— much bigger than the other young stallions. When the big young one matures, the king stallion challenges him to a battle, with the winner inseminating the females of the herd. Darwin cited this as an example of nature's way of arranging to keep the strongest strains going.

I am sure that amongst the early human beings occupying our planet Earth, every once in a while a man was born much bigger than other men. He did not ask to be big; he just found himself to be born so. He found himself continually asked for favors. "Mister, I can't reach the bananas. Could you get some bananas for me?" Being good-natured, he would oblige. And then all the little humans around him said, "Mister, the people over there have lost all their bananas. They're dying of starvation. They're going to come over and kill us to get our bananas. You're big. You get out front and protect us."

So, the big man found his bigness being continually exploited. He said, "All right, people, you've got me out there fighting for you time after time, but between battles I'd like you to help me get ready for the next battle. I need weapons and walls."

The people said, "Okay, we'll make you king and you tell us what to do."

So, the big one became king. Another big stranger came along and said, "Mr. King, you have a soft job here. I'm going to take this away from you." The two battled. The king licked the stranger. The king had his opponent down on the ground and said, "You were going to kill me so you could have my kingdom, weren't you? You understand I can kill you right now, don't you? Okay, you're a very good fighter and I need a lot of good fighters around here, so, if you will promise to always fight for me, I'll release you." The stranger acquiesced. The king found himself to be an institution—a power structure.

The king then said spontaneously to himself, "Don't let two big men come at me at once. I can handle them, but only one by one."

From this instinct there gradually emerged a number-one grand strategy for all power structures: Divide to conquer. To keep conquered, keep divided.

The king said to himself, "I want more of these big men. I'll make one the Duke of Hill A and the other the Duke of Hill B. Then I'll keep my spies watching to see that they don't gang up on me."

Next, a whole lot of little people made trouble for the king by not obeying him. There were some very bright little people around. They refused to do what the king wanted done. The king had one or two of his big men bring in the little offenders. The king said, "Mister, I'm going to cut your head off. You're a nuisance around here." The man replied, "Mr. King, you're making a very great mistake cutting my head off." The king said, "Why?" "Well, Mr. King, I understand the language of your enemy over the hill and you don't. I heard him say what he's going to do to you and when he's going to do it."

The king said, "Young man, you may have a good idea. You let me know every day what my enemy over the hill thinks he is going to do to me, and your head is going to stay on. Then you're going to do something you never did ever before—you're going to eat regularly, right here in the castle. We're going to put purple and gold on you so I can keep track of you."

Then some other physically small character made trouble for the king, and it turned out that he could make better swords than anybody else. He was a great metallurgist. The king made him court armorer and had him live in the castle and wear purple and gold.

Somebody else made trouble and said, "Mr. King, the reason I'm able to steal from you is because you don't understand arithmetic. Now, if I do the arithmetic here in the castle, people won't be able to steal from you." He, too, got the victuals, purple, and gold.

Speaking to all his little "experts," the king said, "You mind your business. And you mind your business. Is it clear to each and all of you that I'm the only one who minds everybody's business?"

The king now had all the great fighters, all the right intelligence, the right arms, the right logistics. His kingdom was getting very big. He wanted to leave it to his grandson. After years of success the king said to each of his experts, "You're getting pretty old. I want you to teach somebody about that mathematics. I want you to teach somebody about that metallurgy of yours," and so on.

Ultimately all the foregoing led to the founding of the educational-category scheming, as manifest in the organization of Oxford University and all other education institutions.

In spite of all humans' innate interest in the interrelatedness of all experience, long ago these world-power-structure builders learned to shunt all the bright intellectuals and the physically creative into specialist careers. The powerful reserved for themselves the far easier, because innate, comprehensive functioning. All one needs to do to discover how self-perpetuating is this disease of specialization is to witness the interdepartmental battling for educational funds and the concomitant jealous guarding of the various specializations assigned to a department's salaried experts on each subject in any university.

In the early 1950s, attending the American Association for the Advancement of Science annual congress in Philadelphia, I happened to find two papers that were presented in different parts of the symposium. One was in anthropology and the other was in biology. A team of anthropologists had for a number of years been examining all the known case histories of human tribes that have become extinct, and a team of biologists had been examining all the known cases of biological species that had become extinct. Both of the papers determined extinction to be the consequence of overspecialization.

How might this be? We know that we can inbreed ever-faster-running horses by mating two very fast-running horses—the mathematical probability of concentrating the fast-running genes is high. When you inbreed special ability, however, you outbreed general adaptability.

Its total energy being fixed and nonamplifiable, physical Universe uses that energy only rarely to do very big things—hurricanes, for example. Nature does smaller tasks more frequently and very small tasks very frequently. As human masters of highly bred racehorses inbreed the high-frequency everyday performance characteristics, they outbreed the rarely used survival capabilities. When the rare big-energy event occurs, the species, having lost its general adaptability to cope with unusual environmental conditions, often perishes. Quantum mechanics is the operating principle.

The energy of Universe may be divided into a few, very infrequent major events or into many, very frequent minor events. The energy of Universe is finite, and since multiplication is accomplished only by division, we recognize that quantum mechanics is the operating principle. Therefore, when humanity today is presented with the option of across-the-board success, it is so specialized as to be unable to recognize this generalized, only comprehensively discoverable and comprehendible course of action to be implemented by an invisible technology.

Humanity was given an enormous range of resources with which to discover that our minds are everything and our muscles are relatively nothing. We note that hydrogen does not have to "earn a living" before behaving like a hydrogen atom. Humans, in fact, are the only phenomenon upon which the power structure has been able to impose the everyday obligation of satisfactorily "earning a living."

Because of high technology's capability to take care of the needs of everybody on the planet, we now know that the prerequisite of having to earn a living is obsolete. Only by virtue of invisible technology's implementation of a revolution in producing constantly the greater performance per unit of invested-resource-accomplished tasks has it come about that there are now adequate resources to take care of, and sustain, everybody at a high standard of living. Such a realization will swiftly alter the fundamental assumptions and activities of our daily lives in a very great way.

A preponderance of fear has long operated in the academic world amongst professional educators working toward, or holding tenuously onto, tenure. A great many teachers would gladly become research professors. If they were assured by some authority that they would be given the income they want, they would prefer to do much of their research and writing at home. Some home-conducted research and telecommuting among academics and other workers would save immense quantities of the irreplaceable fossil-fuel gasoline now used to commute daily to the workplace—especially in the United States.[1]

We must realize that we have all reached a turning point where we can no longer afford to make money rather than good sense.

Every child has an enormous drive to demonstrate competence. If humans are not required to earn a living to be provided survival needs, many are going to want very much to be productive, but not at those tasks they did not choose to do but were forced to accept in order to earn money. Instead, humans will spontaneously take upon themselves those tasks that world society really needs to have done.

If humanity realizes its potential in time to exercise this vital option, we shall witness strong competition among individuals to be allowed to serve on humanity's research, development, and production teams. Never again will what one does creatively, productively, and unselfishly be equated with earning a living. People's sense of accomplishment will derive from showing their peers and demonstrating to the great intellectual integrity of Universe, which we speak of as God, that they vastly enjoy doing their best in the unselfish production of service for others rather than just for the survival needs of themselves and their families.

I think all humanity has crossed the threshold to enter upon its "final examination." It is not the political systems or the economic systems but the human individuals themselves who are in final examination.

How much courage and integrity does each of us have individually to steer a life course according to what our minds have learned through experimental evidence to be the relevant principles governing our situation? How much ingenuity do we have to solve the larger problems of society through anticipatory design rather than through outmoded institutions based on misinformation and the maintenance of the status quo for the vested interests?

I have discovered that we have just such an option. How much courage does each of us have to take the first active step leading to the exercise of that option? What is it that each of us must do? How much willpower must we gather to cast aside deeply ingrained patterns and prejudices? How far must we go to make consciously considerate decisions based on intellectual in-

---

[1]A petroleum geologist friend of mine, François de Chadenedes, once calculated that each gallon of gasoline produced by nature would cost $1 million to produce at the time and energy rates currently charged by utility companies.

tegrity? How much faith must we have in our ability to recognize that intellectual integrity? Or, by default, will the unconscious crowd-following mass psychology of the Dark Ages reign supreme for another aeon?

When nature has an all-important function to be performed by any of her bioinventions and the chances of that biological invention surviving are poor, nature invents many alternative circuits to provide the same results. Nature is not depending solely on the intellectual courage and integrity of this one relatively minor team of human minds on our planet Earth to perform all the local Universe's information gathering and local problem solving. The intellectual integrity of Universe has myriads of alternate fail-safe ways of carrying on. Nature never puts all her eggs in one basket. This gives me reason to surmise that this particular Earthian team is in final examination and that its track record has been far from exemplary.

The human condition today has much improved from when I was young. Illiteracy was then overwhelming. The Soviet Union after the 1917 revolution, for instance, was 95 percent illiterate. For industrialization to work, that condition needed to be reversed, and it was.

The people I worked with on my first pre–World War I jobs were expert craftsmen and very kind human beings, but their on-the-job vocabularies consisted of no more than one hundred English words, almost half of them blasphemous or obscene. Today, the average six-year-old American child has a vocabulary of five thousand words.

The whole communication and information environment of humanity has undergone a revolution. Everybody world-round has a workable vocabulary today. This communications change has taken place at an incredibly rapid rate. In the last twenty-five years I have been around the world forty-eight times, and I am able to communicate wherever I go.

Nature has brought us to the communicating capability where we have 74,000 words in the *Concise Oxford Dictionary* and 150,000 words in most American "college" dictionaries. This proves that we have need of descriptive words for a great many unique experiences. That we could agree on the meaning of 150,000 words is extraordinary. We have reached the point where we are now possessed of sufficient information for each individual human to dare to exercise the option to "make it" rather than having to depend on the decisions of an educated elite.

In astrophysics we can access an omnidirectional 11.5-billion-light-year-radius reach for information. We have photographed the atom. We are at an evolutionary point where we should break out of our Dark Ages eggshell to act in a completely new and unexpected kind of way. A new emergent worldview provides us with clues about our wonderful new metaphysical environment.

Evolution may be classed into two types—what I call "number-one evolution" and "number-two evolution." Number-two evolution is operative wherever and whenever human beings think they are running the world. Number-one evolution is that in which nature is entirely responsible for the evoluting. Number-one evolution suggested in my lifetime that fallout from the comprehensive employment of the doing-ever-more-with-ever-less-resources-perfunction invisible revolution by the military was entirely and unwittingly responsible for the fact that since 1900 we have gone from less than 1 percent to more than 65 percent of humanity enjoying a higher standard of living than had been ever experienced by any potenante when I was young. During that time we have also doubled the population, so we have actually increased by a factor of 130 the number of those so benefiting from this inadvertent technological fallout.

This fallout from political-economics doing the right things for the wrong reasons is what I mean by number-one evolution. There was no planning by any nations or enterprises that sought to alter the lives of all humanity in this historically unprecedented manner. Einstein loved humanity, but was dismayed at the lack of efficient planning to make everyone a success. Official planning was highly biased. There was no organized effort to improve the standard of living across the board. As we discovered earlier, it was assumed you could not, or must not, do so.

In 1938, I predicted in *Nine Chains to the Moon* that by the year 2000 the fundamental needs of everybody could be taken care of. I think Earthlings' final examination has been incrementally advanced and that there may not be that much time. The time remaining to switch over to a winning life strategy is less than a decade, possibly as short as three years. Every day and in every way humanity feels this crisis deeply. Talk of nuclear disarmament and dealing with environmental and social catastrophe is in the air.

Historically, females carrying the young in the womb could not cover as much geography as could the males. Females tended then to stay around a hearth, where they kept a fire going to cook the meals while the male hunted. Because he covered more territory and could report what he saw, man was also the news bearer. Dad could tell his children what he saw from the top of the nearby mountain. He could tell his children what the chieftain over the mountain was saying or doing.

All through history children, starting naked, helpless, and ignorant, have had Dad and Mom telling them what they could eat and what would and would not poison them. Parents told their children what they could and could not get away with in the power system. Dad and Mom were the authorities on how to get on. But Dad was also the authority who brought home the news. Dad's language was local and somewhat esoteric. The kids immediately emulated the way Dad spoke. He was the communication authority. New dialect after dialect was spawned.

Suddenly thrust into my world at age three was the invisible electron. No one took notice. When I was twenty-three, by virtue of that electron, we heard the first human voice on the radio. When I was twenty-seven, the first broadcasting station was licensed. In 1927, when I was thirty-two, all the dads around the country came home one evening to the kids' excited imperative, "Hurry, Daddy, listen to the radio! A man is trying to fly across the Atlantic." Dad said, "What!" and never again was the one to bring home the news.

Nobody thought about this event as a *number-one* anthropological evolution event. The kids knew that Dad and Mom were their private-home authority all right, but quite clearly, Dad and Mom ran across the hallway and got the neighbors to tune in the radio because the radio was going to tell them something important. The children observed for themselves that the radio was more of an authority than was either Dad or Mom. These greater authorities—the radio people—got their jobs on the radio by virtue of the commonality of their diction rather than the esoteric way that Dad said things. The radio people also got their jobs by virtue of the size of their vocabulary and versatility in employing it. To hold their jobs, they had to make their programming ever more popularly understandable, so they developed ever more precise vocabularies and ever-clearer enunciation.

As Dad and Mom accepted the radio-amplified authority, the kids emulated the speech styles of the people on the air. Noting this, many parents also adopted the radio speech, not wishing to be belittled in their children's estimation. This is what overnight changed the speech pattern of humanity the world around, even in the tiniest of hamlets. This was number-one evolution—not planned by humans but altering human interrelations nonetheless.

The speed of sound is approximately 700 miles per hour, given an average temperature. The speed of electromagnetic radiation is 700 *million* miles per hour. Sound waves go no farther than the atmosphere. Radiation goes on and on (without atmosphere) in the Universe, giving us the infinite television views of distant planets remotely transmitted by solar-system-traversing satellites. The amount of information we can get with our eyes is a millionfold greater than what we can get with our ears.

In the mid-1960s, students at the University of California at Berkeley made the world news as the first dissidents in the university educational system. The Berkeley students asked to meet with me. That same year, I was also asked to speak to many of their contemporaries at other universities. In the last half century I have been invited to speak at over 550 universities and colleges around the world. At Berkeley I discovered that the 1965 dissidents were born the year television came into the American home.

The students said, "Dad and Mom love me to pieces. I love them to pieces, but they don't know what's going on." That was exactly the opposite of the way things were when I was young.

My father died when I was very young. My mother said very often, "Darling, never mind what you think. Listen, we're trying to teach you." At school they said, "Never mind what you think. Listen, we're trying to teach you." It was the assumption on the part of the pre–World War I older people that young people's thinking was utterly unreliable.

In 1965, I was fascinated to hear the young world suddenly saying, "Dad and Mom come home from the shoe store and have a beer. Then they watch television, but they have little interest or connection with humans going to Korea or Vietnam or to the Moon. They obviously don't have anything to do with anything. We can see that the people around the world are in great turmoil. We are going to have to do something effective in eliminating that trouble, since Dad and Mom have no understanding of, or concern with, the world's problems."

That 1965 young world's compassion was suddenly of worldwide scope. It could never again be reduced to concern with only themselves and local issues.

The young world was saying, "Dad and Mom don't understand what's going on, so I've got to do my own thinking." Nobody said to them anymore, "Never mind what you think." They begin to think—earnestly, cautiously, individually—and then to test that thinking collectively. Because they did so, they became highly idealistic. They had no experience at taking the thinking-initiative, so they necessarily made mistakes.

The Soviet Union and the United States today spend over $400 billion a year to ready themselves for war. Of that amount approximately $20 billion a year goes for psychoguerrilla warfare—how to break down each other's (and third-party countries') economy and morale before arriving at the point of war by distribution of narcotics, social engineering, political movements, electronically amplified brainwashing, wheeling and dealing of various sorts. Young people's spontaneous thinking is idealistic. In the 1960s, that idealism was sometimes exploited. Quite often the gentle, angry young people discovered that their heads were sometimes used as battering rams rather than for thinking. Then, over the next fifteen years, through experience, they gradually matured, developing an immunity to political and social exploitation.

As I see it now, every child is born successively in the presence of a little less misinformation and in the presence of a great deal more reliable information. The young world is enormously advantaged.

I asked a young man in Pennsylvania who had written an extraordinary book on the Three Mile Island incident to visit me at my Philadelphia office. He was a high school dropout. I said, "How did you get to writing?" I have never read anything more interesting and sustaining than his book on Three Mile Island. It was well informed on all the bureaucratic decisions in Washington, all the mechanisms of the power structure. It was incredibly well done. He said, "Well, I love reading. I liked particularly Shakespeare, Walt Whitman, and Mark Twain." He had quite a range of inspirers—he just loved them—and he had learned how to express himself well. He also had learned how to put relevant information together. He was typical of a young world that is progressing relentlessly. At twenty-one, he was neither misinformed nor misled. I was astonished. He seemed to me to be a heartening manifest of number-one evolution.

Each year, I get letters from children born after humans landed on the Moon. How these young ones find me to be somebody to write to, I do not know, but they do. They say that they understand that I may empathize with *their* concern. The letters are written in superb English. They are familiar with all the tasks that were necessary to get humans to the Moon and back safely. They are familiar with the Apollo Project's critical path. They know that humanity can do anything it needs to do. They wonder, "Why can't we set about to make this planet Earth work?" The young world gives increasing evidence of this level of concern. The after-the-Moon-landing young people will, within a few years, be able to take over the course-setting tasks of humanity as local Universe information gatherers and local Universe problem solvers in support of the integrity of an eternally regenerative Universe.

In 1979 a newspaperman in Los Angeles, Richard Brenneman, arranged for me to meet with a group of very young people to discuss the subjects I have dealt with in this book.

After six months of reading my books, each had prepared a set of questions about my thoughts and statements. They had lively interest in what I had to say. I asked them their ages. The oldest, a boy, was twelve. He said he was interested in learning the tricks of magicians. The next-oldest, an eleven-year-old boy, said that he was interested in electromagnetics. The third member of the group was a little girl who was then ten years old and the only one of these three born after humans had reached the Moon. I asked her in what she was interested. She answered, "I am a comprehensivist, like you. I am interested in everything." All youth born since the 1969 Moon landing are deeply familiar with the appropriation of billions of dollars for the complex technology of the Apollo Project. The Moon that for three million years had represented the unreachable had been successfully reached. The post-Moon-landers say, "Humanity can do anything it sets out to do. We need to make the world work for everybody on the planet. Let's get going." When they find out that I have discovered what can be technologically accomplished, they perk up their ears and roll up their sleeves.

The passion to understand engenders the passion to demonstrate competence. This is about to be demonstrated by that emergent young world.

An unprecedented transformation of all of our affairs is on the horizon. We are about to see all of the more than 150 nations of the world almost imperceptibly vanish, their function outmoded, their selfish and short-term pursuits no longer welcome or workable in an increasingly interdependent world economy. These nations represent more than 150 blood clots impeding the free circulation of the world's metals and the technological advantaging that they implement. When we engage the economics of recirculating all the metals as scrap, the entrenched mining interests will no longer be able to block that free flow. With the vast uncensorable network of communication media, obstacles to the free flow of vital information will become progressively more difficult to erect and enforce. Traditional human power structures and their reign of darkness are about to be rendered obsolete.

Revolutionary changes in every sphere of life must happen, and there is a young world very glad to realize them. I see clearly that intellectual integrity is trying to make humanity a success.

When I first began doing my own thinking in 1927, I said that I was going thenceforward to completely and irretrievably abandon everything I had ever been taught to believe—and, from that time forward, base my decisions only upon my own experimental evidence. It should be the vow of every scientist.

It is a prominent part of everyone's experience that enormous numbers of humanity are deeply moved by some religious credo or another. People manifest a deep sense that something is everywhere operative which is mysteriously greater than that which is negotiable by the knowledge and will of humans.

I constantly ask myself, "Do you have any experientially evidenced reason to assume a greater intellect to be operating in Universe than that of humans?" I answer myself, "The only-by-mind-discovered generalized principles of science that can only be expressed mathematically and mathematics are inherently intellectual." I found that I was overwhelmed by the experiential evidence of a cosmic intellectual integrity at work in the design of Universe. Thus, when I said in 1927 that I was going to try to find out and support what the great cosmic intellectual integrity was trying to do, I committed myself as completely as humans can to absolute faith in the wisdom of the eternal intellectual integrity we speak of as God. In 1930 Einstein's publication of his "Cosmic Religious Sense," which described his "nonanthropomorphic concept of God," told me that the most profoundly thinking human on our planet was also so committed.

At the outset of my 1927 commitment, I realized that my exploration for comprehension of God's design of eternally regenerative Universe might well mean that I could develop some very powerful insights. I asked myself if I could trust never to turn the power of such insights to personal advantage. Never to consider myself special vis-à-vis God. Never to develop a cult. Never to exploit for selfish reasons the insights I was sure to experience in operating an enterprise backed only by intellectual integrity. My answer to myself was, "Yes, I can trust myself not to selfishly exploit the power of cosmic insights." I have kept my promise, which brings me back to my opening statement about myself: I am an average, healthy human—no less, no more. But all average human beings are magnificently endowed with creativity, and mysteriously capable of vastly more than any of us has ever assumed to be possible.

While it is possible to recognize that humanity is still comprehensively locked in by the Earthian power structures' Dark Ages conspiracy, it is as yet not possible to assess exactly how powerful that imprisonment is. The fact that a vast number of humans still assume that it is within the power of their political leadership and the military might they command to resolve our problems is a reasonable manifest of the continued imprisonment of all humanity.

To this author, the dilemma is so great that in 1983 he found himself writing the following paragraphs, which he titled "Integrity":

A very large number of Earthians, possibly the majority, sense the increasing imminence of total extinction of humanity by the more than 50,000 poised-for-delivery atomic bombs. Apparently no one of the 4.5 billion humans on our planet knows what to do about it, including the world's most powerful political leaders.

Humans did not invent atoms. Humans discovered atoms, together with some of the mathematically incisive laws governing their behavior.

In 1928, humans first discovered the existence of a galaxy other than our own Milky Way. Since then we have discovered 100 billion more galaxies, each averaging over 100 billion stars. Each star is an all-out chain-reacting atomic energy plant.

Humans did not invent the gravity cohering the macrocosm and microcosm of eternally regenerative Universe.

Humans did not invent humans or the boiling and freezing points of water. Humans are 60 percent water.

Humans did not invent the ninety-two regenerative chemical elements of the planet Earth with its unique biological life-supporting and protecting conditions.

Humans did not invent the radiation received from our atomic energy generator, the Sun, around which we designedly orbit at a distance of 93 million miles.

The farther away from its source, the less intense the radiation. With all the space of Universe to work with, nature found 93 million miles to be the minimum safe remoteness of biological protoplasm from atomic radiation generators.

Humans did not invent the vast, distance-spanning photosynthetic process by which the vegetation on our planet can transceive the radiation from the 93-million-mile-away Sun and transform it into the complex hydrocarbon molecules structuring and nurturing all life on planet Earth.

Design is both subjective and objective, an exclusively intellectual, mathematical conceptioning of the orderliness of interrelationships.

Since all the cosmic-scale inventing and designing is accomplishable only by intellect, and since it is not by the intellect of humans, it is obviously that of the eternal intellectual integrity we call God.

All living creatures, including humans, have always been designed to be born unclothed, utterly inexperienced, ergo absolutely ignorant. Driven by hunger, thirst, respiration, curiosity, and instincts such as the reproducing urge, all creatures are forced to take speculative initiatives or to "follow the herd," else they perish.

Ecological life is designed to learn only by trial and error.

Common to all creature experience is a cumulative inventory of only-by-trial-and-error-developed problem-solving reflexes.

Unique to human experience is the fact that problem solving leads not only to fresh pastures, but sometimes to ever more intellectually challenging problems. These challenges sometimes prove to be new, more comprehensively advantaging to humanity, mathematically generalizable, cosmic design concepts.

Humans have had to make trillions of mistakes to acquire the little we have thus far learned.

The greatest mistake we have ever made is to assume that the supreme authority governing life and Universe is not God but either luck or the dicta of the humanly constituted and armed most-powerful socioeconomic systems and religions. The combined human power structures—economic, religious, and political—have compounded this primary error by ruling that no one should make mistakes and punishing those who do. This deprives humans of their only-by-trial-and-error method of learning.

The power structure's forbiddance of error-making has fostered cover-ups, self-deceit, egotism, false fronts, hypocrisy, legally enacted or decreed subterfuge, ethical codes, and the economic rewarding of selfishness.

Selfishness has in turn fostered both individual and national bluffing and vastness of armaments. Thus, we have come to the greatest of problems ever to confront humanity: What can the little individual human do about the supranatural corporate power structures and their seemingly ungovernable capability to corrupt?

A successful U.S. presidency campaign requires a minimum of $50 million, senatorships $20 million, representatives $2 million. Through big business's advertising-placement control of the most powerful media, money can buy, and has now bought, control of the U.S. political system once designed for democracy.

Without God, the little individual human can do nothing. Brains of all creatures, including humans, are always and only preoccupied in coordinating the information fed into the brain's imagination—image-I-nation—its scenarioing center, by the physical senses and the brain-remembered previous similar experience patterns and the previous reflexive responses.

Human mind alone has been given access to some of the eternal laws governing physical and metaphysical Universe, such as the laws of leverage, mechanical advantage, mathematics, chemistry, and electricity, and the laws governing gravitational or magnetic interattractiveness, as manifest by the progressive terminal acceleration of Earthward-traveling bodies or by the final "snap" together of two interapproaching magnets.

Employing those principles first in weaponry and subsequently in livingry, humans have been able to illumine the nights with electricity and to intercommunicate with telephones and to integrate the daily lives of the remotest-from-one-another humans with the airplane.

As a consequence of human mind's solving problems with technology, within only the last three-fourths of a century of our multimillions of years' presence on planet Earth, the technical design initiatives have succeeded in advancing the standard of living of the majority of humanity to a level unknown or undreamed of by any pre-twentieth-century potentates.

Within only the last century, humanity has grown from 95 percent illiterate to 65 percent literate. Preponderately literate humanity is capable of self-instruction and self-determination in major degree.

Clearly, humanity is being evolutionarily ejected ever more swiftly from all the yesteryears' group-womb of designedly permitted ignorance.

Regarding the power-structure-supported Scriptures' legend of woman emanating from a man's rib, there is no sustaining experiential evidence. Humanity now knows that only women can conceive, gestate, and bear both male and female humans. Women are the continuum of human life. Like the tension of gravity-cohering, space-islanded galaxies, stars, planets, and atoms, women are continuous. Men are discontinuous space islands. Men, born forth only from the wombs of women, have the function of activating women's reproductivity.

The present evolutionary crisis of humans on planet Earth is that of a final examination for their continuance in Universe. It is not an examination of political, economic, or religious systems, but of the integrity of each and all individual humans' responsible thinking and unselfish response to the acceleration in evolution's ever more unprecedented events.

These evolutionary events are the disconnective events attendant upon the historic termination of all nations. We now have 163 national economic "blood clots" in our planetary production and distribution systems. What is going on is the swift integration in a myriad of ways of all humanity not into a "united nations" but into a united space-planet people.

Always and only employing all the planet's physical and metaphysical resources only for all the people, this evolutionary trend of events will result in an almost immediately higher standard of living for all than has ever been experienced by anyone.

In general, the higher the standard of living, the lower the birthrate. The population-stabilizing higher living standards will be accomplished through conversion of all the high technology now employed in weaponry production being redirected into livingry production, blocked only by political party traditions and individually uncoped-with, obsoletely conditioned reflexes.

A few instances of persistent, misinformedly conditioned reflexes are the failure popularly to recognize the now scientifically proven fact that there are no different races or classes of humans; the failure to recognize technological obsolescence of the world-around politically assumed Malthus-Darwin assumption of an inherent inadequacy of life support, ergo "survival only of the fittest"; the failure to ratify ERA, the equal rights (for women) amendment, by the thus-far-in-history most-crossbred-world-peoples' democracy in the U.S.A.; or, with ample food production for all Earthians, the tolerating of marketing systems which result in millions of humans dying of starvation each year.

Carelessly unchallenged persistence of a myriad of such misinformed brain reflexings of the masses will signal such lack of people's integrity as to call for the disqualification of humanity and its elimination by atomic holocaust.

You may feel helpless about stopping the bomb.

To you, the connection between the equal rights amendment and the atomic holocaust may at first seem remote. I am confident that what I am saying is true. The holocaust can be prevented only by individual humans demonstrating uncompromising integrity in all matters, thus qualifying us for continuance in the semidivine designing initiative bestowed upon us in the gift of our mind.

The best antidote to the powerfully misintentioned sensing and acting reflexes of society is the study of synergetics. The data of synergetics as presented in the two volumes of *Synergetics* and background data in *Critical Path* (1981) are adequate to the task of breaking the Dark Ages stranglehold on the human individual. This book has undertaken to present some of the principal synergetics concepts in a logical sequence. It does not treat the successively acquired realizations in the detailed degree of *Synergetics,* the definitive reference on the subject.

As a guidebook to synergetics in the context of its historic roots, this volume has added new insights and primary concepts to the subject, including some that have accrued since its earlier expositions. Study of synergetics with continued recommitment of human individuals to utter faith in the comprehensive wisdom and absolute power of the intellectual integrity and love governing an eternally regenerative Universe may bring about our ultimate escape from the Dark Ages' race-suicidal obsession with the misconception that cosmic supremacy is vested in little planet Earth's politicians, priests, generals, and monetary-power wielders.

Dear reader, traditional human power structures and their reign of darkness are about to be rendered obsolete.

Bucky Fuller and Thomas T. K. Zung.

Fuller Today: The Legacy of Buckminster Fuller

Thomas T. K. Zung

Since Buckminster Fuller's death, humanity has been experiencing a huge critical lag. Technology has been advancing at warp speed, science has made huge breakthroughs in medicine, and computers and mobile phones have increased our knowledge and abilities, changing our lives exponentially. Yet much of our information—some biased, some prejudiced, some simply unsubstantiated—has us floundering about, lagging behind our own advances intellectually. Fuller's advice would have been to dare to think for yourself and to trust your instincts, using science as a beacon. This new chapter has been added to the 2014 edition of the anthology as an attempt to address this critical lag by presenting a series of vignettes of Fuller's continuing impact and influence after his death. It traces some of the ripples and reverberations from his work and ideas that radiate out into the world to everyday life.

It has been more than thirty years since Buckminster Fuller left Spaceship Earth on July 1, 1983, bound for another intention in the universe. By adding some photographs not included before, this reissue reflects a culture that has changed with time, much of that change predicted by Fuller. Bucky was "like a Messiah of ideas. He was a prophet of things to come," said the late sculptor Isamu Noguchi, who had been Fuller's best friend. Lord Norman Foster, who curated the Bucky Fuller & Spaceship Earth exhibition in 2010–11, said at the exhibition opening in Spain, "Buckminster Fuller is more relevant today than perhaps when he was alive, [with] our Earth in such a fragile state. Fuller provided us with options to make the world work in the shortest time possible with the least disruption to the environment. Bucky is more needed than ever."

One example of the fragile state is the risk associated with the use of nuclear power. Fuller warned about this years ago. Three years after his death, in April 1986, a nuclear disaster occurred at Chernobyl, Ukraine, that shook all of Europe. More recently, in March 2011, a calamity at the nuclear power plant in Fukushima, Japan, rocked the world. In November 2013, Douglas Dasher, a researcher at the Institute of Marine Science at the University of Alaska, Fairbanks, held a press briefing in which he expressed concern about radioactive contamination of fish and wildlife along the west coast of North America. Clearly, a major challenge in our century is achieving sustainability without damage to our environment.

Fuller connected sustainability to humanity. In 2002, a post-Fuller book on sustainability titled *Cradle to Cradle: Remaking the Way We Make Things* was published. In it, the book's coauthors, architect William McDonough and chemist collaborator Michael Braungart, propose that humans should create products based on the principle in nature that "waste equals food." Bill is currently exploring geodesic structures, using waste plastic and reshaping the plastic into useful forms for emergency shelters. His optimization of the full life cycle of products, in which there is no waste, is expounded in his 2003 book, *The Hannover Principles*.

Today architectural students—or any students, for that matter—do not study Fuller's work; he is barely mentioned except in conjunction with the geodesic dome. However, Fuller is more than a patent number. Sixty years ago, when I was one of Fuller's students at age nineteen, my class at Virginia Polytechnic Institute participated in a project studying the design of the Thomas Edison space-frame kite. Using Fuller's concept of tension and compression, known as

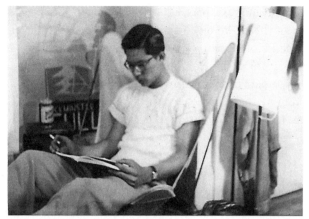
Nineteen-year-old Thomas T. K. Zung working on the Fuller-Edison project

Left to right: Thomas T. K. Zung, William McDonough, and Tommy Bates Zung

tensegrity, we explored the kite's structure and optimized its design. Fuller used this project to encourage us to develop a design science approach to architecture.

When he learned that I had been drafted to serve in the Korean War, Fuller suggested that I spend my off time studying chemistry rather than hanging out in canteens and saloons. As he had said to my class, "chemistry is basic structure, ergo architecture," a prelude to design science, and design science is the basis of architectural thinking as part of the whole, synergetics. I was grateful for his advice, and chemistry prompted my education in the design science application to architecture. After an exciting eight-year stint with his architect friend Edward Durell Stone in New York City, I completed the circle when my road led me back to Fuller.

When the 2013 Nobel Prize in Chemistry was awarded to three U.S.-based scientists, Dr. Martin Karplus, Dr. Michael Levitt, and Dr. Arieh Warshel, who pioneered the use of computers to probe complex chemical reactions, it was not surprising that they had used these technological tools for their scientific research. As Fuller said in his 1963 book *R. Buckminster Fuller on Education*, "Computers as learning tools can take over much of the 'educational metabolics,' freeing us to really put our brains and wisdom to work." He also predicted in that book, "I am quite sure that we are going to get research and development laboratories of education where the faculty [scientists] will become producers of extraordinary moving-picture documentaries." Studying Fuller's writings at Stanford University may yield profound insights and enable scholars, thinkers, and scientists to help solve the world's problems.

To the average reader, some of Fuller's words may not be easily understandable, but taken in its entirety, his body of writing is akin to a Mozartian symphony. Fuller himself said that he did not mind if people did not understand him, but he did not want them to misunderstand him. One of his relatives recently asked me what Fuller's greatest invention was. He was expecting me to say it was the geodesic dome, but to his astonishment, my reply was that Bucky was Fuller's greatest invention. Fuller wrote a prelude for the future of mankind, charting a course with ideas and inventions for generations to pursue.

Fuller Materials at Stanford

The R. Buckminster Fuller Collection, in the Special Collections and University Archives Department of Stanford University Libraries, is undoubtedly the paramount collection of a twentieth-century inventor. Acquired in 1999 from the Estate of R. Buckminster Fuller, the collection is open to all persons who visit the libraries or use the Internet. The Stanford Libraries have made it possible for anyone to access the collection online and freely explore the world of Bucky Fuller at no cost. However, going to the Stanford Libraries in person is even better.

According to Stanford University librarian Michael A. Keller and Special Collection head Robert Trujillo, Fuller's Dymaxion Chronofile, which melds education and technology, is "the most sought-after archive" in Stanford's Special Collections and University Archives Department. There, Keller, who finalized the acquisition of the archives from the Estate of R. Buckminster Fuller, navigates scholars through the vast Fuller files.

The Stanford Guide to the R. Buckminster Fuller Papers with descriptive summary, collection number M1090 in the Manuscripts Division, is amazing and exciting. The papers include Fuller's personal archive, called the Dymaxion Chronofile, as well as manuscripts, drawings, and audio-visual materials. They are broken down into 25 series and many subseries. Series 19 includes Fuller's geometric, architectural, and other models and artifacts, which will

Left to right: Michael A. Keller, Stanford University librarian; Lady Elena Foster of Ivorypress; Robert Trujillo, head of Special Collections; Lord Norman Foster; and Thomas T. K. Zung at Stanford University's Green Library preparing a Buckminster Fuller exhibition on Dymaxion Car No. 4

fill academicians, artists, and other visitors with wonderment at this enigmatic genius. Many of the models in this series can be viewed online.

To inaugurate the acquisition of the Fuller archives in 2001, Keller invited some of those who knew Fuller best to talk about their experiences with Bucky in video interviews at Stanford. These conversations took place over the next two years with E. J. Applewhite, collaborator on *Synergetics* and *Synergetics 2* (2002); Shoji Sadao, a Fuller architectural partner (2002); Stewart Brand, publisher of the *Whole Earth Catalog* (2002); Thomas T. K. Zung, a Fuller architectural partner (2002); Allegra Fuller Snyder, Fuller's daughter (2003); designer Edwin Schlossberg (an off-site interview, 2003); artist Kenneth Snelson, (2003); history professor Theodore Roszak (2003); and architect Lord Norman Foster (2003). All these video interviews are available at the Stanford Libraries Archives.

Exhibitions and Events

During Fuller's lifetime, there were many exhibitions of his work, including in 1959 at the Museum of Modern Art and in 1976 at the Smithsonian's Cooper-Hewitt, National Design Museum, both in New York City. Since his death, interest in Fuller has mushroomed and has led to a flurry of Fuller exhibitions and events, including the following:

1990—Buckminster Fuller Earth, opening on Earth Day as part of an Earth Day celebration, Bard College, Bard, New York

1991—Buckminster Fuller exhibition, Nautilus Foundation Museum, Lloyd, Florida

1999—Your Private Sky, Museum fur Gestaltung, Zurich, Switzerland; Claude Lichtenstein, curator; catalog publication by Joachim Krausse and Claude Lichtenstein

1999—Your Private Sky, Kunsthalle Tirol, Hall, Austria

2000—Your Private Sky, Design Museum, London

2000—Your Private Sky, Stiftung Bauhaus, Dessau-Rosslau, Germany

2000—Your Private Sky, Zeppelin Museum, Friedrichshafen, Germany

2001—Your Private Sky, Museum of Modern Art, Kamakura, Japan

2001—Your Private Sky, Aichi Prefectural Museum of Art, Osaka, Japan

2002—Your Private Sky, Watari Museum of Contemporary Art, Tokyo

Bust of Buckminster Fuller by his best friend, Isamu Noguchi

Front row, left to right: Tsuneko Sadao, Joyce Burke-Jones, Thomas T. K. Zung, and the Swiss New York consulate ambassador to the United States. Back row, left to right: Shoji Sadao, E. J. Applewhite, Claude Lichtenstein, and lady trustee at Zurich museum exhibition Your Private Sky.

Restored Dymaxion house at the Henry Ford Museum in Dearborn, Michigan

2002—*Buckminster Fuller & SIU, Southern Illinois University, Carbondale; Professor Bill Perk, curator*

2006—*Best of Friends: Buckminster Fuller and Isamu Noguchi, Isamu Noguchi Museum, Long Island, New York; Shoji Sadao, curator*

2008—*Best of Friends: Buckminster Fuller and Isamu Noguchi, Henry Ford Museum, Dearborn, Michigan*

2008—*Buckminster Fuller: Starting with the Universe, Whitney Museum of American Art, New York City*

2008—*Buckminster Fuller artifacts, Carl Solway Gallery and Barquet Gallery, New York City*

*2009—Buckminster Fuller: Starting with the Universe, Museum of
     Contemporary Art, Chicago*

*2010—Black Mountain College and Its Legacy, Loretta Howard Gallery,
     New York City*

*2010—Bucky Fuller & Spaceship Earth, with Dymaxion Car No. 4,
     Ivorypress Gallery, Madrid, Spain; Lord Norman Foster, curator;
     Lady Elena Foster, exhibit producer*

Executive Director Dennis Bartels at Exploratorium with Tactile Dome in San Francisco

Left to right: Thomas T. K. Zung; Lady Elena Foster, exhibit producer; and Lord Norman Foster, exhibit curator, at Ivorypress Gallery in Madrid, Spain

Left to right: Lady Elena Foster, Lord Norman Foster, Thomas T. K. Zung, and museum artistic director Roland Nachtigaller at the Bucky Fuller & Spaceship Earth exhibition in the MARTa Herford Museum, Herford, Germany

2010—Fiftieth Anniversary of ASM International (formerly American Society of Metals) geodesic dome installation, in Novelty, Ohio; celebratory event for the landmark by the Cleveland Restoration Society, Kathleen Crowther, director. The ASM dome was the location of Buckminster Fuller's public memorial service in 1983 and was placed on the National Registry of Historic Places in 2009.

2011—Bucky Fuller & Spaceship Earth, with Dymaxion Car No. 4, MARTa Herford Museum, Herford, Germany; Lord Norman Foster, curator; Roland Nachtigaller, museum artistic director

2011—Twenty-six-foot Fly's Eye Dome with collector Craig Robins and the Buckminster Fuller Institute, and Dymaxion Car No. 4 display with Lord Norman Foster, Art Basel Week, Miami Beach, Florida

2012—The Utopian Impulse: Buckminster Fuller and the Bay Area, including Inventions: Twelve around One, San Francisco Museum of Modern Art

2012—Buckminster Fuller, Inventions: Twelve around One, Joel and Lila Harnett Museum of Art, University of Richmond, Virginia; Richard Waller, curator

2013—Twelve around One: Buckminster Fuller. FRAC Provence-Alpes-Cote d'Azur Museum, Marseille, France; Fabienne Clerin, exhibits

2013—*Back to the Future, Toulouse International Art Festival, France, with restored Fuller fifty-foot Fly's Eye Dome, with geometry similar to that of C_{60}, aka buckminsterfullerene or buckyballs; Robert E. Rubin, architectural historian and preservationist*

2013—*The Tactile Dome, Exploratorium, Embarcadero, San Francisco; exhibit encased in a geodesic dome, conceived by architect Carl Day and the late Dr. August Coppola, brother of Francis Ford Coppola and father of actor Nicolas Cage. The science museum was founded by Frank Oppenheimer, brother of atomic bomb scientist J. Robert Oppenheimer. Dr. Dennis M. Bartels became executive director in 2006.*

The Nobel Prize for Buckminsterfullerene

In 1996, thirteen years after Bucky's death, the Nobel Prize in Chemistry was awarded to Sir Harold W. Kroto, Dr. Robert F. Curl Jr., and Dr. Richard E. Smalley for their 1985 discovery of carbon C_{60}, a molecule with sixty atoms, the strongest bonding known in nanotechnology. They named the C_{60} molecule, which resembled a geodesic dome, after Buckminster Fuller, calling it buckminsterfullerene, or buckyballs for short.

Thomas T. K. Zung with Nobel laureate Robert F. Curl at Rice University

Nobel laureate scientist and professor Sir Harry Kroto with students at a Vega educational workshop class

Sir Harry Kroto, currently a professor at the University of Florida and University of Sussex in the United Kingdom, is working with students through his Vega educational organization, providing education based on science and geometry. Using Fuller's Dymaxion maps and geodesic domes, Kroto demonstrates for his students using geometry with hands-on experience. He provides science as a base, and he told me in personal correspondence in 2011 that he is very concerned about youth education, particularly with "political lunatics who misinform our children with information not based on scientific information," and asked, "What else are they teaching the children?"

When I corresponded with Robert Curl, a professor at Rice University, in 2012, he told me that what he said in his Nobel banquet speech rings just as true to him today as it did then. "I will never recover, or am I likely to recover, from the astonishment that these beautiful three dimensional molecules are formed spontaneously when carbon vapors condense under the right conditions, and that we were fortunate enough to discover this amazing fact. Suddenly a door opened into the new world of fullerenes." His feeling of astonishment remains as great today as it was when the molecule was first discovered.

The third Nobel laureate, Richard Smalley, sadly died of cancer in 2005. Prior to his death, he had been working on nanotechnology with material said to be fifty to a hundred times stronger than steel and one-sixth the weight. Smalley had renovated his home near Rice University with a glass skylight shaped like half a buckyball, with precisely proportioned steel struts representing the bonds between atoms of C_{60}. His passing was a great tragedy for science.

U.S. Postage Stamp

On July 12, 2004, on what would have been Fuller's 109th birthday, the U.S. Postal Service released at Stanford University a commemorative Buckminster Fuller stamp, marking the fiftieth anniversary of Fuller's patent for his famed geodesic dome. Bucky's daughter, Allegra Fuller Snyder, helped unveil the stamp, and Stanford University librarian Michael A. Keller welcomed the enthusiastic gathering. "The stamp is a tribute to a man whose keenness for man is capsuled with his 'what one man can do that great nations and international mega-companies cannot do, think for themselves,'" said Keller. "A visionary thinker, though considered 'loony' in previous decades, was proven right with the passage of time." Also on hand for the stamp unveiling was astronaut Col. Edwin "Buzz" Aldrin, the "man on the moon," who had been a friend of Fuller's. Aldrin spoke on space synergetics and his conversations with Fuller on tensegrity and cosmic energy.

The postage stamp image was inspired by a painting of Fuller by artist Boris Artzybasheff that had appeared on the cover of *Time* magazine in January 1964. Today the Fuller 37-cent stamp is a coveted memento for philatelists and Fuller friends to treasure.

Astronaut Buzz Aldrin unveiling the U.S. postage stamp at Stanford University on the fiftieth anniversary of Fuller's geodesic dome patent

Second day of stamp issue event held by Global Energy Network Institute (GENI) in San Diego. *Left to right:* D. W. Jacobs, writer-director of "The Bucky Play"; Peter Meisen, president of GENI; Carl Herrman, art director and designer of the stamp; and Jim Olsen, acting San Diego postmaster.

The Bucky Play

A two-act multimedia play titled *R. Buckminster Fuller: THE HISTORY (and Mystery) OF THE UNIVERSE*, written and directed by D. W. Jacobs, was drawn from Fuller's life, work, and writings. Of the many Fuller plays produced, this one perhaps represents the most captivating aspect of Bucky Fuller and his habit of thinking out loud. It explores how Bucky's persona and his legacy live on in today's world, at this critical moment for Earth. The play premiered at San Diego Repertory Theatre in 2000 and has since been produced and performed in many cities throughout the United States. In 2005, a French version of the play, titled *R. Buckminster Fuller: MEMOIRES (et Mysteres) DE L'UNIVERS*, was performed in Montreal, Quebec, produced by Theatre Alambic and Theatre Denise-Pelletier, translated by Maryse Pelletier, with Jean Boilard playing Bucky, and directed by Bernard Lavoie. The U.S. venues where the play has been performed include the following:

San Diego: San Diego Repertory Theatre, performed by Ron Campbell

San Francisco: Lorraine Hansberry Theatre and others, performed by Ron Campbell

Chicago: Mercury Theater, performed by Ron Campbell

Seattle: Intiman Theatre, performed by Ron Campbell

Ventura, CA: Rubicon Theatre and Z Space Studio, performed by Joe Spano

Portland, OR: Portland Center Stage, performed by Doug Tompos

Washington, DC: Arena Stage, performed by Rick Foucheux

Cambridge, MA: American Repertory Theater (Harvard), performed by Thomas Derrah

San Jose: San Jose Repertory Theatre, performed by Ron Campbell

Carbondale, IL: Southern Illinois University Guyon Auditorium, performed by D. W. Jacobs

R. Buckminster Fuller: The History (and Mystery) of the Universe

Fuller Friends

Early in Fuller's life, he connected with and influenced many artists of the day in New York City; Black Mountain and Raleigh, North Carolina; Ann Arbor, Michigan; Philadelphia; and Carbondale, Illinois. The list of Fuller friends is awesome, including such notables as John Denver, David Rockefeller, John Huston, Ellen Burstyn, Dr. Henry Heimlich, Valerie Harper, and the surrealist painter Salvador Dalí, who used a geodesic dome in his Teatro-Museo Dali in Figueres, Spain. This section showcases Fuller with some of his friends in photos from the collection of Fuller's executive secretary, Shirley Sharkey, and from the Stanford University archives.

Left to right: John Denver, Bucky, and Shirley Sharkey planning Choices for the Future, the Windstar Symposium in Aspen, Colorado, for Denver's concert with Fuller

Left to right: Tommy Zung, John Denver, Bucky, and Peter Kent getting ready for a John Denver concert in Philadelphia. Bucky is wearing a concocted "space cadet" earphone that allows him to hear the music and words of Denver's song "What One Man Can Do."

Left to right: Bill Baird, Shirley Sharkey, Bucky, Bill Sharkey, and Ellen Burstyn discussing a proposed play about Bucky's great-aunt Margaret Fuller Ossoli, a transcendentalist, friend of Ralph Waldo Emerson, and perhaps America's first feminist

Left to right: Bucky, Shoji Sadao, Thomas T. K. Zung, banker Whit Whitlow, and Isamu Noguchi at the David Rockefeller compound in Bar Harbor, Maine

Left to right: Thomas T. K. Zung, Shirley Sharkey, Ed Applewhite, Isamu Noguchi, Bill Sharkey, and Shoji Sadao at Bucky's eighty-fifth birthday party in Maine, hosted by Neva Goodwin, David Rockefeller's daughter

Bucky and David Rockefeller discussing *Critical Path* at the Rockefeller home in Maine

Dr. Henry Heimlich, inventor of the Heimlich maneuver, and Bucky discussing tension and compression at Fuller's Philadelphia office

Left to right: Shirley Sharkey, Bucky, film director John Huston, and Thomas T. K. Zung in Budapest, Hungary. John Huston and Fuller were planning films (now at the Stanford University Archives) involving Huston's son Tony, a noted screenwriter who worked directly with Fuller.

Fuller's Travels in China

During Fuller's extensive travels, he often met with world leaders; the breadth of his interactions is astonishing. Previous publications on Fuller show photos of Indian prime minister Indira Gandhi, Princess Margaret of the United Kingdom, and Bucky in front of the Moscow geodesic dome, with Russian premier Nikita Khrushchev in the distance.

American Nobel laureate Pearl S. Buck and Fuller both wished to visit China. When travel restrictions in China were relaxed in the early 1970s, they applied to the Chinese embassy in Ottawa, Ontario, for visas. Coincidentally, future People's Republic of China ambassador to the United States Zhang Wenjin was en route to Canada at the time. Prior to his foreign appointments, Zhang had been an interpreter to Premier Mao Zedong and a right-hand man to Premier Zhou Enlai. It was Zhang who met secretly with then National Security Advisor Henry Kissinger to plan President Richard Nixon's first visit to China. Zhang took an interest in Fuller's work. Later we became acquainted, and Zhang was kind enough to introduce me to his successor, Ambassador Han Xu, as well as Henry Kissinger. Both Chinese ambassadors visited Fuller's Ohio office to see firsthand his Dymaxion works.

COPYRIGHT © KARSH

Nobel laureate Pearl S. Buck, at her desk with *The Good Earth* manuscript, contacting Fuller about her wish to visit China.

China's ambassador to the United States, Zhang Wenjin, visiting Fuller's Ohio office to learn firsthand about Bucky Fuller, with Thomas T. K. Zung

Fuller visited China, where he was received enthusiastically, but Madame Buck died before she could revisit China, the setting of her book *The Good Earth*. Her son Edgar Walsh, the literary executor, and one of his sisters, Janet Walsh, lamented the fact that her dream remained unfulfilled. It would have meant so much to their mother, as China had been her home, a fact highlighted in a Broadway play about Pearl S. Buck starring Valerie Harper, *All under Heaven*. Fuller noted that Buck had done for the China poor what Dickens had done for London's forgotten; with technology, this would have helped lift their burden.

Fuller's trip to China presented an opportunity for a geodesic dome project, planned to be an international trade center. Sadly, the tragic circumstances of the Tiananmen incident caused the project to be abandoned.

Thomas T. K. Zung with former U.S. secretary of state Henry Kissinger, who was interested in Fuller's work and his visit to China

Thomas T. K. Zung with Han Xu, China's ambassador to the United States and a former neighbor to George H. W. Bush in China, learning about a Fuller dome. Han, a former interpreter to Premier Mao Zedong, expressed hope for a geodesic dome in China.

Proposed China international trade center with a 450-foot geodesic dome, a plan abandoned after the Tiananmen tragedy

Chinese ambassador Zhang Wenjin, former right-hand man to Premier Zhou Enlai, with Ohio governor Richard Celeste in Beijing, China, discussing the trade center dome. Celeste later officiated at Fuller's public memorial service at the ASM dome in Ohio in 1983.

The Campuan Meetings and Bucky in Southeast Asia

Bucky's travels and impact in Southeast Asia merit a fuller account. In 1969, after calling on India prime minister Indira Gandhi in New Delhi, Fuller visited Singapore, then Bali, Indonesia, and Penang, Malaysia. He established friendships in the region, especially with Massachusetts Institute of Technology (MIT) graduate architect Lim Chong Keat, with whom he convened successive meetings with special friends on world issues at Campuan, Bali, beginning in 1976.

Fuller's final Campuan meeting was in 1983, held at Bellevuc in Penang, where he had been collaborating with Lim and Shoji Sadao on the Komtar Dome, perhaps one of his last major works. It was finally completed in 1986. In conjunction with the opening of this geodesic dome, then named Dewan Tunku after Malaysia's first prime minister, an exhibition on Fuller's work was held there and yet another Campuan meeting was convened in his memory. Lim and other friends subsequently continued the series with a meeting in 1995 at MIT and Harvard.

Komtar Dome in Malaysia

Geodesic dome representing the United States, in the international garden at Suan Luang Rama IX Public Park in Bangkok

It was in Penang that Fuller had finalized *Critical Path* and worked on *Cosmography*. He also copublished with Lim *Synergetics Folio*, featuring ten original posters that had been launched by Bucky at an exhibition at Alpha Gallery in Singapore. In 1988, Lim and Thai architect Sumet Jumsai were responsible for a memorial to Bucky in Southeast Asia: a geodesic dome representing the United States, in the international garden at Suan Luang Rama IX Public Park in Bangkok, Thailand.

Noted Geodesic Domes

The U.S. Pavilion at Expo 67 in Montreal, Canada: the Buckminster Fuller "Cathedral." The iconic dome started a trend for every World's Fair. It is interesting that Fuller's first concept was a giant octet-truss space frame, while his second concept was the famed 250-foot geodesic dome. The dome framework is still standing strong as ever in Montreal.

Climatron, at the Missouri Botanical Garden in St. Louis, opened to the public in 1960 and renovated in 1988 to replace the original acrylic sections with glass panels. The geodesic dome was structurally able to support the extra weight of glass, demonstrating the dome's flexibility and strength.

Spaceship Earth at the Epcot Center, at Disney World in Orlando, Florida. This geodesic dome is perhaps the most recognizable to the public and houses an attraction by the same name. Conceived by two of America's foremost visionaries, Walt Disney and Buckminster Fuller, this dome was meant as a symbol of things to come. Disney, who ventured into the future through cartoons, nature, and cities, and Fuller, who cared deeply about the ecology and the environment of man, were two contemporaries who best exemplified original thinking. Although Disney may not have drawn every one of his cartoon characters, nor did Fuller calculate every strut of each dome, history would be much less full without their conceptual ideas. Those of us who worked on their projects felt privileged to practice in their company; we all swam in the original waters of these two geniuses.

Headquarters of ASM International (formerly the American Society of Metals), Materials Park, Ohio. ASM celebrated the fiftieth anniversary of this building in 2010. Fuller's dome concept was a tensegrity symbol, used to highlight the importance of metals to benefit all humankind. The original architect for the building was the late J. T. Kelly, and Synergetics, Inc., was the engineer. The office complex and dome structure were renovated in 2011 by the Chesler Group. Thomas Passek is the CEO of ASM International.

Elongated sports dome in Cleveland, Ohio. This elongated barrel, geodesic vault dome was used as a sports facility for Cleveland State University. Fuller called it the caterpillar dome. Designed as a temporary structure, the students insisted it remain; after forty years, in 2009, it was replaced with a student center.

Close-up of a geodesic space frame with Fuller's grand-godson Tommy Zung. The dome was engineered by Synergetics, Inc., and Sverdrup & Parcel engineers.

Kuwait Towers, a national landmark in Kuwait City. The world may recall seeing images of these towers during the Persian Gulf War in 1990–91. The main sphere contains a restaurant and an observation deck.

Reunion Tower in Dallas. A geodesic sphere perched five hundred feet in the air, this symbol tops a renovated hotel complex framed by the late entrepreneur Lamar Hunt. This Texas-size tower with restaurant, radio station, and observation tower has become a symbol of Dallas.

Shanghai Tower in China. This emerging nation wished to replicate a Fuller geodesic sphere. Fuller and his colleagues Shoji Sadao and Thomas T. K. Zung visited China many times, and perhaps this had an influence on Chinese design students. Nobel laureate Pearl S. Buck told Fuller that his 1969 book *Operating Manual for Spaceship Earth* had been translated into Chinese, a rarity in the late 1970s.

Other Uses of Geodesic Domes

The strength of geodesic domes under severe conditions and the ease of erection of the manufactured parts have led to their use in many different situations. One example was the 415-foot-wide, 130-foot-high geodesic dome that housed Howard Hughes's Spruce Goose. This dome was laid out in sections. Crews placed the parts and units on the ground, and each section was then lifted up to assemble the dome. This dome was manufactured at Temcor with Don Richter, a Fuller student and colleague.

In 1975, the Amundsen-Scott South Pole Station was rebuilt as a geodesic dome. The dome was manufactured by Temcor, whose vice president, Don Richter, had been a student of Fuller's at Black Mountain College and a colleague of Zung's. U.S. Navy Seabee crews had to brave freezing temperatures and high winds to assemble the dome from modular components, shipped by plane. The structure withstood winds of up to 130 miles per hour and freezing temperatures for more than forty years, with huge snows pushing against the dome sides. After a new station was built, the dome was dismantled; it was reassembled in 2011 at the U.S. Navy Seabee museum in Port Hueneme, California.

A geodesic dome was erected in Greenland near the North Pole in 2008 for the North Greenland Eemian Ice Drilling project (NEEM). It is a scientific site dedicated to ice-core drilling and analyzing carbon isotopes from the last three ice ages. This international effort to comprehend the dramatic effects of climate change is supported by the National Science Foundation's Division of Polar Programs. The NEEM dome was designed by Blair Wolfram, a post-Fuller student, and manufactured by Dome, Inc.

South Pole dome, erected by the U.S. Navy to house scientific instruments in one of the world's coldest environments. Buried in snow, the temporary dome lasted nearly half a century.

Geodesic dome erected in 2008 near the North Pole for the North Greenland Eemian Ice Drilling project (NEEM). This dome is used by scientists from fourteen nations for drilling and studying ice cores that date as far back as three ice ages ago. The dome can withstand 150-mile-per-hour winds.

Buckminster Fuller also made a significant contribution to the protection of the United States during the cold war, when geodesic domes were used as radomes to protect Department of Defense dish antennas in locations around the globe as part of the Distant Early Warning (DEW) Line radar system.

Buckminster Fuller Institute (BFI)

After Fuller's death, his daughter Allegra Fuller Snyder and his grandson Jaime founded the Friends of Buckminster Fuller, which morphed into the Buckminster Fuller Institute (BFI). Allegra and Jaime invited E. J. Applewhite, Bill Perk, Don Moore, Martin Leaf, Hans Meyer, Shirley Sharkey, and Thomas T. K. Zung as founding board members in Los Angeles. BFI moved from Los Angeles to Santa Barbara, California, then to its current location in Brooklyn, New York, in 2004. Since BFI's inception, Allegra has served as board chairperson, and for many of those years, Neil Katz served as board president. With recent restructuring, Allegra is now board chair emerita, David McConville is the current board chair, and Elizabeth Thompson is the executive director.

In recent years, the Fuller Challenge was initiated, with a $100,000 award granted each year to aid in developing and implementing the winning idea. The challenge is open to original environmental and social ideas that are "innovative solutions to some of humanity's most pressing problems." For the first five years, an anonymous donor made this possible, and Allegra Fuller Snyder continues the legacy. Following are the Fuller Challenge grant winners thus far:

Thomas T. K. Zung with BFI founders Allegra Fuller Snyder and Jaime Snyder

2008—The Challenge of Appalachia: Dr. John Todd's vision for the design of a new economic model for a carbon-neutral world

2009—Sustainable Urban Mobility and Mobility-on-Demand System: Efficient electric vehicles designed by an interdisciplinary team of students from Massachusetts Institute of Technology Media Lab Smart Cities Group

2010—Operation Hope: Savory Institute's management principles for reversing desertification to provide permanent water and food security for Africa's impoverished millions

2011—Blue Ventures: Development of programs in partnership with communities to conserve threatened marine environments

2012—The Living Building Challenge: Innovations in the built environment so that it will be integrated with its ecosystem

2013—Ecovative: Creation of innovative packaging, building, and other products from biomaterials

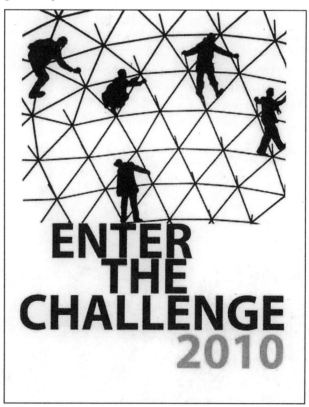

The 2010 Fuller Challenge

Synergetics Collaborative (SNEC)

Shortly after the formation of the BFI, E. J. "Ed" Applewhite, Fuller's collaborator on both the first and second books on *Synergetics: Explorations in the Geometry of Thinking*, convened a meeting of Fuller colleagues at the Isamu Noguchi Museum in New York. The purpose was to explore gathering an informal group of scholarly minds to share synergetic discoveries. In addition to Applewhite, the initial members were Shoji Sadao, Thomas T. K. Zung, Shirley Sharkey, Russell Chu, Joe Clinton, Carl Solway, David Zung, and Tommy Zung. The group dubbed itself SNEC, an unseemly but geographically based name that was the acronym for Synergetic North East Coast.

The organization's mission was furthered by John Belt, a professor at the State University of New York at Oswego, and Joseph Clinton, a former Fuller graduate student at Southern Illinois University. Belt and Clinton held workshops at Oswego illustrating Fuller's "I seem to be a verb" philosophy. SNEC evolved into Synergetics Collaborative, sponsoring international symposia on synergetics and the study of nature's geometry at Oswego State University and the Rhode Island School of Design (RISD).

Don Richter, a Bucky student at Black Mountain College and a founder of the dome-building company Temcor, at a SNEC symposium in Oswego, New York

Joseph Clinton and Thomas T. K. Zung with a dome model for a SNEC workshop in Carbondale, Illinois

Interest has been so keen that SNEC has held workshops and seminars at the Noguchi Museum, American University, George Washington University, the Henry Ford Museum, and RISD, where SNEC programs are now held semiannually. At the Noguchi Museum, where retired Shoji Sadao was past executive director, the current director Jenny Dixon offered the museum facilities to Fullerites for a SNEC symposium about Black Mountain College in 2005. Fuller colleagues who have participated at sessions include Peter Pearce, designer of Biosphere 2 in Arizona, originally built to be a self-contained ecosystem, and Don Richter of the dome-building company Temcor, who encouraged the thousand-foot geodesic dome to Thomas Zung. Dr. Donald Ingber of Harvard University; Estrellita Karsh, widow of Yousuf Karsh; Swiss designer Caspar Schwabe; artist, engineer, and architect Chuck Hoberman; and Japanese architect Edward Suzuki, as well as many other notables, have appeared at SNEC programs.

Keeping the group informed and recording the SNEC work archives is hardworking C. J. Fearnley from Philadelphia. John Belt, the current SNEC president, received the 2012 Chancellor's Award for Excellence in Teaching from the State University of New York.

Thomas T. K. Zung with a thousand-foot Gigundo dome model at Fuller's Ohio studio

SNEC president John Belt and students at a SNEC workshop in Oswego, New York

Rhode Island School of Design (RISD)

At Harvard University, Dr. Arthur Loeb, a Fuller friend, conducted the Synergetics program, a popular curriculum with students, and he accumulated an expansive treasure of teaching artifacts. After his death in 2002, these models, spheres, geometric shapes, and pattern images were too valuable to be discarded, and his widow, Charlotte "Lotje" Loeb, wished to pass on these incredible teaching tools to the next generation of educators. Based on suggestions from some of Loeb's former students, Sandy Johnson, Amy Edmondson, Rachel Reimer, and Karl Thidemann, Lotje contacted me to seek an institution sympathetic to Loeb and Fuller's vision.

Loeb and Fuller had been familiar with the Rhode Island School of Design. Former RISD president Roger Mandle and Yousuf Karsh were both members of a club in New York that Fuller frequented, and in 1975, Fuller had been invited by Brown University–RISD to be the keynote speaker at the National University Conference on Hunger. In 2006, Shoji and I met in New York with John Maeda, president of RISD, hosted by Professor Peter Dean and the director of RISD's Edna W. Lawrence Nature Lab, Karen Idoine. This began the series of events that saved the Harvard-Loeb artifacts and created a new home for them at the nature lab. Lotje Loeb also generously donated an Alexander Calder sculpture along with the artifact collection endowment, and the Loeb Synergetics Design Lab tools were moved from Harvard University to RISD.

SNEC workshop in the Edna W. Lawrence Nature Lab at RISD

Left to right: Thomas T. K. Zung; Estrellita Karsh, widow of Yousuf Karsh; and Jerry Fielder, Karsh curator, donating a Bucky Fuller sculpture artifact and Karsh portraits to the RISD Art Museum

The Edna W. Lawrence Nature Lab offers workshops and Fuller-Loeb design courses. The current director is Neal Overstrom. Others involved in the SNEC collaboration with RISD were Peter Dean, senior critic in furniture design; Carl Fasano, senior critic in foundation studies; Amy Leidtke, in industrial design; Kyna Leski, department head of architecture; and Steve Metcalf, RISD board member.

Estrellita Karsh, widow of the famed photographer Yousuf Karsh, was impressed with RISD, and she consulted with me about donating a noted sculpture piece by Fuller to the RISD Museum. Generously, Estrellita also donated Karsh portraits of Buckminster Fuller, Isamu Noguchi, Mies van der Rohe, and a bevy of well-known artists to the RISD Museum.

SNEC has collaborated with RISD to convene biennial workshops and symposia in Providence for students and design scientists from around the world. The Fourth Biennial Design Science Symposium took place at RISD in early 2014.

Bucky's Dome Home in Carbondale, Illinois

Bucky and his wife, Lady Anne, lived in a dome home in Carbondale, Illinois, for twelve years, from 1959 to 1971. Some describe this period, when Fuller taught at Southern Illinois University (SIU), as one of his golden eras, though some assert his eleven years in Philadelphia were his defining years for his first and second books on *Synergetics*. It was while living in Carbondale that he wrote *Operating Manual for Spaceship Earth*, *Education Automation*, and *I Seem to Be a Verb*, among many other manuscripts.

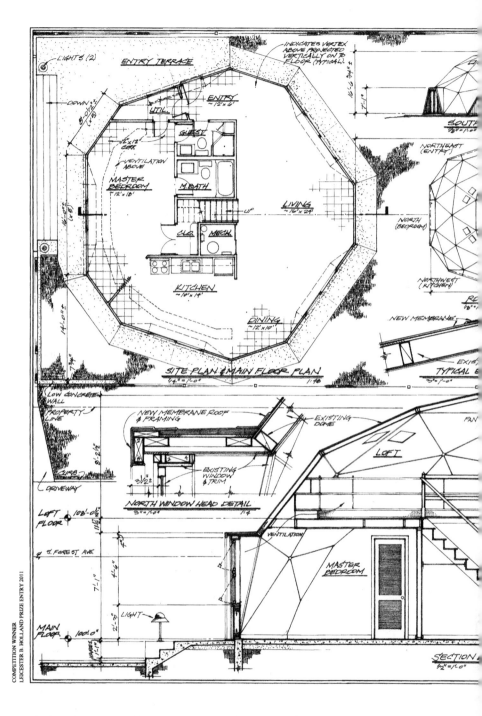

Thad Heckman's award-winning architectural submission for the Holland Prize 2011. The RBF Dome NFP received a Save America's Treasures grant to restore Bucky's dome home in Carbondale, Illinois. The final product is intended to be an educational and tourist resource and an international think tank facility.

R. BUCKMINSTER FULLER DOME
CARBONDALE, ILLINOIS

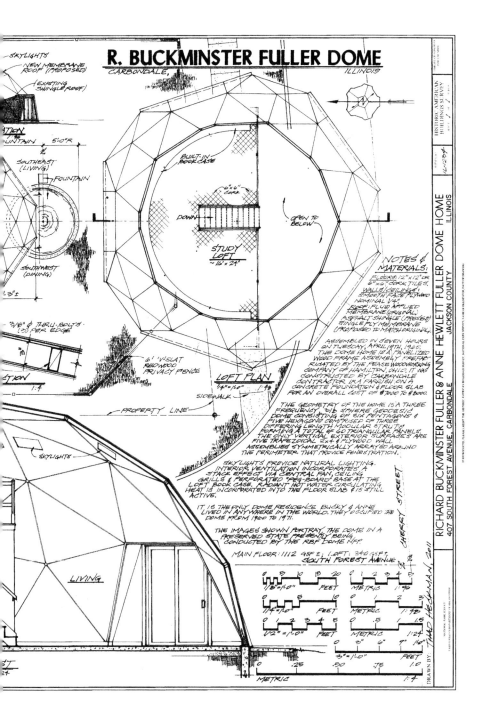

SKYLIGHTS

NEW MEMBRANE ROOF (PROPOSED)

(EXISTING SHINGLE ROOF)

...TION FOUNTAIN — 5'-0"R

SOUTHEAST (LIVING)

FOUNTAIN

SOUTHWEST (DINING)

...3"±

3/8" Ø THRU-BOLTS (3) PER EDGE

...TION 1:4

BUILT-IN BOOK CASE

6"x6" CORK

DOWN

STUDY LOFT ~ 16' x 24'

OPEN TO BELOW

LOFT PLAN 1/4"=1'-0" 1:48

6' 'V'-SLAT REDWOOD PRIVACY FENCE

SIDEWALK

PROPERTY LINE

SKYLIGHTS

LIVING

CHERRY STREET

SOUTH FOREST AVENUE

NOTES & MATERIALS:

FLOORS: 12"x12" OR 6"x6" CORK TILES.
WALLS/CEILINGS: SMOOTH FACE PLYWOOD NOMINAL 1/4".
ROOF: FLUID APPLIED MEMBRANE (ORIGINAL) ASPHALT SHINGLE (PRESENT) SINGLE PLY MEMBRANE (PROPOSED TO MATCH ORIGINAL).

ASSEMBLED IN SEVEN HOURS ON TUESDAY, APRIL 19TH, 1960. THE DOME HOME IS A PANELIZED WOOD FRAME ASSEMBLY, PREFABRICATED BY THE PEASE WOODWORKING COMPANY OF HAMILTON, OHIO. IT WAS CONSTRUCTED BY CARBONDALE CONTRACTOR IRA PARRISH ON A CONCRETE FOUNDATION & FLOOR SLAB FOR AN OVERALL COST OF $7000 TO $8000.

THE GEOMETRY OF THE HOME IS A THREE FREQUENCY 3/8 SPHERE GEODESIC DOME CONSISTING OF SIX PENTAGONS & FIVE HEXAGONS COMPRISED OF THREE DIFFERING LENGTH MODULAR STRUTS FORMING A TOTAL OF 60 TRIANGULAR PANELS. THE ONLY VERTICAL EXTERIOR SURFACES ARE FIVE TRAPEZOIDAL 2x4 & PLYWOOD WALL ASSEMBLIES SYMMETRICALLY ARRAYED AROUND THE PERIMETER THAT PROVIDE FENESTRATION.

SKYLIGHTS PROVIDE NATURAL LIGHTING. INTERIOR VENTILATION INCORPORATES A STACK EFFECT VIA CENTRAL FAN, CEILING GRILLS & PERFORATED "PEG-BOARD" BASE AT THE LOFT BOOK CASE. RADIANT HOT WATER CIRCULATING HEAT IS INCORPORATED INTO THE FLOOR SLAB & IS STILL ACTIVE.

IT IS THE ONLY DOME RESIDENCE BUCKY & ANNE LIVED IN ANYWHERE IN THE WORLD. THEY OCCUPIED THE DOME FROM 1960 TO 1971.

THE IMAGES SHOWN PORTRAY THE DOME IN A PRESERVED STATE PRESENTLY BEING CONDUCTED BY THE RBF DOME NFP.

MAIN FLOOR: 1112 GSF ±; LOFT: 340 GSF±.

HISTORIC AMERICAN BUILDINGS SURVEY

RICHARD BUCKMINSTER FULLER & ANNE HEWLETT FULLER DOME HOME
407 SOUTH FOREST AVENUE, CARBONDALE, JACKSON COUNTY, ILLINOIS

DRAWN BY: THAD HECKMAN, 2011

1/8"=1'-0" FEET METRIC 1:96

1/4"=1'-0" FEET METRIC 1:48

1/2"=1'-0" FEET METRIC 1:24

3"=1'-0" FEET METRIC 1:4

The Fullers' dome home in Carbondale, Illinois, a Pease dome designed by Alvin E. Miller. For twelve years, this was Buckminster Fuller and Lady Anne's residence, the only dome he called home.

After years of neglect, retired SIU design professor Bill Perk purchased the R. Buckminster Fuller dome home in 2001. The following year, a group of SIU faculty members, alumni, and Fuller friends formed a not-for-profit organization called the RBF Dome NFP, and Perk donated the property to the NFP. In February 2011, the RBF Dome NFP secured a Save America's Treasures grant to restore the Fuller home that will match up to $125,000 raised by the NFP. Much credit goes to those involved with the fund-raising efforts, especially Perk for spearheading the project and Peter Bahn for his financial support of the project.

In 2011, architect Thad Heckman, who is also the vice president of the RBF Dome NFP, won the inaugural Leicester B. Holland Prize, an architectural competition administered by the Heritage Documentation Programs of the National Park Service, for his plans to restore the Fuller dome home. According to Heckman, the plans are now housed in the Library of Congress, acknowledging Fuller's place at the SIU campus. With funding under way, a groundbreaking celebration took place on April 19, 2014, to start the restoration, which is being done by Minnesota-based Dome, Inc. Once the Fuller dome home has been historically preserved, it will be made available to SIU and the local Carbondale community. It is envisioned as an educational and tourist resource, a world think tank facility, and a meeting place for the synergetic comprehension of ideas for humankind, all in the spirit of Bucky Fuller.

Recent Geodesic Design Use: Oil Recovery and Containment Domes

Inspired by what Fuller called anticipatory design science, a new dome to mitigate oil spills was recently patented. The Oil Recovery and Containment Geodesic Dome (ORCoD) is a giant metal geodesic dome for deep-water use to protect our seas from environmental harm.

Humankind is facing a dilemma regarding oil exploration. The 2010 Deepwater Horizon offshore oil drilling accident in the Gulf of Mexico killed eleven people, disrupted millions of lives, and resulted in over $40 billion in litigation to date. Since then, there have been other offshore oil spills, including off the coasts of China and Brazil and in the North Sea. In October 2012, the *Guardian* reported that since 2000, there had been 4,123 oil spills in the North Sea alone. But despite the hazards of deep-sea oil exploration and drilling, it is necessary to meet current and future energy demand levels. The inherently dangerous process demands backup systems. The ORCoD geodesic dome is a proposal for safer deep-water oil exploration. ORCoD was described as "a sphere [that] limits the effects of any spill in deep seas" in an article in the March 2013 issue of *Oil Magazine* titled "Prophylactic Design for Minimizing Risk."

The idea took root in 2010, when Lord Norman Foster invited Shoji Sadao and me to the Hearst Corporation reception for the New York premier of his film *How Much Does Your Building Weigh, Mr. Foster?*, a documentary on his architecture. As we discussed the idea of an oil containment dome as envisioned thus far, Norman was encouraging and offered his New York office for the first meeting to brainstorm the concept further. Attending were

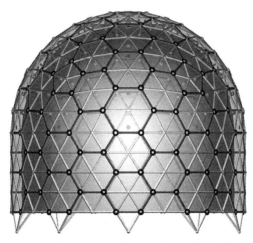

Oil Recovery and Containment Geodesic Dome (ORCoD) elevation

Shoji, Shirley Sharkey, Joe Clinton, Blair Wolfram, Tommy Zung, Foster staff member Brandon Haw, Peter Han, and me.

The first patent application was registered in June 2011, and we received encouraging feedback from the Bureau of Ocean Energy Management, Regulation and Enforcement. I had the opportunity to introduce the concept of ORCoD's design at a speaking engagement at the Cleveland City Club by former U.S. senator and past U.S. Navy secretary John W. Warner and the CEO and president of the American Council on Renewable Energy (ACORE), retired U.S. Navy vice admiral Dennis McGinn, and interest was immediate. With the assistance of John Warner's able executive secretary, Pauline French (whose father was an architect), ongoing discussions followed, and both Warner and McGinn, who is currently the assistant secretary of the navy for energy, installations and environment, thought the oil dome merited further study.

The dome's design involves the co-implementation of science and architectural engineering. The 150- to 200-foot ORCoD geodesic dome can collect between 340,000 and 700,000 barrels of oil in integrated interior containment bags, or bladders. This oil will then be vacuum pumped to nearby surface vessels. The structural integrity of ORCoD makes it the strongest manufactured structure ever invented to date: a geodesic dome incorporating an upper section with a lower barrel section. Importantly, 15 to 20 percent of the dome is open at the base to ensure pressure stabilization at deep-sea depths. Constructed of materials that will withstand saline waters with virtually no moving parts, the oil dome will outperform any comparable structure. The investment in this reusable dome will both protect the environment and ensure significant savings in life and insurance.

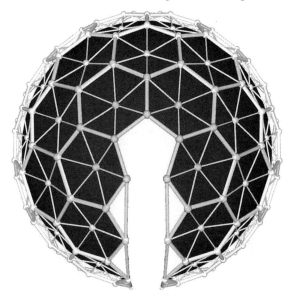

Split Emergency Geodesic Dome (SED) plan

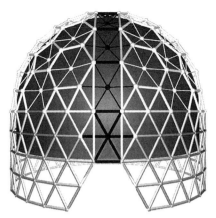

Split Emergency Geodesic Dome (SED)

Manufactured in parts that can be shipped to any site adjacent to a seaport, the ORCoD geodesic dome is assembled and then transported to the drilling site aboard a semi-submersible heavy lift vessel like the *Blue Marlin*, an airship, barges or a catamaran. The oil dome is secured on the sea bottom with tie-down anchors similar to those used to secure floating oil rigs. Buoyant flotation bags are integrally attached to the dome to control the distribution weight. Once in place, the geometry of the structure allows sensors at the connection nodes to detect sea currents and seismic movements, with the exploration process monitored by cameras and other devices controlled from above. The exterior of the dome is covered with mesh to ward off inquisitive sea creatures. At the base of the dome are several large openings to allow robotic operating vehicles (ROVs) to enter for any purpose.

The dome is equipped with two different bladder options. The first option is folded four-ply containment bags or bladders that remain collapsed at a specific dome level until activated from above by the pull of a cable. This solution is used when it is determined that the ocean currents are so severe and unpredictable that under normal circumstances, an open cage structure would subject the dome to less turbulence. The second option is a fixed polymer surface material attached to the interior dome to contain the oil and gas. Both options may heat the containment bag to minimize hydrates. Spill repair commences with the opening and closing of the top dome access cap, which we named the Bucky cap.

Another patent configuration is the Split Emergency Geodesic Dome (SED), designed to be installed and slide around any existing vertical riser after an incident to capture the oil leak. This is accomplished with a continuous vertical split opening. The doors operate similarly to elevator doors and open and close magnetically. After closure, to ensure the structural integrity of the dome, the doors are further secured by a series of knife latches.

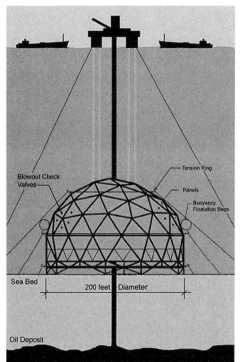

Blowout Check Valves

Tension Ring

Panels

Buoyancy Floatation Bags

Sea Bed

200 feet | Diameter

Oil Deposit

Dome lowered to seafloor

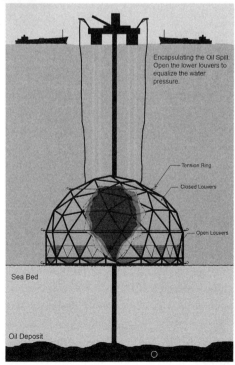

Encapsulating the Oil Spill: Open the lower louvers to equalize the water pressure.

Tension Ring

Closed Louvers

Open Louvers

Sea Bed

Oil Deposit

Dome capturing an oil spill

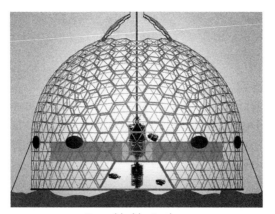
Dome bladder in place

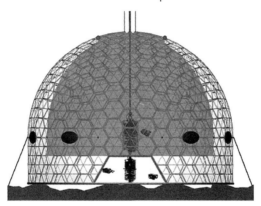
Dome bladder activated

Another of the many challenges of deep-sea oil drilling is methane hydrate buildup, a culprit in the Deepwater Horizon disaster. In an article titled "Pipe-cleaner," published in *The Economist* on April 17, 2012, MIT professor Dr. Kripa Varanasi proposed making the interior of pipes carrying oil so slippery that the methane hydrate will be unable to stick. Our dome idea is to coat the pipes vacuuming the oil to surface vessels with an eighty-twenty mixture of poly(ethyl methacrylate) and fluorodecyl polyhedral oligomeric silsesquioxane solution as another option for the hydrate dilemma.

Upon submission of the first ORCoD patent application, my Fuller architectural partner Shoji Sadao shared his belief with me that "of all the post-Fuller ideas, the oil dome will best express the years with Fuller; this invention could be a powerful, useful contribution to society for environmental design."

Oil Dome Credits: ORCoD and SED are two oil dome patent inventions by Thomas T. K. Zung of Buckminster Fuller, Sadao & Zung Architects (BFSZ). The BFSZ team includes the principal members Shoji Sadao and Thomas T. K.

Zung; senior associate Joseph D. Clinton, design scientist; Shirley Sharkey, former executive secretary to Buckminster Fuller; Blair Wolfram of Dome, Inc.; Tommy B. Zung, designer; Tony Huston, researcher and editor; John Warren, designer; and Joyce Burke-Jones, planner. The key advisors are Steven Percy, former CEO of BP America; Sir Harold W. Kroto, Nobel laureate in chemistry; and Michael A. Keller, university librarian at Stanford University. Additional advisors are Robert D. Storey, former board overseer at Harvard University; Thomas Rayburn of EnSafe environmental consultants; Richard Alt of Carnegie Capital; Dr. Paul Schroeder, Case Western Reserve University professor, PEI, Pagnotta Engineering, Inc.; Thomas Passek, CEO of ASM International; and retired U.S. Coast Guard captain Edwin M. Stanton, incident commander of the Deepwater Horizon oil spill. Distinguished special honorary advisor is former U.S. senator and past U.S. Navy secretary John W. Warner, currently a senior advisor at Hogan Lovells, US LLP, an international law firm.

An Epilever

The title of this section is borrowed from Fuller friend Ed Schlossberg, who used the term instead of *epilogue* for the section he provided in *Tetrascroll: Goldilocks and the Three Bears, a Cosmic Fairytale* (1975), a lithographic book written and illustrated by Buckminster Fuller with the aid of print artist Tatyana "Tanya" Grosman. As Amei Wallach explains in the introduction to that book, "Instead of acting as an ending, the 'log' is made into a lever; an epilever brings the whole experience once more around to the beginning."

Reviewing the Fuller years through documents, inventions, artifacts, and prognostications, we may ask ourselves, what lessons have we learned? What might we glean for the future by studying the past, as suggested by Churchill? Not for some distant dystopian future, but earnestly reflecting, "OK, we got some of it, and now what"? The leap of knowledge as predicted by Fuller might give us cause to accelerate our minds to use synergy, which, as Bucky said in his 1976 book *And It Came to Pass—Not to Stay*, "means behavior of whole systems unpredicted by the behavior of their parts."

Fuller's 1933 Dymaxion car, recently re-created as Dymaxion Car No. 4 by Lord Norman Foster, was Bucky's way to examine the entire car industry in a holistic review. With the car modeled by his sculptor friend Isamu Noguchi, its sleek boat-shaped body guided by naval architect Starling Burgess, Fuller captured the science of wind energy, the physics lift of dynamics, with a car so advanced that it resembles a modern-day jet ready to leap into the skies. Little wonder the Dymaxion car was so amazing in fuel efficiency, and it remains an artifact so advanced in style and performance that it defies comparison.

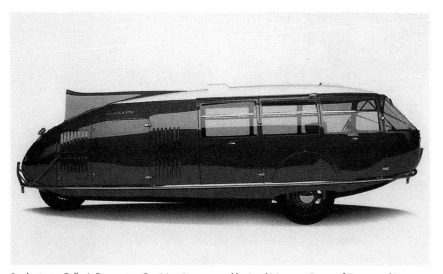

Buckminster Fuller's Dymaxion Car No. 4, re-created by Lord Norman Foster of Foster and Partners, a timeless design from yesterday catapulted to today. Exhibited at Lady Elena Foster's Ivorypress Gallery, Madrid, Spain, and at the MARTa Herford Museum, Germany.

As Foster said at the opening of his Bucky Fuller & Spaceship Earth, with Dymaxion Car No. 4, exhibition at the Ivorypress Gallery in Madrid, using Bucky's optimistic belief in the future and his intelligent use of intellect, "There is much more to learn."

Fuller created the Dymaxion Map in the 1940s after the Wright brothers heralded a new paradigm with air flight. The popular Mercator map distorts the size and relationship of the continents and oceans of Earth. Fuller concluded that we no longer lived in an east-west world, but in a north-south world. Now that man could fly, the shortest distance between two points was a geodesic line. The Dymaxion map filled the need for a twentieth-century map, yet we still cling to the old Mercator map, giving our youths a distorted view of planet Earth.

At the end of World War II, when housing was at a premium, Fuller designed the Dymaxion house, an attempt at retooling a building industry mired in high costs and one-of-a kind-design waste. Produced by Beech Aircraft using the Henry Ford concept of high-speed production, the house revealed Fuller's intention to redo the housing industry from the bottom up. With a one-piece bathroom, revolving shelves, and natural air-conditioning, everything made with factory precision and at lower cost, the unit cost made the Dymaxion house as affordable as the Ford Model T auto. After several years of restoration, the Dymaxion house is now on exhibit at the Henry Ford Museum in Dearborn, Michigan, overseen by the chief curator of industry and design, J. Marc Greuther, and conservator Clara Deck.

Twelve-foot geodesic sphere used for the New Year's Eve ball drop in New York's Times Square. The ball was constructed by Waterford Crystal, with lighting by Phillips.

Model of Fuller's Dymaxion house. A full-size house has been restored and is on exhibit at the Henry Ford Museum in Dearborn, Michigan.

The ubiquitous geodesic dome by Fuller is considered by most to be Bucky's biggest legacy. But architect partner Shoji Sadao and I believe that a hundred years from now, C_{60}, aka buckminsterfullerene or buckyballs, nanotechnology and his visionary ideas will be the best references to Fuller, the Leonardo da Vinci of our time. In a June 2006 *National Geographic* article titled "Nano's Big Future," Jennifer Kahn wrote, "Nanotechnology has been around for two decades, but the first wave of application is now taking hold. As it does, it will affect everything from the batteries we use to the pants we wear to the way we treat cancer." As Nobel laureate Sir Harry Kroto said to me in a recent telephone conversation, "Science has [not] yet unfolded all that C_{60} has in store."

In the 1950s, Fuller proposed that a two-hundred-foot geodesic sphere be suspended over the East River in New York. Called the Geoscope, its design would encompass a miniature Earth. Recently, Nina Colosi of Streaming Museum, an international educational, cultural arts organization, joined with Marnie Muller of the Universe Story Journey to propose that a twenty-foot Fuller Geoscope—a living Dymaxion map, a vision accessible everywhere, for everyone—be placed at the Dag Hammarskjold Plaza, a park in Manhattan near the United Nations New York headquarters. Sherrill Kazan, president of Friends of Dag Hammarskjold Plaza, and Shimina de Gonzaga, executive director of World Council of Peoples for the United Nations, are both championing this idea for the site.

Each year on New Year's Eve in New York City's Times Square, the ball dramatically drops. Since 2009, this ball has been a twelve-foot-diameter geodesic sphere that weighs 11,875 pounds. As we cheer in the New Year, we should reflect on our "critical path." We are currently undergoing a third technological revolution. The first revolution in technology took us from wood to coal, the second from coal to oil and gas, and ours, perhaps the most challenging, is clean, sustainable renewable energy for all humanity.

As we face the rest of this millennium, with the relatively recent phenomenon of the Internet and the multitude of possibilities arising from the ready availability of high-speed digital information, we have high hopes. Will the computer age lead to the changes we need for the "comprehensive" world? Since Fuller's 1936 discussion with Dr. Albert Einstein, documented in *Nine Chains to the Moon*, the human population on Earth has tripled. The phrase "doing more with less" is an ever more critical path for planet survival, and the word for dynamic, maximum, and tension is *Dymaxion*.

Auspiciously, some passengers from Earth made some spectacular comments about our third planet from the Sun. In June 1983, Sally Ride became the first American woman in space. She looked back at Earth and, as she later described it, "I saw the blackness of space, and then the bright blue Earth. . . . It looked as if someone had taken a royal blue crayon and just traced along Earth's horizon. And then I realized that that blue line, that really thin royal blue line, was Earth's atmosphere, and that was all there was of it. And it's so clear from that perspective how fragile our existence is." The late Apollo astronaut Wally Schirra said so succinctly, "I left Earth three times. I found no place else to go. Please take care of Spaceship Earth."

On February 14, 1983, St. Valentine's Day and Chinese New Year, in a letter I received from Bucky, the grand strategist for Spaceship Earth, he asked, "Are humans a worthwhile to Universe invention?" Fuller had asked two decades earlier in *On Education*, "Where will the world be in 2025? . . . Nature is trying very hard to make humans successful. If we do make it, we're going to make it by virtue of that young world [in his lectures, Fuller urged the young to change the environment and not try to change man] and its determination to learn the truth and the synergetic intersignificance of all the truths. Once you give the young world a synergetic clue, they will find they can really understand technology and their Universe. Then, knowledge is going to proliferate very rapidly."

In that book, he had concluded, "Seeing much of our young all around the world, I would say there is a good chance we can make it, as we now have the means through technology to achieve for the world's people the highest standard of living ever known to man." As with all things, it will be touch and go: "to continue will depend entirely on the integrity of the human individual, available in each of us, and not on any political or economic systems."

It is an extraordinary occasion to add accounting to this polymath thinker; his life was a "metaphor, an organic artifact for the instruction of others," said his friend E. J. Applewhite. It is hoped this revised edition will lead the reader to delve into Fuller's own writings. In reading Bucky's own words, one will gain the keen privilege that some of us who worked with him experienced, rare among contemporaries, the enigmatic R. Buckminster Fuller, a synergetic man, the Spaceship Earth's friendly genie.

In closing this reissue, as Bucky would say after one of his absorbing talks, "Thank you . . . you darlin' people!"

Respectfully and faithfully yours,
Thomas Tse Kwai Zung, adjuvant to Buckminster Fuller

Fuller as a Poet of Technology

An appreciation presented on the occasion of the opening of the retrospective exhibit Your Private Sky—R. Buckminster Fuller, Museum für Gestaltung, Zürich, June 1999, and reprinted by permission of the author.

E. J. Applewhite

Throughout his career, Buckminster Fuller—not without reason—was despairing of being misunderstood. He said he could endure that people might not understand what he was trying to say, as long as they did not misunderstand him. The risk prevailed, even when Fuller was using the plainest colloquial diction. His constant refrain of exhorting young people not to feel they had to "earn a living" is an example. What he wanted them to do instead was "doing what needs to be done." The corollary was often overlooked, and many people in the generation of hippies (and their parents) misunderstood the point.

Some of the simple metaphors he introduced, notably "Spaceship Earth," were powerful and unambiguous. But when he ventured into engineering and geometrical abstraction, he became more demanding. His coinage of "tensegrity" (for structural design derived from tension rather than compression) was challenging to the reader and listener. And when it came to the writing style of his synoptic geometrical philosophy in *Synergetics*, he abandoned convention altogether. Its unfamiliar, polysyllabic vocabulary was deliberately oblique in its challenge to rectangularity. Some of his contemporaries and legatees are still waiting for *Synergetics* to be translated into plain English. I suggest that putting it all into an orthogonal vocabulary cannot be done—not without great loss to its subversive intent. *Synergetics* is a work without genre: it will never be admitted into the scientific canon; if it has a home anywhere it is as a work of literature.

Fuller had a penchant for mythic projection of events and artifacts—real or imagined—for largely heuristic purposes. For example, his protestations that he was just a "normal ordinary human being" invited the audience to share his daring personal epiphany. Just as Fuller abandoned all established academic disciplines in favor of his own heroic self-discipline, he had to employ a new vocabulary—very unwelcome, even threatening—because he wanted to force the reader to venture into new pathways of mental reception: pathways subversive of the conventions of philosophy and mathematics of (what he regarded as our culturally imprisoned) intellectual establishment. The price has been a heavy one: few academics— except for Joachim Krausse and Claude Lichtenstein (and Arthur Loeb and Hugh Kenner)—have been willing to stop what they have always been doing in order to master his peculiar idiom.

I submit that the license Fuller invokes is that of a poet—a calling that he never arrogated to himself. He said it was not a profession. It was not something that you could proclaim or put on a calling card; it could only be bestowed by others, an accolade. (Anyone can say she's a painter; only others can say she's an artist.) As Fuller put it to an audience of UNESCO delegates in Tiflis (Tbilisi, Georgia) in 1968, "The term 'artist' is not really professable. It is a term which society alone can bestow and then only retrospectively."

In 1962, Fuller was appointed to the Charles Eliot Norton Chair of Poetry at Harvard. He liked that, and he wore the cap and gown of his thirty-some honorary degrees with pride and dignity. But the very notion of a *professor* of philosophy (or *a fortiori* poetry) to him was hilarious. How could you pay such a compliment to a mere institution?

His idea of poetry was devoid of any precious aesthetic or literary taint. Who but Fuller could describe Henry Ford and Einstein as poets? "Ford," he proclaimed, "was a great conceiver. His logistics were like conducting a great orchestra." And Einstein was "the prototype scientist-artist . . . of the twentieth century." He defined language as the first industrial tool, to be reinforced by emerging technology as a worldwide integrator of human activities.

For Fuller the vast world patterns of industrial energy and the geometrical relationships of the events of experience—these invisible trails of our manufacturing processes and these polyhedral forms of our consciousness, not mere words—are the stuff of poetry. He is a poet of geometry. He defined thinking itself as the dismissal of irrelevancies, a process he saw as inherently tetrahedral. His favorite line of poetry was $E = mc^2$. And for him the most beautiful event in nature was the hydrogen atom in its minimal orbital simplicity. His use of "trim tab" shows the dynamic propensities of any mechanical device. In his script even chemical bonding becomes geometrically metaphorical.

This kind of apprehension is not accessible by nouns, and barely by verbs.

He was a poet of energetic patterns mediated beyond all language by the artifacts of technology we behold in this exhibit.

He disdained the Bauhaus aesthetic as an irrelevant phase of superficial design sensibility, oblivious of "performance per pound" and the invisible tensile strength of materials. Modernism per se was almost as bad as the modernistic, doomed as a retrograde cultural backwater. The aim of the comprehensive designer should be to reform the environment in order to release humans from the doctrine of scarcity. Here we have a technological ethic by which engineering design has an imperative to turn the energies of the universe to human advantage.

Fuller rejected the relentless critical search for the source of artistic "creativity." He deplored the pedantic sociological inquiry into the supposed conditions attracting the cohort of young talented artists to Black Mountain College in North Carolina in the 1930s—as if there were something in the water there. To him it was like stalking the Swiss patent office in Berne to look for another Einstein.

He did not use the word "creativity" in the ordinary sense. "I would not suggest that it is the role of the individual to add something to the universe. Individuals can only discover the principles and then employ them to move forward to greater understanding." In matters of artistic imagination, Fuller—like Coleridge—believed the role of the artist is limited: there is only one creator.

Indeed, his use of geometry and mechanical devices is programmatic and didactic. The ethical imperative is inescapable. He has been called the first poet of industrialization. And, more extravagantly—by Arthur C. Clarke—the first engineer-saint.

Writing these words on the occasion of the auspicious June 1999 opening of the Museum für Gestaltung Zürich exhibition of Fuller's artifacts, I recall several times when critics said that his *Synergetics* reads as if it had been translated from the German. If their intent was disparaging, I dismiss it: I think the suggestion is a tribute to the compound-word-making facility of *die deutsche Sprache.*

A Fuller Family

Some readers will be interested in Bucky's immediate family. Fuller was raised in a New England family. His father, a businessman, died unexpectedly when Bucky was still a child. This was a big blow to Bucky and the family. There were four children: Leslie, Bucky, Wolcott, and Rosie. Bucky was always an attentive, curious child with good table manners and kind words for everyone. I recall that even after his painful hip transplant, and while still mending, Bucky would jump up out of his chair whenever a lady entered the room. In my forty years with Bucky, I never heard him even once utter a cuss word. On one occasion, after we left a meeting where obscene words were being used for emphasis, Bucky leaned over and whispered to me, "I do not approve of using barnyard language, and I have no *right* to correct them; but only the ignorant and fools use gutter words." Bucky was always a gentleman.

After being thrown out of Harvard for the second time, Bucky Fuller met Anne Hewlett in New York. Anne was the eldest daughter of a family of ten. Her father was James Monroe Hewlett, a prominent architect; he was vice president of the American Institute of Architects and director of the American Academy in Rome. It was Anne's father who introduced Bucky to architecture, the arts, and a new way of thinking that would influence his entire life.

In 1927, at age thirty-two, Bucky made a pact with himself "to do my own thinking, and see what the individual, starting without any money or credit—in fact considerable discredit, but with a whole lot of experience—to see what the individual, with a wife and a newborn child, could produce on behalf of his fellow man." He believed that "by doing more with less, we could take care of everybody and there need not be suffering around the world." Bucky's experiment as "Guinea Pig B" presented a challenge to his family. Fuller was convinced that "money is absolutely irrelevant" and if he carried on this experiment, his family would understand, and somehow "society would take care of their needs if he was operating properly for the benefit of all mankind."

Lady Anne, as many of us would address her, was creative, courageous, and able to adjust to the family affairs. Always supportive of her husband, Lady Anne noted that her life with this idiosyncratic man was not always easy, but she would have it no other way. With fortitude and always with grace, she managed beautifully. Her and Bucky's marriage lasted just eleven days shy of sixty-six years. Anne passed away thirty-six hours after Bucky.

Allegra, Bucky's daughter, followed her own interests and is a noted authority in dance. She is a graduate of Bennington College, and she holds an M.A. in dance from UCLA. Allegra was Professor of Dance and Dance Ethnology at UCLA, where she chaired both the Department of Dance and the pioneering World Arts and Culture Program. The recipient of numerous awards as well as a Fulbright research fellowship, she has many publications and films to her credit. Allegra was cofounder of the Buckminster Fuller Institute and is currently the chairwoman of its board. Bucky Fuller had asked Allegra and her son, Jaime, to serve as co-executors of his estate, a role they continue to fulfill.

Robert Snyder is Fuller's son-in-law. A filmmaker, Bob received an Academy Award for *The Titan: Story of Michelangelo* and a nomination for *The Hidden World.* Other prize-winning films include *Michelangelo, Self-Portrait,* several films on Fuller, and films on Henry Miller, Anaïs Nin, Pablo Casals, and Willem de Kooning. Bob edited the Bucky Fuller book *Autobiographical Monologue Scenario* and others on Miller and Nin. Bob and Allegra have two children, Alexandra and Jaime.

Jaime Snyder, a graduate in composition from the UCLA School of Music, is a producer and director of educational media. He was cofounder of the Buckminster Fuller Institute, is an active member of its board, and for many years was its executive director, instituting a highly successful series of educational programs on Fuller's work. With Masters and Masterworks Production, and under his father's wing, Jaime produced and directed two award-winning films on Bucky Fuller, and he now serves as the company's CEO. Jaime and his wife, Cheryl, a teacher of meditation, reside with Cheryl's daughter, Mira, on the West Coast.

Alexandra Snyder May was named for Bucky and Anne's first daughter, Alexandra Willets, who died as a child. Alexandra was the associate director of the Isamu Noguchi Museum and Gardens and the initiator of its educational department. Alexandra and her husband, Sam May, a stock analyst in wireless communications, live on the West Coast. Alexandra and Sam have two children, Olivia and Rowan. Alexandra serves as a board member of World Game.

The R. Buckminster Fuller public memorial service was held at Fuller's geodesic dome at the American Society of Metals headquarters in Cleveland, Ohio. The governor of Ohio, Richard F. Celeste, former Peace Corps director and currently the United States Ambassador to India, hosted the ceremonies in 1983. Alexandra followed Governor Celeste's comments and read this poem by grandfather Bucky as a tribute to Anne Hewlett Fuller.

Why "Yours Truly"

I'm not "yours," you're "not" mine.
My years of life are seventy-nine.
Mysteries deepen. I opine
"curvaceous, nymphacious, sylphaceous you"
I'm nigh inefficacious. What may we do?
I can't eat you and have you too.
Let's enjoy laughter! And wisdom too.
You're eternally lovely . . . The truly you.

You can't "see" me, I can't "see" you.
But we may know one another
and sometimes do . . .
Then learn that we both love only all that's true
Wherefore we both love the truly you
I'll love you forever
The *truly* you.

B. Fuller
1974

Frank Lloyd Wright and *Nine Chains to the Moon*

R. Buckminster Fuller's first major publication was *Nine Chains to the Moon* (1938), and the book was reviewed by Frank Lloyd Wright in the *Saturday Review of Literature*. Wright and Fuller remained friends throughout their lives. Wright once described Fuller as an engineer who loved architecture and himself as an architect who loved engineering.

I recall Anthony Puttnam, now vice president of Taliesin Architects, relating to me how sensitive Bucky was to the Frank Lloyd Wright apprentice fellows after the death of Frank Lloyd Wright. One evening at Taliesin West in 1960, a small exhibit was on display of Wright's original drawings. Bucky took in the drawings as if seeing them fresh, each a living presence, looking directly at the "creator" showing through Wright. Later that evening, Bucky was sitting on a long bench that faced Mrs. Wright's swimming pool terrace in the house's garden. Overhead was a bright full moon. He gestured and asked Tony Puttnam to sit by him. He wanted Tony to see, from his vantage point, the moonlight filtering through the cloth awning above. Tony told me that Bucky's kindness to him and the Taliesin fellows and his sharing with them of his poetic vision left an indelible impression and made a connection he will remember always.

Posthumously, R. Buckminster Fuller received the Frank Lloyd Wright Creativity Award from Wright's widow, Olgivanna Hinzenberg Wright, in a ceremony attended by Fuller's grandson, Jaime Snyder, and myself. The Buckminster Fuller Institute reciprocated by presenting to Frank Lloyd Wright via Mrs. Wright the R. Buckminster Fuller Dymaxion Award.

Buckminster Fuller—you are the most sensible man in New York, truly sensitive. Nature gave you antennae, long-range finders you have learned to use. I find almost all your prognosticating nearly right—much of it dead right, and I love you for the way you prognosticate. To address you directly will be a hell of a way of reviewing your book—I know. I should write all around you, take you apart, and put you together again to show—between the lines—how much bigger my own mind is than yours and how much smarter than you I can be with it and leave the essence of your thought untouched.

But I couldn't do it if I would and I wouldn't if I could. To say that you have now a good style of your own in saying very important things is only admitting something unexpected. To say you are the most sensible man in New York isn't saying much for you—in that pack of caged fools. And everybody who knows you knows you are extraordinarily sensitive. . . .

Faithfully, your admirer and friend, more power to you—you valuable "unit."

FRANK LLOYD WRIGHT
Taliesin
Spring Green, Wisconsin,
August 8th, 1938.

Excerpt from a review by Wright of Fuller's book *Nine Chains to the Moon* (Lippincott, 1938). The passage quoted was published in the *Saturday Review of Literature,* September 17, 1938.

Thomas T. K. Zung, Mrs. Frank Lloyd Wright, and Jaime Snyder (Bucky's grandson) at Taliesin West

Isamu Noguchi and $E = mc^2$

R. Buckminster Fuller's lifelong and best friend was the sculptor Isamu Noguchi. "I first met Mr. Fuller, as I used to call him," said Noguchi, at Romany Marie's restaurant in 1929. "Some time later, I found an old laundry room on top of a building on Madison Avenue and 29th Street with windows all around. Under Bucky's sway I painted the whole place silver so that one was almost blinded by the lack of shadows. There I made his portrait head in chrome-plated bronze—also form without shadow.

"Bucky was in a continuous state of dialectic creativity, giving talks in any situation before any kind of audience. . . . He would talk to me as though to a throng; walking and talking everywhere—over the Brooklyn Bridge, over innumerable cups of coffee. Bucky drank everything—tea, coffee, liquor—with equal gusto and would often be in a state of wide-awake euphoria for three days straight. Drink did not seem to affect him otherwise.

"He used to drink like a fish. He had become a God-possessed man, like a Messiah of ideas. He was a prophet of things to come. Bucky didn't take care of himself, but he had amazing strength. He often went without sleep for several days, and he didn't always eat either. Bucky's zest for life is part and parcel of his creativity. However, he has the capacity and resolution to come to grips in unknown hours and retreats of the mind to fathom new secrets from the universe."

Once Noguchi, who was then in Mexico City, asked Bucky, in New York, to "please wire me rush Einstein's formula and explanation thereof." With his last ten dollars, Bucky sent the Western Union reply on page 369 to his friend.

EINSTEIN'S FORMULA DETERMINATION INDIVIDUAL SPECIFICS RELATIVITY READS "ENERGY EQUALS MASS TIMES THE SPEED OF LIGHT SQUARED" SPEED OF LIGHT IDENTICAL SPEED ALL RADIATION COSMIC GAMMA X ULTRA VIOLET INFRA RED RAYS ETC. ONE HUNDRED EIGHTY SEVEN THOUSAND MILES PER SECOND WHICH SQUARED IS TOP OR PERFECT SPEED GIVING SCIENCE A FINITE VALUE FOR BASIC FACTOR IN MOTION UNIVERSE. SPEED OF RADIANT ENERGY BEING DIRECTIONAL OUTWARD ALL DIRECTIONS EX-PANDING WAVE SURFACE DIAMETRIC POLAR SPEED AWAY FROM SELF IS TWICE SPEED IN ONE DIRECTION AND SPEED OF VOLUME INCREASE IS SQUARE OF SPEED IN ONE DIRECTION APPROXIMATELY THIRTY FIVE BILLION VOLUMETRIC MILES PER SECOND. FOR-MULA IS WRITTEN LETTER E FOLLOWED BY EQUA-TION MARK FOLLOWED BY LETTER M FOLLOWED BY LETTER C FOLLOWED CLOSELY BY ELEVATED SMALL FIGURE TWO SYMBOL OF SQUARING. ONLY VARIABLE IN FORMULA IS SPECIFIC MASS. SPEED IS UNIT OF RATE WHICH IS AN INTEGRATED RATIO OF BOTH TIME AND SPACE AND NO GREATER RATE OF SPEED THAN THAT PROVIDED BY ITS CAUSE WHICH IS PURE ENERGY LATENT OR RADIANT IS ATTAINABLE. THE FORMULA THEREFORE PROVIDES A UNIT AND A RATE OF PERFECTION TO WHICH THE RELATIVE IMPER-FECTION OR INEFFICIENCY OF ENERGY RELEASE IN RADIANT OR CONFINED DIRECTION OF ALL TEM-PORAL SPACE PHENOMENA MAY BE COMPARED BY ACTUAL CALCULATION. SIGNIFICANCE: SPECIFIC QUALITY OF ANIMATES IS CONTROL WILLFUL OR OTHERWISE OF RATE AND DIRECTION ENERGY RE-LEASE AND APPLICATION NOT ONLY TO SELF MECHANISM BUT OF FROM-SELF-MACHINE DIVIDED MECHANISMS. RELATIVITY OF ALL ANIMATES AND INANIMATES IS POTENTIAL OF ESTABLISHMENT THROUGH EINSTEIN FORMULA.

Isamu Noguchi and Bucky Fuller

The United States Postage Stamp Campaign

As would be expected, Bucky attracted a cadre of students and supporters. He had a way of introducing each idea within a small group and would set the group to work to solve a unique problem. It was like casting pebbles into a pond and noting the water's patterns of integrity, each pebble making its own ripples. Tony Huston worked with Bucky, and they shared their own patterns of integrity. They became friends, and Bucky was also a longtime friend of Tony's father, film director John Huston. Tony, a screenwriter, has served as a member of the board of the Buckminster Fuller Institute.

In 1993, many of us began a campaign for the issuance of a United States postage stamp to commemorate the centenary of Bucky's birth. Sadly, the proposal was rejected. In 1998, urged on by Tony, we tried again, and again we were turned down. Although we were disappointed, I print Tony's letter, as it is typical of the tremendous support Fuller received.

Undaunted, we shall try to have the Citizen's Stamp Advisory Committee consider a stamp to honor R. Buckminster Fuller on the occasion of the fiftieth anniversary of the geodesic dome patent, which will be in year 2004. Reproduced on page 372 is a proposed stamp design from our previous attempt.

Citizen's Stamp Advisory Committee
U.S. Postal Service
475 L'Enfant Plaza SW Room 4474E
Washington, DC 20250-2437

September 29, 1998

Dear Sirs,

I should like to add my voice to the chorus of those who are asking that the memory of R. Buckminster Fuller be commemorated by a U.S. postage stamp.

It would be redundant of me to repeat all the public reasons why Bucky should be so honored. The trouble with awards and honors is that, in a case such as his, they are so much smaller than the man.

Frequently one gets the impression that the highest aspiration of our society is success in commerce. This was not Bucky. Until the day of his death, his mission was nothing less than to make humanity a success. Everything he gained he ploughed back into this mission—he never became complacent or wealthy.

I was fortunate to travel with him for some months nearly thirty years ago and to remain his friend until death. He is the only famous person I have met whose stature increased as a result of close proximity. Indeed Bucky is the only great man I have ever met.

Putting him on a stamp will encourage his memory to affect our society and inspire our children to make the world a better place. In other words, honoring him in this way is not just for his sake, but also for ours.

Yours sincerely,
Tony Huston

Proposed U.S. postage stamp

Fuller's Last Diary Entry

As R. Buckminster Fuller became better known, more time was demanded of him. His office work, his studio time, his publications, and his invitations soon got out of hand. There simply was not enough time in one day. Always the New England gentleman, and always accommodating, Bucky would fall into the trap of trying to be in three places at one time. Because of his constant traveling, Bucky kept three watches. One watch was set to the time zone of his office time, which was also Anne's time. The second watch was set to the time zone where he currently was, and the third watch was set to the zone where he was to appear next—his thought was always anticipatory.

After his move to Pennsylvania, the dynamic and chaotic Philadelphia office needed order. His Chronofile, the basis of Bucky Fuller's archives, demanded it. The chore fell to Shirley Swansen Sharkey, one of his students from Southern Illinois University, who had been helping his Carbondale secretary, Naomi Smith Wallace. Shirley helped move Bucky's office to the University of Pennsylvania's Design Science Center. With a staff of able assistants—Kiyoshi Kuromiya, Peter Kent, Ann Mintz, Amy Edmundson, Rob Grip, and Chris Kitrick, among many—Bucky appointed Shirley his executive administrator. Shirley has also served as a member of the board of the Buckminster Fuller Institute.

Shirley Sharkey kept Bucky's schedule, which he sometimes called a diary. Reproduced on the following pages is the last diary entry of R. Buckminster Fuller.

On June 23, 1983, I had flown to Chicago to rendezvous with Bucky, to be interviewed with him for the hanging library, as I was the Design Adjuvant for Bucky's "Hang It All" invention. The unit was to be produced by the Thonet Company, and *Time* magazine wanted a story on Bucky's latest patent.

June 24 was to be the last time that I was to be in Bucky's company. I hugged Bucky as he boarded the airplane. He died seven days later, on July 1, 1983, in California, at Lady Anne's bedside. Anne died thirty-six hours later.

Jun.14 Tue.   10:00am Temcor Board of Director's Meeting,
                       2825 Toledo Street, Torrance, CA

                       Stay: Pacific Palisades

Jun.15 Wed.

                       Stay: Pacific Palisades

Jun.16 Thu.   12:00n. Fly from Los Angeles, CA   United #1110

               1:12pm Arrive San Francisco, CA

                       Stay: Mark Hopkins Hotel

Jun.17 Fri.   11:00am- Speak at opening of new North Face
               6:00pm Headquarters, 999 Harrison St., Berkeley, CA

                       Contact: Bruce Hamilton

              10:00pm Fly from San Francisco, CA   TWA # 50

Jun.18 Sat.   2:00pm Arrive Philadelphia, PA (met)
                       (note: the plane was re-routed to New York
                       and BF arrived in Philadelphia via bus from
                       NYC)

                       work in Philadelphia office

               6:40pm Fly from Philadelphia, PA   TWA #756
                       (with Shirley Sharkey)

Jun.19 Sun.   6:20am Arrive London, ENGLAND   (met)

               9:00am- Breakfast with Mr. and Mrs. Jones and
              12:00 n Joyce Burke-Jones of Cleveland office at
                       Stafford

                       Datuk Lim Chong Keat arrives Stafford Hotel

                       Stay: Stafford Hotel
                             16 St.James Place, London, England
                       ----
Jun.20 Mon.   9:00 am-Breakfast with Lim Chong Keat at Stafford

              11:30 am-Meet with Norman and Wendy Foster, 172-182
               4:30 pm Great Portland Street, telephone

               6:30 pm Tea at Stafford with Lim Chong Keat

               8:30 pm Dine with Lim Chong Keat, Shirley, Cedric
                       Price at Etoille Restaurant

                       Stay: Stafford Hotel

Jun.21 Tues. 9:00 am Breakfast meeting at Stafford w/ Aida Svenson
                     re: London Integrity Day

            10:00 am Interviewed by Chris Nicolson of BBC-Radio
                     re: skyscrapers

            12:00 n.-Meet with Norman and Wendy Foster

             6:00pm Her majesty the Queen of England has given
                     her consent for the award of this year's
                     Royal Gold Medal for Architecture to be made
                     to Norman Foster. Formal presentation
                     ceremony to be held at Royal Institute of
                     British Architects, 66 Portland Place, London
                     BF to act as sponsor. (BF gives 10 min. talk)

                     dinner follows

                     Stay: Stafford Hotel
            _____

Jun.22 Wed. 9:00 am Meet with Louis Hellman re: BF cartoon book

            10:00 am-Meet with Paul Matthews, Lim Chong Keat,
             4:30 pm Norman and Wendy Foster, Loren Butt re:
                     autonomous dwelling project

             4:30 pm Interviewed by Richard Boston for
                     The Guardian newspaper, Stafford Hotel
                     re: CRITICAL PATH

                     Contact: Hutchinson Publishers 011-44-1-2811

             5:00 pm Picked up at Stafford by Oliver Caldecott of
                     Hutchinson Publishers

             5:30 pm Interviewed by BBC's Kaleidoscope program
                     at Broadcast House

             6:30 pm Meet with Michael Glickman

             7:30 pm Dine with Norman and Wendy Foster, Shirley,
                     Jmes and Jill Meller, at Odins Restaurant
                     27 Devonshire Street

                     Stay: Stafford Hotel

R. BUCKMINSTER FULLER   DIARY   1983

Jun.23 Thur. 9:00 am Meet with Lim Chong Keat, James Meller

            10:00 am Meet with James Meller, Raymond Foulk of
                     Milton-Keynes

            11:00 am View completed model of Garden of Eden dome.
                     Foster's office

             1:00pm Fly from London, ENGLAND   Concord
                     (with Shirley Sharkey)

            12:20pm Arrive Washington, DC (Dulles)
                     See John H. Lee in customs

                     Helicopter to Nat'l Airport
                     (pilot John Nielsen)

                     Limo to White House

             1:00pm BF & Ann Lewin present proposal for
                     "Scenario Universe" Exhibit for Capital
                     Children's Museum, to members of U.S.Counsil
                     for World Communications Year 83

             2:30 pm Meet EJA at Nat'l Airport

             2:45pm Fly from Washington, DC (Natl.) Midway #53
                     (with S.S.)
             3:30pm Arrive Chicago, IL (Midway) (met by limo)

             5:00pm BF & SS attend TIME Magazine's Chicago
                     Jubilee Dinner Dance to honor men & women
                     with connections to Chicago who have been
                     subjects of TIME cover stories. Museum of
                     Science & Industry; 57th St.& Lake Shore Dr.
                     (black tie)

                     Contact: Jean Gussenhouen

                     Stay: Park Hyatt Hotel
                          (Water Tower Square, 800 No.Michigan Ave.)
                     ——

Jun.24 Fri. 10:00 am Photographed at Thonet Showroom, Merchandise
            12:00 n. Mart, 11th floor, suite 100, with Hanging
                     Library for TIME Magazine with Thomas Zung

                     Contact: Martha Wadsworth   (312) 644-5765

             1:10pm Fly from Chicago, IL (O'Hare)   United #107

             3:15pm Arrive Los Angeles, CA   (met)

                     Stay: Pacific Palisades

Jun.25 Sat.   10:00am— Integrity Day, presented by Friends
(Full Moon)   5:00pm   of Buckminster Fuller Foundation, sponsored
                        by Church of Religious Science, Huntington
                        Bch., CA

                        Contact: Ron Landsman

                        Stay: Pacific Palisades

Jun.26 Sun.   2:00pm   1st of 3 interviews with Heather Williams,
                        who's doing portrait of BF (at PAC PAL)

                        Stay: Pacific Palisades

Jun.27 Mon.   Meet with Anwar Dil

                        Stay: Pacific Palisades

Jun.28 Tue.   7:40am   Picked up at Pac Pal

              8:30am   Attend construction of 3 office spaces —
                        "Inner Space Odyssey"

              10:00am   Keynote speaker for Building Owners &
                        Managers Assoc., Grand Ballroom, Century
                        Plaza Hotel (1 hour), Los Angeles, CA

                        Contact: George Brown

                        Stay: Pacific Palisades

Jun.29 Wed.

                        Stay: Pacific Palisades

Jun.30 Thu.

                        Stay: Pacific Palisades

JULY

July 1 Fri.   morning   Meet with Anwar Dil

              2:45 pm   Buckminster Fuller suffers massive heart
                        attack at bedside of Anne Hewlett Fuller,
                        room 446, Good Samaritan Hospital, Wilmer &
                        Wilshire Blvd, Los Angeles.

              4:45 pm   Buckminster Fuller pronounced dead

July 3   Anne Hewlett Fuller dies, 36 hours later

Epilogue

....

On June 16 and 17, 2000, a symposium was held in London, on *R. Buckminster Fuller: The Art of Design Science* at the Royal Institute of British Architects. The sponsor of the event was CargoLifter A.G. in Germany, specializing in Airocean transportation by Zepplin, not unlike one of Fuller's concepts in 1927.

The symposium was held in conjunction with the largest array of Fuller's work exhibited in any one place, ever, by the Design Museum of London. Produced in Zürich by Claude Lichtenstein and Professor Joachim Krausse of the Museum für Gestaltung Zürich, with Laurence Mauderli coordinating the London event, the exhibit will travel to ten cities over the next three years.

The symposium challenged and inspired in ways that one must ponder while reading this book. The significance of Bucky Fuller to the twenty-first century is sometimes direct, sometimes oblique, but always omnidirectional.

The program began with my laying a visual foundation presentation for the two-day symposium with an overview of the artifacts of R. Buckminster Fuller. Shoji Sadao, USA, shared his insights into the development and the importance of the Dymaxion Airocean world map. E. J. Applewhite, USA, commented on Fuller's experiential approach to Geometry and Structure with Synergetics. Hinrich Schliephack, Germany, spoke on the Genesis of an industry, CargoLifter, made for the twenty-first century. Jaime Snyder, USA, spoke about Humanity's Option for Success. James Meller, UK, Fuller's first foreign student in the United States, gave a wider view of Fuller, and Caspar Schwabe, Switzerland, demonstrated his unique Kinematic Phenomena.

The second day began with Donald E. Ingber, M.D., USA, on the Architecture of Life, who discussed discoveries through bio-medicine that one day may lead via microbiology to medical applications for living cells and tissues, "buckyballs" to cure diseases by linking medication directly to the source without damaging the healthy area. He has also written, "Fuller was himself a miracle of Nature. Starting with a terribly nearsighted child, you end up with one of the most insightful men of the twentieth century." He continued, "Cell biologists, such as myself, literally have observed molecular geodesic domes within the skeletal framework ('cytoskeleton') of living cells. We have discovered that the basic architectural principle of tensegrity that Fuller explored in theoretical terms—and his student, Kenneth Snelson, explored through sculpture—guides biological organization, from the simplest carbon molecules ('Bucky Balls') to the most complex living organisms. In other words, Fuller's vision of Nature and the Universe was crystal clear; it was everyone else who needed inch-thick eyeglasses."

Dr. Jonathan Hare, UK, associate to Nobel chemist Harry Kroto, described working to discover new carbon molecules, fullerenes, and Nanotubes. Application of these new discoveries with micro-activity could easily benefit mankind. Professor Joe Clinton, USA, described Geodesics: before and after Fuller. Engineer and Professor Ing Jorg Schlaich, Germany, offered his thoughts on structural efficiency through double curvature and showed experiments of alternative units of energy. Martin Pawley, UK, author and editor, spoke of Fuller as a historic figure. Professor Dr. Joachim Krausse, Germany, mentioned the origins of Fuller's key concepts. The symposium ended with Steve Lacy, USA, on the saxophone with Irene Aebi, France, a soprano and improviser of jazz, performing a geodesic jazz duo.

This London symposium would have made Bucky Fuller smile. I submit

R. Buckminster Fuller
Fifty Years of Design Science
1933 Dymaxion Car   1983 Fly's Eye Dome
(Photo taken on Bucky's 85th Birthday)

from the variety of participation in the symposium that Fuller's influence may well be described in the new millennium with micro and macro discoveries leading to the adaptation of tensegrity from molecules to cosmography. It would be Synergetics at its faculae.

In my mind it recalls a question that R. Buckminster Fuller posed to all of us in one of his last communications. Only he could write this perfect epilogue.

"Human integrity is the uncompromising courage of self determining whether or not to take initiatives, support or cooperate with others in accord with 'all the truth and nothing but the truth' as it is conceived in each individual.

"Whether humanity is to continue and comprehensively prosper on Spaceship Earth depends entirely on the integrity of the human individuals and not on the political and economic systems.

"The cosmic question has been asked—ARE HUMANS A WORTH-WHILE TO UNIVERSE INVENTION?"

R. BUCKMINSTER FULLER
Penang, Malaysia
February 14, 1983

Bucky Fuller, Shoji Sadao, and Thomas Zung

Books by R. Buckminster Fuller

4D Timelock, privately published, 1928

Nine Chains to the Moon, 1938

The Dymaxion World of Buckminster Fuller, 1960

Education Automation, 1962

Untitled Epic Poem on the History of Industrialization, Simon & Schuster, 1962

Ideas and Integrities, 1963

No More Secondhand God, 1963

R. Buckminster Fuller on Education, 1963

World Design Science Decade Documents, privately published, 1965–1975

What I Have Learned "How Little I Know," Simon & Schuster, 1968

Utopia or Oblivion, 1969

Operating Manual for Spaceship Earth, 1969

The Buckminster Fuller Reader, Penguin Books, 1970

Buckminster Fuller to Children of Earth, 1972

Intuition, 1972

Earth, Inc., 1973

Tetrascroll: Goldilocks and the Three Bears, 1975

Synergetics: Explorations in the Geometry of Thinking, 1975

And It Came to Pass—Not to Stay, 1976

Synergetics Folio, privately published, 1979

Synergetics 2: Further Explorations in the Geometry of Thinking, 1979

Buckminster Fuller Sketchbook, University City Science Center, 1980

Critical Path, 1981

Synergetic Stew, Buckminster Fuller Institute, 1982

Grunch of Giants, 1983

Inventions: The Patented Works of R. Buckminster Fuller, 1983

The Artifacts of R. Buckminster Fuller, 1984

Synergetics Dictionary: The Mind of Buckminster Fuller, 1986

Cosmography: A Posthumous Scenario for the Future of Humanity, Kiyoshi Kuromiya, adjuvant, Macmillan, 1992

Suggested Reading List

....

R. Buckminster Fuller. John McHale. Braziller, 1962.

Cover piece on R. Buckminster Fuller, *Time*, Jan. 10, 1964.

"Profile" of Buckminster Fuller by Calvin Tomkins, *The New Yorker*, Jan. 8, 1966.

Bucky: A Guided Tour of Buckminster Fuller. Hugh Kenner. Morrow, 1973.

The Mind's Eye of Richard Buckminster Fuller. Donald W. Roberston. Vantage Press, 1974.

Geodesic Math and How to Use It. Hugh Kenner. Morrow, 1976.

Ho-Ping: Food for Everyone. Medard Gabel. Anchor/Doubleday, 1979.

Buckminster Fuller: An Autobiographical Monologue Scenario. Robert Snyder. St. Martin's Press, 1980.

Buckminster Fuller: Inventions, Twelve Around One. Catalog. Carl Solway Gallery, Cincinnati, Ohio, 1981.

Cosmic Fishing. E. J. Applewhite. Macmillan, 1985.

A Fuller Explanation: The Synergetic Geometry of R. Buckminster Fuller. Amy C. Edmondson. Birkhauser, 1987.

Buckminster Fuller's Universe. Lloyd Steven Seiden. Basic Books, 1989.

Concepts and Images: Visual Mathematics. Arthur L. Loeb. Birkhauser, 1993.

Bucky Works. Jay Baldwin. Wiley, 1997.

Your Private Sky: R. Buckminster Fuller, the Art of Design Science. Joachim Krausse and Claude Lichtenstein. Lars Muller Publishers, 1999.

New Views on R. Buckminster Fuller. Hsiao-Yun Chu and Roberto Trujillo. Stanford University Press, 2009.

Buckminster Fuller: Starting with the Universe. K. Michael Hays and Dana Miller. Whitney Museum of American Art, 2009.

Dymaxion Car: Buckminster Fuller. Norman Foster. Architecture Ivorypress, 2010.

Buckminster Fuller and Isamu Noguchi: Best of Friends. Shoji Sadao. Isamu Noguchi Foundation and Gardens, 2011.

Black Mountain College. Robert S. Mattison and Loretta Howard. Loretta Howard Gallery, 2011.

Index

....

Italicized page numbers indicate figures. An italicized letter *n* refers to an endnote.

"THERE *is nothing in a* CATERPILLAR THAT TELLS YOU IT'S GOING *to be a* BUTTERFLY"

— BUCKY FULLER